CAMBRIDGE LIBRARY COLLECTION

Books of enduring scholarly value

Monographs of the Palaeontographical Society

The Palaeontographical Society was established in 1847, and is the oldest Society devoted to study of palaeontology worldwide. Its primary role is to promote the description and illustration of the British fossil flora and fauna, via publication of an authoritative monograph series. These monographs cover a wide range of taxonomic groups, from microfossils, trilobites and ammonites through to Coal Measure plants, mammals and reptiles, and from all ages from Cambrian to Pleistocene. They form a benchmark for understanding the past life of the British Isles and many include the original descriptions of numerous key species. The first monograph (on the Crag Mollusca) was published in March 1848 and the Society still continues this work today. Notable authors in the series include Charles Darwin (fossil barnacles) and Richard Owen (dinosaurs and other extinct reptiles). Beginning in 2014, the Cambridge Library Collection and the Society are collaborating to reissue the earlier publications, focusing on monographs completed between 1848 and 1918.

A Monograph of the British Fossil Corals

In this supplementary monograph, originally published 1866–72, the English geologist Peter Martin Duncan (1824–91) treats only those British fossil corals that were unknown to Edwards and Haime in their preceding monograph (also reissued in this series), and those that needed revision. The two works were meant to be used together, and Duncan includes combined indexes. Featuring similar white-on-black engravings, the work starts with the Eocene and proceeds stratigraphically downwards, ending with corals from around the Triassic–Jurassic boundary. The intention to cover Palaeozoic corals was never fulfilled, so this monograph concerns scleractinians. Duncan includes an introductory treatment of the anatomy of living scleractinians, in support of his claim to offer a new taxonomical approach. He also draws attention to 'the entirely new [Eocene (Bartonian)] Coral-fauna of Brockenhurst'. This was auspicious, since recent study of this fauna sheds light on the origin of the modern Indo-Pacific coral hotspot.

A Monograph of the British Fossil Corals

Second Series

P. Martin Duncan

CAMBRIDGE
UNIVERSITY PRESS

University Printing House, Cambridge, CB2 8BS, United Kingdom

Cambridge University Press is part of the University of Cambridge.
It furthers the University's mission by disseminating knowledge in the pursuit of education, learning and research at the highest international levels of excellence.

www.cambridge.org
Information on this title: www.cambridge.org/9781108080866

This edition first published 1866–1872
This digitally printed version 2015

ISBN 978-1-108-08086-6 Paperback

THE

PALÆONTOGRAPHICAL SOCIETY.

INSTITUTED MDCCCXLVII.

LONDON:

MDCCCLXVI—MDCCCLXXII.

THE SUPPLEMENT

TO THE

BRITISH FOSSIL CORALS OF THE TERTIARY, CRETACEOUS, OOLITIC, AND LIASSIC FORMATIONS.

DIRECTIONS TO THE BINDER.

This Supplement will be found in the Volumes of the Palæontographical Society issued for the years 1865, 1866, 1867, 1868, 1869, and 1872.

Cancel **the title-pages and table of contents given in the Volumes for the years 1866, 1867, 1868, 1869, and 1872, and substitute the accompanying title-pages and tables of contents, and place the sheets and plates in the order indicated below. The plates of the Tertiary Corals to follow their pages; the plates of the Cretaceous Corals to follow their pages; the plates of the Oolitic Corals to follow their pages, and the plates of the Liassic Corals to follow their pages.**

ORDER OF BINDING AND DATES OF PUBLICATION.

PAGES	PLATES	ISSUED IN VOL. FOR YEAR	PUBLISHED
General Title-page	—	1890	April, 1891.
PART I, Tertiary			
Title-page, Table of Contents	—	1865	December, 1866.
i—iii; 1—66	—	,,	,,
	I—X	,,	,,
PART II, Cretaceous			
Title-page, Table of Contents	—	1890	April, 1891.
1—26	I—IX	1868	February, 1869.
27—46	X—XV	1869	January, 1870.
PART III, Oolitic			
Title-page, Table of Contents	—	1890	April, 1891.
1—24	I—VII	1872	October, 1872.
PART IV, Liassic			
Title-page, Table of Contents	—	1890	April, 1891.
i, ii, 1—43	I—XI	1866	June, 1867.
45—73	XII—XVII	1867	June, 1868.
Index to Tertiary Species			
Title-page, 3—6	—	1872	October, 1872.
Index to Secondary Species			
Title-page, 3—12	—	1872	October, 1872.

A MONOGRAPH

OF THE

BRITISH FOSSIL CORALS.

SECOND SERIES.

BY

P. MARTIN DUNCAN, M.B.Lond., F.R.S., F.G.S.,

PROFESSOR OF GEOLOGY TO, AND HONORARY FELLOW OF, KING'S COLLEGE, LONDON.

Being a Supplement to the

'Monograph of the British Fossil Corals,' by MM. Milne-Edwards *and* Jules Haime.

LONDON:

PRINTED FOR THE PALÆONTOGRAPHICAL SOCIETY.

1866—1872.

PRINTED BY
ADLARD AND SON, BARTHOLOMEW CLOSE.

A MONOGRAPH

OF THE

BRITISH FOSSIL CORALS.

SECOND SERIES.

PREFACE.

Twelve years have elapsed since MM. Milne-Edwards and Jules Haime completed their great 'Monograph of the British Fossil Corals' for the Palæontographical Society.

During this period Geology and Palæontology have been very carefully studied, with the aid of all the accessories of modern scientific research. Many strata which had been considered almost unfossiliferous have been discovered to contain both known and unknown species, and those beds which yielded the specimens so admirably described and figured by the great French Zoophytologists have been successfully searched for others.

The interest in the study of the Madreporaria has been greatly increased since the publication of the "Introduction" in the 'Monograph of the British Fossil Corals,' for the list and the description of the genera which it contained facilitated the diagnosis of species. In 1857 the authors of that "Introduction" commenced a work which has remained the best and, in fact, the only text-book for the student of recent and fossil Corals. The 'Histoire Naturelle des Coralliaires' was completed in 1860, by Milne-Edwards, after the death of his able and amiable colleague, M. J. Haime. The anatomy, physiology, and classification of the Zoantharia are admirably given in this work, and the classification is, with slight modifications, adopted by all Zoophytologists.

The distinguished authors modified many of their genera and introduced others, consequently the "Introduction" in the First Part of the 'British Fossil Corals' is incomplete and behind the day.

Many authors have added to the general knowledge of the comparative anatomy of recent Corals, and a few have given elaborate descriptions of fossil species [1] since the publi-

[1] Reuss, in his 'Beiträge zur Charakteristik der Kreideschichten in den Ostalpen,' &c. De Fromentel, 'Polypiers Fossiles.' Laube, 'Corals of St. Cassian, in Die Fauna der Schichten von St. Cassian,' 1 Abtheil. Michelotti et Duchassaing, 'Mem. Acad. Turin,' 2nd series, vol. xix, p. 279, 1861. Seguenza, 'Disquisizioni Paleon. intorno di Corall. Foss. di Messina.' There are also memoirs on the West Indian, Australian, Sindian, Maltese, and Javanese fossil Corals, by myself.

cation of the Monograph already alluded to, whilst the majority of Palæontologists have gradually learned to appreciate the value of the evidence afforded by Corals in many of the most important geological inquiries.

MM. Milne-Edwards and Jules Haime had not the advantages of the inspection of many collections made by private individuals and provincial Geological Societies, from the lower members of the Lias and from the Mountain-limestone; they had not an opportunity of studying the Coral-fauna of Brockenhurst; and time as well as some unintentional difficulties prevented their examining many of the most interesting forms from some of our Museums.

It has been felt, moreover, that although the "Introduction" in the First Part of the 'Monograph on the British Fossil Corals' was a great advance on all that had been done before, still the absence of those anatomical details which were so elaborately given in the Histoire Naturelle des Coralliaires' rendered the Monograph of no very great practical value.

No one could comprehend the minute details which distinguish species, by the study of the "Introduction" alone, but a very superficial examination of the 'Histoire Naturelle des Coralliaires' renders the anatomy of Corals, and the principles of their classification, easy of comprehension.

It is of very little use having detailed descriptions of species unless the anatomy of the whole class to which they belong is understood, and the publications of a Society like this should be instructive as well as recording.

A Supplement, or a Second Series, to the Monograph by MM. Milne-Edwards and Jules Haime is thought to be required. It might introduce the anatomy and physiology of recent Corals, the new genera, with descriptions of new species, and it might embody a general scheme of classification.

Following the plan adopted for the Brachiopoda in Mr. Davidson's Monograph, the relation between the hard and soft parts of the Corals will be considered, and their anatomy will be explained as correctly and as briefly as is possible. The earlier pages of this Second Series will refer to the fossil Corals of the Tertiary and Secondary rocks, and the classification of the species found in them will be given at once; that of the Palæozoic species will not be attempted until after the completion of the description of the Secondary Coral-fauna.

There will be some irregularity in the succession of the parts of the Second Series, for it is necessary to describe those large collections which can be had at once, and which might be scattered after a short period. Thus, the entirely new Coral-fauna of Brockenhurst, and many new species from Bracklesham, Barton, and Sheppey, will appear first of all; their description will be followed by that of the hitherto neglected Liassic Coral-fauna; and the Cretaceous species will be then considered, or the Oolitic, if necessary.

At the end of the description of the species from every formation, the forms already described by MM. Milne-Edwards and Jules Haime, or others, will be placed in a catalogue, and their last synonyms will be given, the name of the first describer of the species being attached; alterations in the generic names and specific determinations by the authors of the Monograph, subsequently to its completion, will be noticed, and also whatever fresh information may be requisite about previously described species.

It is hoped that after the description of all the new species has been finished there will be an opportunity for noticing the geographical distribution of Corals, and the peculiarities of the palæontological evidence offered by them.

NOTE.—In writing this Supplement, or, as I have termed it, "Second Series," I am most anxious to acknowledge that the foundation of all my knowledge upon the anatomy, physiology, and classification of the Zoantharia was derived from the writings of MM. Milne-Edwards and Jules Haime. It will be found that the greater part of the following Introduction is taken, if not in exact words, still in ideas, from those writings; and if any palæontologist or naturalist should think that I have neglected other works, it may, perhaps, be an excuse, that it is right, in following such distinguished men as those who wrote the "First Series," to carry on their train of thought, and to choose the results of their labours in preference to those of others in compiling the "Second Series."

A MONOGRAPH

OF THE

BRITISH FOSSIL CORALS.

(SECOND SERIES.)

INTRODUCTION.

I.—General Anatomy of Recent Corals.[1]

Madreporaria, or Sclerodermic Zoantharia.

When a simple or solitary Coral is living in pure and well aerated sea-water its superficial soft tissues are noticed to form a *disc,* marked with a central depression and more or less covered by *tentacules,* as well as a covering to the general external surface.[2]

The disc is superior, and the other soft tissues are inferior to it.[3]

The tentacles[4] surround the central *mouth* at varying distances; and the mouth is capable of being elevated above the level of the disc by the protrusion of a conical process.[5] Certain ridges or radiating lines mark the sides of the mouth (the *lips*), and extend outwards amongst the tentacules to the margin of the disc.

The margin of the disc gives origin to those soft tissues which are visible on the outside of the coral.

When any unusual stimulus is applied to the tentacules they contract, become smaller, and the conical mouth usually projects more than before.[6] If the irritating influence persists, the mouth is retracted, the disc sinks, the tentacules disappear, and finally the hard parts of the *calice* come into view, covered simply by the flaccid and transparent soft parts. At the same time much water escapes through openings at the end of the tentacules, and the tissues covering the outside appear to lose their colour.

[1] The Introduction is illustrated by Plates I, II, III, IV, as well as by reference to some of the figures in those plates which refer more especially to species.

[2] Plate II, figs. 4, 9, 11, 12, 13, 16.

[3] Plate II, figs. 12, 13, 16.

[4] Plate II, figs. 4, 9, 11.

[5] Plate II, fig. 10.

[6] Plate II, fig. 10.

The relation of the soft to the hard parts can then be well seen, and it will be at once comprehended that there is a correspondence between the disc and the star-like upper opening of the hard parts, which is called the *calice*.[1]

On examining a dried coral, or a well-preserved fossil specimen, certain plates will be seen projecting inwards from the edge of the calice like the spokes of a wheel; these are the *septa*,[2] and each is usually composed of two *laminæ*, but their union is so exact that it often requires microscopic sections for its determination.

On the edge of the calice, and running down the outside of the coral, are some projections, not so long as the septa, but corresponding generally with them, which are called *costæ*.[3]

The rim or edge of the calice, although it appears to be made up to a great extent by the bases of the *septa and costæ*, still presents a structure which unites their bases laterally; or, in other words, if the septa and costæ were all planed off, there would remain a more or less cup-shaped structure, called the *theca* or *wall*.[4]

The wall determines the shape of the coral; and it may be even horizontal, or more or less turbinate, cup-shaped, &c. The lowest part of the wall is called the *base* of the coral, and it may be broad or pedunculated.

The outside of the base, and more or less of the outside of the coral, are occasionally covered by a calcareous investment, which results from a soft tissue, called by Dana "foot-secretion."

The inside of the base forms the floor of a cavity, whose superior termination is the calice. This cavity is divided off by the septa, and its axis is usually filled up by a structure called the *columella*,[5] which, in transverse sections of corals, occupies the relative position of the axle to the spokes and tire of a wheel. The upper end of the columella is free, and usually forms centrally the bottom of the calice.

In some corals[6] there are thin processes, which are more or less oblique or even horizontal in their direction; they are situated between the *septa*, and they separate the cavity into compartments, the upper or calicular being the newest. In other forms these dissepiments (*dissepimenta*) are nearly vertical; and in one great series they simply connect the septa laterally, without dividing or restricting the cavity. These latter processes are called *synapticulæ*.[7] Horizontal dissepiments are termed *tabulæ*.

There are corresponding processes between the costæ in many corals, and they are often so fully developed as to project beyond and over them. The processes which are inside the wall and between the septa compose the *endotheca*,[8] whilst those without the wall and in relation to the costæ are termed *exotheca*.[9]

The "foot-secretion" is an *epitheca*.[10]

[1] Plate II, fig. 11.
[2] Plate I, figs. 1, 3, 14, 18.
[3] Plate I, figs. 2, 7, 11.
[4] Plate I, figs. 3, 4, 14, 15, 17, 18.
[5] Plate I, figs. 5, 6, 8, 10, 12, 14, 18.
[6] Plate I, figs. 15, 13, 18.
[7] Plate III, figs. 1, 2.
[8] Plate I, figs. 13, 15, 18.
[9] Plate I, figs. 11, 18.
[10] Plate I, fig. 16.

On looking into a calice and down the internal cavity the vacant spots between the septa become apparent; these are the *interseptal loculi;* they are restricted in depth when dissepiments exist, and extend from the bottom to the calice when there is no endotheca.[1]

The septa vary in size, and may or may not reach from the wall to the columella, and all the space left between them, restricted or not by dissepiments or *tabulæ*[2] (horizontal dissepiments), forms in living corals part of the *visceral cavity.* When there is no columella there is a central space, into which the interseptal loculi open; the visceral cavity is then all the larger, but the depth of its inferior boundary always depends upon the existence of the endotheca. The septa are frequently raised in an arched form[3] above the level of the top of the *wall* (theca); and a line carried across their tops over the calice would bound a cavity whose base is the top of the columella and the internal ends of the septa. This cavity is the *calicular fossa;* the interseptal loculi open into it, and it is very variable in size and depth. When the columella is very prominent the calicular fossa is all the more restricted in depth; but when the wall is high, the columella absent, and the septa not exsert, the fossa is deeper.

It will now be evident that the hard parts of a coral form the boundaries to a system of cavities (the interseptal loculi), and to the calicular fossa, into which they open.

The disc, in living corals, elevated very slightly above the tips of the septa, closes the calicular fossa above, and opens into it over the columella, so that when the mouth is widely open the markings on the free surface of this structure can be seen faintly covered by the tissue which lines all the hard parts of the coral above the newest dissepiment or the base, as the case may be.

The septa, dissepiments, and the columella, being covered with a soft tissue, which is continuous with the margin of the disc, it is evident that there is a cavity in the soft parts of the coral which corresponds with that already mentioned as being within the calcareous portion.

Thus, the *interseptal loculi, calicular fossa,* and the space between the tops of the septa and the disc, all lined by continuous soft tissue, form the whole *visceral cavity.*

The *mouth,* seen on the upper surface of the disc, opens into a short stomach, which in its turn opens into the visceral cavity by means of a *pyloric orifice* situated above the level of the top of the columella (or junction of the inner ends of the septa when there is no columella).

The *stomach* is an inversion of the membranes of the disc, is tubular, ridged longi-

[1] Plate I, figs. 5, 14. [2] Plate III, figs. 9, 11, 16. [3] Plate I, figs. 4, 14, 15.

tudinally, and very short. It is bounded above by the lips with their ridges,[1] below by the pyloric constriction, and its outside is free in the visceral cavity.

The ridges correspond with *mesenteric folds,* which are attached to the under surface of the disc and to the outer or visceral surface of the stomach. Where the *mesenteric folds* are attached to the lower margin of the stomach (the pyloric constriction), some *tubular prolongations*[2] arise which float in the visceral cavity. There is an intimate relation between the mesenteric folds, the septa, the interseptal loculi, and the tentacules. These last open inferiorly into the visceral cavity between the mesenteric folds; and, being hollow and also perforated at their free extremity, they connect the visceral cavity with the outside. The septa are developed between the mesenteric folds, and correspond with the *subtentacular spaces.*

There are, in some species, processes which are internal and accessory to certain septa; they arise from the base internally, and pass upwards in the form of thin plates, and are attached to the columella. These are the *pali.*[3]

The costæ and the exotheca are covered by, and, like all the other hard parts, are developed by, soft tissues.

The coloration of the soft tissues is very varied and beautiful; they are, of course, not preserved in the fossil state, but they occasionally leave behind them the chemical proofs of their former existence.

The soft tissues are—

1. The disc and its accessories.
2. Membranes of the visceral cavity.
3. Stomach.
4. External membranes.

The disc supports the tentacules and forms the lips. The external membrane covering the costæ arises from its external margin. It is marked by radiating ridges.

The membranes of the visceral cavity line the interseptal loculi, and cover the septa, wall, pali, and columella; they form also the mesenteric folds and the tubular processes.

The stomach, formed by membranes continuous above with those of the disc and below with those of the visceral cavity, is bounded above by the mouth with its lips, which are capable of being extended above the level of the disc.

The foot-secretion or *epitheca* has its especial membrane.

The membranes or tissues of these cavities of the disc and tentacules consist of three layers.

The *Sclerenchyma,* skeleton, or calcareous polypary—the hard parts, as they may be more simply called—consist of the *wall* or *theca, septa, costæ, columella, pali, endotheca, exotheca,* and *epitheca.*

[1] Plate II, figs. 11, 13. [2] Plate II, fig. 2. [3] Plate I, figs. 8, 9, 10, 14, 18.

The base, sides, calice, calicular edge or margin, are self-explanative terms. The terms calicular fossa, and interseptal loculi, have been noticed.

These are the usual structures observed, and they are modified in every way to produce the various shapes of corals.

The word *corallum* is used to individual corals when solitary in their growth; but when aggregated to form a compound mass each individual of the mass is called a *corallite*, the aggregation retaining the name of *corallum*.

The *corallites* of a compound *corallum* may be united together by the fusion of their walls, no costæ existing, or they may be united by a great development of the costæ and the exothecal dissepiments. Sometimes the exotheca is so developed as to form a very distinct tissue between the corallites; it is then more or less cellular, and is termed *cœnenchyma* and *peritheca*.

Some simple and many compound corals extend by a process of lateral calicular growth, so that there is not a circular or ovoid calice, but a long, and often gyrate assemblage of septa; such a calice is called "*serial*." The shape of compound corals is determined, to a great extent, by their method of *gemmation*,[1] and by the existence of *fissiparous*[2] and *serial*[3] *calices*.

II.—Anatomy of the Sclerenchymatous Structures.

Calice, Wall, Septa, Pali, Columella, Costæ, Endotheca, Exotheca, Epitheca, Peritheca, Cœnenchyma.

Calice.—The upper and open extremity of a corallum is called its *calice*.[4] Its outline is formed by the upper or marginal part of the *wall*, and is very various in its form. The superior boundary is determined by the greater or less exsertness of the *septa*, and its depth by the greater or less prominence of the structures forming the floor of the *fossa*.

The periphery of the calice is called its margin, and its floor is formed by the septa, the interloculi, the top of the columella, and, when that structure does not exist, by the axial space.

Every variety of form may be noticed in the outlines of calices; they may be circular, circular and slightly compressed, oval, elliptical, elliptical and slightly angular at the end of the long axis, ovoid and compressed from side to side, ovoid at one end, linear or leaf-

[1] Plate III, fig. 15; Plate IV, figs. 10, 11, 17, 18.

[2] Plate IV, figs. 12, 13.

[3] Plate IV, figs. 14, 15.

[4] Plate I, figs. 1, 11, 6; Plate II, figs. 11, 13, 14; Plate III, figs. 15, 17, 18, 19, 20; Plate IV, figs. 8, 11, 12.

shaped, wavy in their outline, nipped in centrally or in the figure of eight, more or less square, pentagonal, hexagonal, polygonal, polygonal and elongated, linear or serial, serpentine, &c.

The margin is not always on the same plane throughout. It may be ridged, so as to form an ornamental series of projecting angles; the plane of the minor axis may be much higher than that of the major, and *vice versâ*. In corals which are simple and horizontal the wall is covered completely by the calice, and the septa are necessarily very exsert.

The calice may be prominent, and even placed at the end of a cone, or may be depressed below the surface, as in many compound corals. Calices may be distant or connected together by their walls, or they may form series by a succession of calices running one into the other in a linear or radiating direction.

The opening of the calice may be very wide and everted or contracted and inverted; the calice may be deep, shallow, wide, narrow, and widely open; its margin may be broad, flat, or narrow, and sharp; moreover, it may be below or above the bend of the top of the septa. Deformed calices are produced by the pressure incident to the growth of crowded corallites in a compound corallum, and a great number of calices are more or less altered in outline by the phenomena of fissiparous and calicinal reproduction.

The calices vary in size on different parts of the same corallum.

In some genera one half of the calicular margin may be lip-shaped or more elevated than the other, and in a few the distinction between the calicular fossa and the general surface is by no means easy.

Wall.—The wall gives support to the costæ externally and to the septa internally, and it can be seen in the most complicated corals between the costæ at the bottom of the intercostal spaces and between the septa, where it bounds externally the septal interloculi. It determines the shape of the corallum and the amount of its solidity; moreover, it has intimate relations with the columella and endotheca, as well as with the exotheca.

The hardness and thickness of some walls[1] is as remarkable as the porosity, reticulate character, and fragility of others, and the so-called perforate[2] condition of the last is always noticed in an important section of the *Madreporaria*. Every possible variety of thickness and solidity may be noticed, as well as of fragility, thinness, and porosity; moreover, these opposite conditions are brought together by the existence of perforations in comparatively solid walls.

Usually the wall is a very prominent feature in the corallum;[3] but it may become so united to exothecal structures or to the cœnenchyma as to be indistinguishable from them; and in some large simple corals, where the epitheca is strongly developed, the wall is either rudimentary or has become absorbed. In these species the coral is kept together by the enormous development of the dissepiments or tabulæ.

[1] Plate I, figs. 3, 14. [2] Plate III, figs. 3, 4; Plate IV, fig. 18. [3] Plate I, figs. 3, 14.
[4] Plate IV, fig. 6.

Some simple forms have walls which are moderately stout superiorly and excessively thick and hard inferiorly, so as to encroach on the visceral cavity; this filling up of the lower part of the corallites is observed in some compound corals. It is very evident that the thickness and the hardness of the wall are determined by the nutrition of the coral; but no defect in this will produce the perforate condition.

Two series of wall-shapes are noticed,—one more or less horizontal and the other ranging from a shallow cup to a long cylinder in shape; the square, polygonal, and compressed outlines of some walls are either the result of pressure or are characteristic of the species.

The horizontal wall produces shallow, disc-shaped corals; the septa arise from its upper and the costæ from its lower surface. In some species the under surface is concave, so that the cup-shape is seen reversed.

The second and commonest form may be slightly horizontal at first, and with growth the edges turn up and enclose the calicular cavity; then any height, width, and contortion may result; the turbinate, subturbinate, conical, conico-cylindrical, tubular, and other forms, may thus arise.

The wall forms the most important part in some corals, but only a secondary in others; it may be uncovered externally by costæ or by epitheca, or it may be in such close contact with neighbouring walls, in compound corals, as to become fused.[1] The upper termination or margin of the wall is very visible when the septa are not exsert; and in compound corals, when the walls have become united, this margin may be sharp or broad, and variously marked. Usually the walls of neighbouring corallites (not fused together) are separated by a dense tissue, which is ornamented superiorly, and often traversed by costæ.

The wall occasionally gives out processes, and is often marked by growth-rings, constrictions, and ridges. It is rarely symmetrical; for most simple corals are curved, twisted, or more or less compressed; and this is equally true as regards the compound. The base of the wall is often attached to foreign substances, and may be broad, even concave from rupture, or very delicate and pedunculate. The epitheca, where it exists, is generally more strongly developed over the base; the inner base is the floor of the visceral cavity.

Septa.—The septa have been already noticed in a general manner; and it has been mentioned that they are developed between the mesenteric folds, and that they are localized in the intermesenteric or subtentacular spaces.

The number of tentacules has a direct ratio to that of the septa and pali.

The septa, in their simplest condition, are spiniform agglomerations of nodules, projecting slightly into the calice from the wall,[2] and there is every imaginable variety

[1] Plate III, figs. 3—16; Plate IV, fig. 11.

[2] Plate III, figs. 5, 6.

of structure between these and the highly developed septa of some Tertiary corals, where the laminæ composing the septa are distinct, very long, broad, and imperforate, very much arched and exsert, beautifully dentate on their free upper margin, and magnificently ornamented with granules in regular series.[1]

The number of septa in a calice varies in many species, and there is great diversity in their arrangement. The number and arrangement of the septa differ according to the age and development of the individual, to a certain extent. Many species have six septa, never more and never less; others have a second series, and a new septum is introduced between each of the old. Thus twelve septa and no more are found in a species of *Alveopora.* (Plate III, fig. 5.)

The six septa which appear first of all, are termed the *primary*, and they constitute a *cycle* or *order;* the next six, which are developed between the primary, are termed the *secondary*, and constitute a second cycle. The *Alveopora* has, then, two cycles of septa, or six primary and six secondary. In very many species other septa are developed, which are always found regularly distributed, one occupying each interseptal loculus. That is to say, in every interseptal loculus between the original primary and the after-coming secondary septum a third arises from the wall. There are, therefore, twelve of these tertiary septa, and the twelve form the *third cycle* or *order*. The three cycles, first, second, and third, combined, form twenty-four septa.[2] That is to say, between two primary septa there is one secondary and two tertiary septa. These septa between the two primary constitute a *system;* and when the primary septa are six in number there are six systems. If there be twelve septa, there are six systems of two cycles; and if there be twenty-four, there are three cycles in six systems.

There are interlocular spaces between the first septa and the tertiary, and between the tertiary and the secondary; any more septa must be developed one by one in these spaces. The additional septa are, in fact, developed in the space which intervenes between the first and the third septa, and simultaneously others come in between the second and the third septa, so that in each system four more septa arise. Those between the primary and tertiary constitute the *fourth order* of the *fourth cycle*, and those between the secondary and the tertiary the *fifth order* of the *fourth cycle*. The septa arise simultaneously in all the systems in this manner.[3]

The number of the septa in the last instance is forty-eight, or five orders of four cycles in six systems. Each system contains the following orders:

$$\underbrace{\text{1st. 4th. 3rd. 5th. 2nd. 5th. 3rd. 4th}}_{8}\text{—1st.}$$

Any other septa are introduced between the primary and the fourth septa, then

[1] Plate I, fig. 15. [2] Plate IX, figs. 5, 6. [3] Plate V, figs. 3, 4, 9.

between the secondary and the fifth; then others between the third and fourth and third and fifth. This regular cyclical arrangement multiplies the septa rapidly and regularly, and determines the symmetry of the calice and of the tentacular disc. When the fifth cycle is complete, there are ninety-six septa, or sixteen in each system.[1]

When six cycles are developed, no less than 192 septa result; and seven cycles, when perfect, produce 384.

It is rare for these higher cycles to be complete and the septa are aborted in many of the interlocular spaces.

The primary septa are usually larger, more exsert, and extend further inwards than the others; but, as the cycles become complicated, the secondary and even the tertiary septa often resemble the primary. Nevertheless, in the majority of instances, it is easy to determine the orders of the septa. The development of six systems of septa is seen in the majority of corals, but there are some very curious and important exceptions to its universality. Some species have four, five, eight or ten systems, and a corresponding number of large or primary septa. Moreover, monstrosities often occur, and produce an extra system, with a normal cyclical arrangement.

The pentameral, octomeral, and decemeral[2] arrangements are accounted for either by the abortion or duplication of a system or by their being natural and normal types.

The palæozoic corals belong generally to species in which there are four primary septa, or in which vacant spaces produced by aborted large septa are counted with the other large septa. But even this generalization is not free from great exceptions, and there are many genera where no trace of the quaternary septal arrangement is to be made out.

It must be acknowledged that septa do not always exist, and in the genus *Axopora* there is a proof of this.[3]

The septa thus elaborated as regards their succession and number present many peculiarities in their direction, size, length, breadth, height, exsertness, ornamentation, and in the structure of their lamellæ and margins. They usually pass directly inwards from the wall towards the columella or the centre of the calicular fossa and middle of the visceral cavity; occasionally they vary in this course; and it is by no means uncommon for the smaller septa to turn towards and even to join their larger neighbours. In calices where there is fissiparous growth, or the development termed serial, the septa pass inwards almost at right angles to the wall.

[1] Plate V, fig. 16.

[2] De Fromentel, 'Introduct. Polyp. Foss.,' may be consulted concerning these unusual types; and see my "Memoirs on Maltese and Australian Tertiary Corals," 'Ann. Nat. Hist.,' Sept., 1865.

[3] Plate VII, figs. 11, 12, 13, 14.

There is every possible variety in the size of the septa; but, as has already been mentioned, the primary are the largest, and the members of the higher orders the smallest.

The same observation holds good, as a general rule, with regard to the height. The exsertness of septa varies greatly; some are arched and extend far higher than the top of the wall, and others do not extend upwards above the wall at all. The longer septa in some species meet and are twisted centrally, whilst those of the higher orders only just project within the calice.

The breadth of the septa depends very much on the habit and size of the corallum; the bi-laminate arrangement is very distinct in some species, whilst in others it cannot be seen, and the septa are thin, delicate, and very fragile. The genus *Dasmia* has a tri-plated arrangement of the septa.

The thickness of the septa varies in corals of the same genus, and it becomes of some importance in a diagnostic sense.

Usually all the septa are thickest at their origin from the wall, they then thin off towards their inner edge, but very often there is an increase of their bulk near the columella and midway.

The ornamentation consists of ridges, papillæ, spines, and granules, which are variously arranged in radiating, parallel, or irregular series.

The structure of the laminæ differs in many species. The laminæ may be dense and imperforate, or more or less perforate generally or only in certain parts. In some corals the septa are mere spiny processes, in others they are spongy in appearance, and in the other extreme they are very dense and solid.

The upper or superior margin of the septa is free, and the inner margin or end is towards the columella or long axis of the corallum. The upper margin may be smooth or incised, lobed or entire, granular or largely dentate, serrate and spined; it may be arched, or may be directed downwards and inwards, and it may be enlarged at any part.

The inner end or margin may be free, may join a columella by processes of dense or of lax hard tissue,—may send off processes to form a columella, with others from other septa,—may be attached to pali,—and it is often very ragged, twisted, clubbed, and perforate.

The inner ends of small septa may become attached to the sides of the larger.

Finally, the sides of the septa are marked more or less by the *dissepiments*[1] and *tabulæ*,[2] and they give origin to these structures as well as to the *synapticulæ*.[3]

Note.—The description of the septa of the *Rugosa* is omitted until the introduction to the palæozoic corals is commenced. For an exhaustive essay on the septa, see Milne-Edwards and J. Haime, 'Hist. Nat. des Corall.,' vol. i, p. 40. M. E. de Fromentel's criticisms on it, and his own able descriptions, may be found in his 'Introduction à l'étude des Polypiers Fossiles,' p. 18.

[1] Plate I, figs. 12, 13, 15; Plate IV, fig. 4. [2] Plate III, fig. 16; Plate IV, fig. 2. [3] Plate III, fig. 2.

Pali.—The pali[1] are the small processes which exist between certain septa and the columella. They generally arise from the base of the visceral cavity, or close to it, and pass upwards, united by one edge to the columella, and by the other to the inner end or margin of the septa. When there is no columella they are adherent to the septa, present a free edge to the cavity in the axis of the corallum, and arise with the septa.

The upper or free margin of the pali is usually lobed, and is thicker than the end of the septum to which it corresponds; it may project higher than the end of the septum, and may form a very marked feature in the calicular fossa.

The sides are more or less broad, and are usually ornamented differently to the septa.

The inner and outer edges are united to the septa and columella, either by processes or by a perfect fusion.

The number and size of the pali vary in different species. The pali may exist before (or rather internally to) one or several orders of septa, and they are then said to form one or more crowns.

The development of the pali has often, but not invariably, a very singular relation to that of certain septa. Thus, when there is but one row or crown of pali[2] they are placed at the inner edge of the penultimate septa; and when there are two rows, or crowns, they may be seen at the inner edge of the penultimate and antepenultimate cycles of septa. In some corals with numerous septa pali are found in contact with all the septa, except those which have been developed the last—the last cycle. In others the pali are found in relation to all the cycles. One genus has the pali attached to all the septa except those of the cycle which precedes the last, and a genus well marked in the West Indian fossil coral-fauna has no pali before the principal septa, but they exist before the penultimate and antepenultimate cycles.

There is a curious relation between the perfection of the septal development and the presence of pali. Milne-Edwards and J. Haime have proved that, if in one half of a system the cycles of septa are not complete, there is a corresponding absence of pali; thus, a coral with four cycles may have pali before the secondary and tertiary septa; but if one or both of the orders of the fourth cycle are wanting in one half of a system there would be no palulus before the tertiary septum of that half-system.

When there is a columella the appearance of the pali is generally very distinct, at the same time they may be confounded with its papillæ; but when the calice has been worn away, the attachment of the pali to the columella is often so distinct that they may be mistaken for the ends of large septa.

The large spines on the inner end of some septa, or some enlargement of the laminæ at that spot, may be mistaken for pali, and the terms paliform tooth and swelling are very

[1] Plate I, figs. 8, 9, 10, 14, 18.

[2] For an exception, see pali in *Porites panicea*, Lons.

commonly met with. In the genus *Acervularia* the distinguishing of so-called paliform lobes or enlargements and teeth is sufficiently difficult.

The number of genera without pali is very considerable.

Columella.—This structure is in the axis of the coral, and may be noticed in the centre of the calice or of transverse sections of corallites, whilst in longitudinal sections it is to be seen passing from the base upwards, having the pali or septa on either side.[1]

The columella is not invariably present, but in some species it forms the most important part of the calicular apparatus.[2] The most highly developed columellæ spring from the centre of the base of the young corallum, increase in height with the growth of the septa, and always appear as prominent organs in the calice. These columellæ grow independently of the septa, and are not formed by their internal and free terminations. For this reason they are called "*essential*" or "*propria;*" they generally assume the styliform, the fasciculate, or the lamellar character, and may or may not have pali attached to them.[3]

The second kind of columella is termed "*septal,*" and is produced by the inner ends of the septa dividing into longitudinal "poutrelles." They have a fascicular arrangement. These "septal" columellæ are rare, and may, for all practical purposes, be considered with the next kind.

The third kind of columella is formed by the septa dividing into numerous processes before they approach closely; the processes unite centrally, and throw out lateral growths, so that a more or less dense, spongy, or cellular structure results. This columella is termed *parietal,* and may be very highly developed or may be rudimentary. In the latter instance the columella may only be recognised by a slight bifurcation of the inner ends of the septa, with a sparely developed cross tissue.

False columellæ are formed by the soldering together of the inner ends of two or more septa, by the twisting of the inner ends of several septa, and by the presence of endotheca close to the septal inner margin.

Rudimentary columellæ are often observed, which cannot be classified with any of the above; they may be formed by a lateral junction of the inner ends of the larger septa, by processes connecting them, and by the inner ends becoming clubbed in outline, and more or less irregular in their direction.

There are many modifications of these varieties of columellæ, but their division into essential, septal, parietal, and false, is of great practical value, and they can always be distinguished with care. The calicular terminations of the columellæ vary in size, projection, outline, and arrangement.

[1] Plate I, figs. 5, 6, 10, 18.

[2] Plate VII, fig. 12.

[3] Plate I, figs. 5, 6, 8, 10, 12; Plate IX, figs. 3, 6, 10.

Amongst essential columellæ the styliform may end in a cylindrical and pointed process, or in a more or less compressed and blunt, which may project even higher than the septa, or in a bulbous termination marked by ridges corresponding with large septa; or the organ may be angular in transverse outline, and project but slightly above the bottom of the calicular fossa. The styliform columellæ may be studied in the genera *Turbinolia*, *Synhelia*, *Stylophora*, *Axosmilia*, *Stylosmilia*, *Stylina*, *Holocœnia*, *Stylocœnia*, *Astrocœnia*, *Stephanocœnia*, *Holocystes*, *Cyathoxonia*, *Syringophyllum*, and *Phillipsastræa*. They are nearly solid, spring from the base, and may or may not be attached by processes to the septa. Very visible in well-preserved specimens, these columellæ are readily destroyed by rolling, and cannot then be distinguished except by sections. In many species, especially in the *Astrocœniæ*, the columella appears to be very large in certain fossil conditions; but this appearance arises from a mechanical adhesion of calcareous particles to the outside of the columella and between the inner ends of the septa. There are examples of styliform columellæ (Plate IX, figs. 3, 6, 10).

The lamellar form of essential columella (Plate I, fig. 6; Plate IV, fig. 14) may occur in circular, elliptical, or in elongated calices. It is seen as a sharp edge, generally at the bottom of the calicular fossa, and may be in contact both with septa and pali. Its sides are occasionally ornamented with granules. In the genus *Madrepora*, and in some species of *Solenastræa* this lamellar columella does not really exist, but is simulated either by the junction of opposite septa or by the irregular development of neighbouring septal ends. The true lamellar columella is not formed by septa, but springs from the base of the corallum.

The fascicular columella is a very complicated organ. In its simplest form it is a bundle of rods coalesced laterally, adherent below, and rounded at the free calicular surface.

This structure is well seen in the genus *Axopora* (Plate VII, fig. 14), and in an Australian fossil, the *Conosmilia anomala*[1] (nobis). Here are two riband-shaped processes arising from the base, and projecting in the calicula fossa; each is simply twisted five or six times, so that the riband's edge takes on a spiral form; this is the simplest form of the common fascicular columella, and in Plate I, fig. 13, several processes, really riband-shaped, but much twisted, are seen in lateral contact, the whole forming the columella. The number of the processes varies in different species, and it is tolerably constant in certain forms; the processes, were they untwisted, would form a number of flattened and lamellar columellæ in lateral apposition. The septa and pali do not contribute to their formation. The calicular surface of the fascicular columellæ may be papillary, or even twisted; and it most frequently resembles the arrangement of the central portions of the flowers of certain Compositæ; hence the term "chicoracé," which is most significant and explanatory of the appearance of the calicular surface of the columella in the genus *Caryophyllia*.

[1] 'Ann. Mag. Nat. Hist.,' ser. 3, vol. xvi, pl. viii, fig. 4*e*.

The septal columellæ may be mistaken for the fascicular and essential; but a longitudinal section will show that the inner edges of the septa forms the organ, and that it does not arise from the base.

The parietal columellæ are very common, and their structure is illustrated (Plate IV, fig. 13; Plate VII, fig. 9).

The calicular surface of the columella may be prominent or depressed, papillary or spongy; and the organ may be very dense or consist of very lax tissue.

The columellæ of the following genera may be studied with regard to this variety:—*Parasmilia, Eusmilia, Dendrosmilia, Lithophyllia, Circophyllia, Rhabdophyllia, Mæandrina, Manicina, Diploria, Heliastræa, Solenastræa,* &c.

As a general rule, when pali exist, they are in close contact with the columella, and as they spring from the base they often look like lateral processes of essential columellæ. It will be observed, in the descriptions of living corals, that the columella fills up much of the visceral cavity, and is developed by the inner layer of the soft tissues. Playing a very important part in the economy, and being in relation both with the septa and pali, the columellæ are structures whose variations in form are of generic import.

Costæ.—The costæ may be considered in a general sense to be the continuations of the septa beyond the wall.[1]

In some *Turbinoliæ* the continuity between the costæ and the exsert septa is very evident, and both of the structures are much higher than the upper margin of the wall.[2] But it is very probable that this exsert condition of the septa and costæ is to be referred to the corallum having attained its full development as regards height; the further upward growth of the wall was arrested, and only the combined costo-septal apparatus grew on. For when the costæ of the same specimens are broken off low down, it is tolerably evident that the wall intervened between their bases and those of the corresponding septa.

It would appear that the costæ and septa are not developed by the same parts of the soft tissues except when they are exsert and above the wall; and the want of correspondence between the septa and costæ about to be mentioned is in consequence of this.

It is probably quite correct to give the costæ an origin independent of the septa, and to assert that they are frequently separated by the thickness of the wall from the septal laminæ.

The costæ are developed by the inner layer of the tissue which covers the wall externally, and the outer surface of the wall and the exothecal structures are also formed by it. The costæ follow, as a rule, the cyclical development of the septa, and are called primary, secondary, &c.

[1] Plate I, figs. 2, 6, 7, 11, 15, 18.

[2] Plate I, figs. 6, 14, 15, 18.

All the varieties of length, thickness, porosity, solidity, and ornamentation, observed on the septa are represented in the costal structures. As a rule, the costæ are shorter than the septa in transverse section, but there are many exceptions to it, and it is very common to find a rudimentary septum of a high cycle with a corresponding well-developed costa.[1] The projection of the costæ from the wall and the size of the space between them (intercostal space) vary greatly; in some species the costæ are close and form simple prominent ridges, whilst in others they are wide apart, project greatly, and may be covered with great spines, dentations, or serrations. The greater projection of certain costæ, the ornamentation of others, and their correspondence with the cyclical arrangement of the septa, are readily studied in different species.

The costæ do not always project at right angles from the wall, and those that are very long often curve and twist. Whatever may be their form or length, they have sides and a free surface. The sides of neighbouring costæ are frequently joined by the dissepiments of the exotheca, or they may be simply marked by dissepiments which do not stretch across the intercostal space.[2] The sides are often spined or granulated, and are even perforated in certain species. The variety in the ornamentation of the different cycles of costæ in the same individual is very interesting, and its study is of great use as a secondary method of specific diagnosis.

In many compound corals the costæ of one corallite run into and join those of the neighbouring corallites,[3] whilst in others, where the walls are fused,[4] the costæ abort altogether. There are many species where the costæ are simply rows of granules; in others the rows of granules[5] become lines of slight elevation, and finally well-developed costæ. The reverse occurs, and well-developed costæ on the outside of a calice often become granular or even become aborted on the wall.[6]

The exothecal dissepiments extend beyond the costæ in some instances, and, as a rule, the costæ are then feebly developed.[7] The following are some of the most important variations in the structure of costæ. They may be absent or rudimentary, and they may arise on the corallum at various heights from the base. They may be recognised under the following aspects:—Small, large, finely granulated, indistinct, generally indistinct inferiorly, prominent, prominent near the calice only, prominent inferiorly, sub-equal, equal, alternately large and small. As faint ridges, as striæ, moniliform, very thin, perforate, wedge-shaped, flexuous, broad, flat; formed by a series of globules, spines, and granules; wide apart, close, rounded, cristæform, tubercular, largely spined, dentate, alæform, crenulated, striated, verrucose, folded in zigzag, echinulate, long, dichotomous, inclined, &c.

The costæ do not invariably correspond to septa, and are not constantly continuous

[1] Plate IX, fig. 11.
[2] Plate IX, fig. 7.
[3] See 'Descriptions of the "*Thamnastrææ*," in 'Brit. Foss. Corals,' MM. Milne-Edwards and J. Haime.
[4] Plate IV, fig. 11.
[5] Plate V, fig. 6.
[6] Plate I, fig. 4.
[7] Plate V, fig. 2.

with them. It will be noticed that in some species of *Cyathophyllidæ*, and in many Tertiary[1] simple corals, that the external edges of the septa correspond with the intervals between the costæ, and not with those organs themselves. This is not an accidental variation in growth, but is constant in several species.

In some species there are small costæ which do not correspond to any septa; the large costæ are continuous with septa; but these so-called rudimentary costæ simply project externally, and correspond internally with an interseptal space.[2]

In some corals the epitheca, whilst covering the costæ and hiding them from view, appears to have produced their partial absorption, for above the limit of the epithecal structures the costæ may be seen to be prominent and to be greatly ornamented.[3] It may be inferred that in young specimens whose epitheca is not fully developed the costæ would command more attention in the specific diagnosis than is proper, and this has taken place in more than one instance. The costæ may, however, retain all their ornamentation when covered by a very dense and membraniform epitheca, and this peculiarity is generally constant. Occasionally the long spines on the costæ of some *Lithophyllaceæ* project through the epitheca, but in the majority of instances they are included. It is evident that the costæ were well developed before they were covered by the epitheca.

The more prominent the costæ, the more they are exposed to the destructive influences of rolling and of wear and tear; it happens, therefore, that the large cristæform costæ, the long delicate spines on their edges, and the finely granulated dentations, are rarely distinguishable in many fossil species, and their former existence can only be suggested in consequence of scars and raggedness on the surface, or by the preservation of an ornament here and there.

In examining the costal structures the specimen should be placed in several positions and in different lights, for small structural peculiarities are often hidden in the shadows.

Endotheca.—The structure which, stretching from one septum to another, closes more or less the interseptal loculi,[4]—the horizontal processes which, extending from side to side in a corallite, shut out all beneath from communication with above,[5] and certain exaggerated septal papillæ, which meet in the interlocular spaces and form a system of joistwork,[6] constitute the *Endothecal Sclerenchyma.*

The first variety, termed by Milne-Edwards and Haime "*Traverses*" or *Endothecal dissepiments*,[7] characterises many genera; whilst the second, termed by these authors "*Planchers*" or *Tabulæ*,[7] serves to distinguish a great series of *Madreporaria*. The third variety is seen in the family *Fungidæ*, which it characterises, and the name *Synapticula* is given to it.

[1] 'Ann. Mag. Nat. Hist.,' loc. cit. [2] In *Turbinolia Forbesi*, Dunc. [3] Plate I, fig. 16.
[4] Plate I, figs. 16, 18; Plate IV, figs. 2, 4, 6, 8. [5] Plate III, fig. 8, 9, 10, 11.
[6] Plate III, figs. 1, 2. [7] Plate V, fig. 3; Plate I, figs. 15, 18. [8] Plate I, figs. 3, 5, 14.

The endothecal dissepiments, greatly developed in some genera,[1] are either rudimentary or quite absent in others;[2] they are nearly horizontal, inclined and nearly vertical in different species, and they may be concave or convex upwards; moreover, they may either be very numerous in each interlocular space or but one or two only may exist.

As a rule, there is no exact correspondence in all the interloculi as regards the distance of the last dissepiment from the upper septal margin. In some species the distance is considerable, whilst in others the dissepiments fill in the interloculi close up to the bottom of the calicular base.

The dissepiment is attached to the septum on either side of the interlocular space and to the inside of the wall. Its inner edge is either free or joins another dissepiment, which, not reaching the wall, is carried inward in its growing course, and so with other dissepiments in succession. It results that, according to the convexity and size of the dissepiments, they produce more or less *cellular* or *vesicular divisions*[3] in the interloculi.

The dissepiments may be very coarse or the reverse, and in some species they are found of several sizes. The distance between the dissepiments varies, and the cellular condition of the outer part of the interloculi is often very marked. The straight dissepiments do not produce the vesicular appearance. Dissepiments often form a vesicular tissue when tabulæ exist. There are some important genera without dissepiments, and whose species contain individuals whose internal base forms the lower margin of the visceral and interlocular cavities.

In some species, a filling-up of the interior of the corallum by a process of thickening of the lower part of the wall and base supplies the place of the endotheca.[4]

The second variety of endotheca, the *tabular*, is recognised by the horizontal direction of the processes,[5] and by each process being on the same level with regard to the interseptal loculi. In fact, the *tabulæ* give the idea of passing through septa and everything else in their horizontal course, for they appear to shut out all the space beneath them most perfectly. Their extent varies with the diameter of the corallite, and is influenced by the occasional presence of *vesicular endotheca*[6] near the *wall;* but, as a rule, they are attached to the inside of the wall and to the septa: they may be distant or very close, very delicate or very strong, and they are often marked either by depressions or elevations on their upper surfaces. Some tabulæ are not quite horizontal, but curve upwards in the long axis of the corallite, and others are inclined between horizontal series.

In *Axopora Fisheri* (nobis) the great fasciculate columella clearly passes through the tabulæ, and in the genus *Columnaria* the large tabulæ may be broken off the septa, in longitudinal sections, and it may be readily observed that the septa are continuous and that the tabulæ are not their foundation.

1 Plate V, fig. 3; Plate I, figs. 15, 18.

2 Plate I, figs. 3, 5, 14.

3 Plate I, fig. 15.

4 Noticed in many West Indian Tertiary corals.

5 Plate III, figs. 9, 10, 11.

6 Plate IV, fig. 2; Plate III, fig. 16.

The *synapticulæ* are not considered to be endothecal structures by MM. Milne-Edwards and Haime, but their development in some species renders their present classification necessary. In their feeblest development they are papillæ (on opposite septal laminæ), which have coalesced, and thus form a bar across the interlocular space, whilst in their greatest they form long ridges between the septa, and they cannot be distinguished from very vertical dissepiments except that they do not tend to close a cavity.

Exotheca.—There are structures resembling endothecal dissepiments between the costæ of some species;[1] in others these sclerenchymatous laminæ—the *exotheca*—extend beyond the costæ and form a more or less cellular envelope to the corallite, by which it is joined to its fellows to form a compound corallum.[2]

The simplest exothecal dissepiments are stretched horizontally across the intercostal spaces, they generally reach the free edge of the larger costæ, and now and then hide the smaller. They may be inclined or not.

The highest dissepiment, or that nearest the calice, bounds the lowest reflection of the soft tissues, just as the highest endothecal dissepiments bound and form the base of the soft tissues of the visceral cavity.

In some species there are dissepiments between the costæ very high up, and in others much lower down. The distance between the dissepiments, their arched or plane course, their vesicular character, and the presence of vertical laminæ dividing the space between dissepiments into cells, are all seen to vary greatly in different species.

The dissepiments are very feebly developed in most simple corals, and they may be noticed as simple fold-like elevations on the sides of costæ and as forming dimple-shaped depressions on the wall at the bottom of the intercostal spaces in some of the *Turbinoliæ*.

In *Solenastræa* they may be distinguished as forming cells on the wall and between the costæ and as a tissue which extends around each corallite.

The upper surface of the dissepiments is often marked with elevations resembling blunt papillæ.

The genus *Galaxea* has this exothecal cell-growth in excess; it is termed in such an instance *Peritheca.*[3]

Cœnenchyma.[4]—Some corallites in many compound corals are separated by a very dense sclerenchyma, which is variously ornamented on its free or intercalicular surface. In some species the walls of the corallites are evidently independent of this structure, but in others this is not the case. It would appear that this tissue, which is very cellular in its simplest development and hard and solid in its greatest, is really an *exothecal* structure, and that it is formed by the lowest and reflected layer of the external soft tissues. The costal markings, the granules, spines, monticules, ridges, and depressions, on the surface of the cœnenchyma differ greatly in many species.

[1] Plate I, figs. 11, 18.

[2] Plate V, figs. 2—5.

[3] Plate I, fig. 19.

[4] Plate IV, figs. 7, 12, 17, 18.

Epitheca.[1]—This structure is occasionally seen both in simple and in compound corals; it is the " foot-secretion " of Dana,[2] and may either be closely applied to the wall of the corallite or may simply cover the costæ, leaving them more or less perfect in their ornamentation. In some simple corals it covers the wall so closely as to resemble a coating of varnish, in others its texture is rough and marked with concentric or encircling ridges, and in a few instances it is marked by chevron-shaped lines. The epitheca may be very thin or very dense, and it may simply cover the base or only reach a short distance upwards from it; or it may cover all the external surface as far as the calicular margin. The dense *epitheca* of some *Montlivaltiæ* is accompanied by a great diminution in the strength of the wall; this is seen also in many *Rugose* corals. The epitheca of compound corals is rarely ornamented, but is laminate and often readily destroyed. Its preservation in fossils is comparatively rare, and it should therefore not be made of very great classificatory value.

The epitheca developes processes in certain species and only covers the base of others; it is porcellanous in some, as in *Flabellum*, and pellicular in others, as in *Balanophyllia.* It is membranous, striated, verrucose, marked by growth-rings, shining, rough and partial, in different species.

It is a structure evidently formed after the development of the costæ, and results from a tissue which is a continuation of that which determines the agglutination of the bases and peduncles of certain corals to their supporting earth, stone or rock, or foreign organism.

III.—Anatomy of the Soft Tissues.

The membranous surface which covers the calice, supports the tentacules, and is perforated by the mouth, is called the *Tentaculiferous Disc.*[3]

The opening of the mouth is central, and is either circular or elliptical in outline; it is at the top of a truncated cone[4] whose base is continuous with the disc and whose height varies according to circumstances. The margin of the opening—the lip—is usually marked by radiating ridges, is very prehensile, and can be moved in different directions. The cone, whose upper extremity is the mouth, varies in its power of protrusion in different species; this is especially great when the tentacules are small and are only arranged at the margin of the disc; and, as a rule, when the tentacular development is considerable the labial protrusion is slight. In some species, such as *Heliastræa cavernosa* and *Lithophyllia Cubensis,* there is a considerable space between the mouth and the tentacules, and these last are feebly developed; consequently the mouth can be so protruded as to form a hollow between its cone and the base of the tentacules.

[1] Plate I, fig. 16.

[2] Plate IV, fig. 6.

[3] Plate II, figs. 4, 9, 11, 10, 13, 14, 16, 17.

[4] Plate II, figs. 10, 11.

The *ridges* which mark the lips are continued on to this vacant space, and radiate towards the bases of the tentacules.

In some species the moveable mouth and the hollow between it and the tentacules are of more use in obtaining food than the tentacules themselves.[1]

The contrary is very evident in *Caryophyllia clavus*[2] (the *Caryophyllia borealis* of British zoophytologists), and in *Cladocora cæspitosa.*[3] In these species the tentacules are greatly developed and extend close up to the base of the cone which is surmounted by the mouth and lips; there is but little of the disc unoccupied, and the power of protrusion on the part of the cone is comparatively slight. Yet it must be observed that when the tentacules are withdrawn, the mouth is capable of being projected further than when they are in full extension.

The lips, the external surface of the cone, and the disc, are covered with cilia. At the marginal extremity of the disc in some species, and scattered over more or less of the whole disc and extending even very close to the labial orifice, in others, are the tentacules.[4] These organs vary in length and thickness in different species, but each has a base continuous with the tissues of the disc and opening into the upper part of the visceral cavity. Generally terminated by a bulbous swelling, the tentacules are perforated throughout by a delicate canal, and consist of tissues which render them very mobile, contractile, extensile, and more or less prehensile. The external margin of the disc corresponds with the calicular margin; it is separated from it by a very small space, is continuous with the tissues covering the outside of the coral, and in some species has a small fold which covers in the tentacules.

The opening of the mouth, when fully expanded, admits of the columellary surface being seen at the bottom of a shallow cavity; and the sides of this cavity, marked by the continuation of the ridges noticed on the lips and disc, are often protruded through the lips.[5] The cavity is the stomach, and it is separated from the visceral cavity, which is below or at about the level of a prominent columella, by a faint constriction—the *pylorus.* The stomach is very short and very extensile.

The sides of the cavity are continuous, by means of the lips, with the outside of the disc; they are formed by the same tissues, but the tegumentary layer of the disc is altered and becomes the superficial layer of the mucous membrane of the stomach.

The ridges already noticed on the lips, disc, and stomach, correspond on the under side of the disc and outside of the stomach with *mesenteric folds.*

The pylorus opens into the visceral cavity, whose upper boundary is the lower surface of the tentaculiferous disc, and it therefore is clear that the stomachal membranes con-

[1] Plate II, fig. 10. [2] Plate II, figs. 7—11. [3] Plate II, fig. 4.

[4] Plate II, figs. 4, 7, 11, 12, 13, 14, 17, 18, 19, 20.

[5] The ridges are seen in Plate II, figs. 11, 13, 14.

tinued over the pylorus are reflected, upwards again, outside the stomach to cover the lower surface of the disc. Here, moreover, they form the *mesenteric folds*, upper attachment is to the under surface of the disc, and whose inner is in part to the ridges of the lips and the corresponding structures on the outside of the stomach. There are openings between these mesenteric folds corresponding with the bases and canals of the tentacules. The pylorus exists more in name than in reality, for the passage into the visceral cavity is large and easily passed. Around the lower margin of the pylorus, and attached where the ridges already alluded to end, are the free edges of the mesenteric folds and a *tubular structure.*[1] There is a distinct numerical relation between the development of the ridges, mesenteric folds, tentacules, septa, and pali.

If the disc were removed from the subjacent corallum by cutting the membrane which is continued from below upwards to its margin, and the pylorus were pulled upwards, the septa, pali, columella, wall, and dissepiments, would be exposed to view, covered by soft tissue; in other words, all the boundaries of the visceral cavity except the upper would be seen.

The upper boundary—the under surface of the excised disc—presents a series of radiating soft folds, separated *by intermesenteric spaces*, which are perforated by foramina, continuous with the tentacular canals. The pali and septa are developed in these spaces, and hence it is that the tentacules over these hard parts appear to grasp them by their bases.

The visceral cavity is bounded below and externally by the tissues covering the inside base, the wall, and the dissepiments which close in the calicular fossa, as the case may be.

The cavity is divided by the septa and mesenteric folds into a series of *radiating fissures*, which may be recognised in the dead specimen by means of the interseptal loculi.

The absence of the columella and of endothecal dissepiments infers a large visceral cavity, and it may be readily understood that a coral developing endothecal dissepiments rapidly will have a short visceral cavity, for the newest dissepiment bounds the calicular fossa inferiorly.

The sea-water and its minute organisms would pass into the mouth, through the stomach and pylorus, and would enter between the mesenteric folds into one of the perivisceral fissures of the great visceral cavity, and the water passes out again through the tentacular canals.

The under surface of the disc is continuous with the soft tissues covering the septa and wall (internally) by their direct continuation upwards. The contiguity of the tissues covering the costæ and outer part of the *wall* with the outer rim of the *disc* has been noticed.

The *disc* thus constituted is, when the polype is well nourished and lively, slightly

[1] Plate II, fig. 2.

elevated above the calicular margin; its tentacules are stretched out and overlap the hard parts, whilst the conical mouth is barely visible. Under other circumstances the disc is contracted, the mouth open, the tentacules more or less retracted, and the outer part of all the septa is visible through the translucent tissues.

In certain "*serial*" corals, such as *Diploria cerebriformis*,[1] the edge of the disc gives exit to *prehensile cirrhi*, and these organs are to be seen projecting from the rim of the disc in *Caryophyllia clavus*.[2] They are very thread-like, and have prehensile powers. The microscopic anatomy of these cirrhi has not been studied.

The tubular structures, "*cordons pelotonnés*," which are attached to the juncture of the mesenteric folds with the pylorus,[3] float about in the visceral cavity, and especially near the inner margin of the smaller septa; their lower end is unattached and often rises on to the top of the columella. These tubular structures are very much twisted, hollow, and contractile, and are covered with cilia. They often contain ova. The relation between the mesenteric folds and these tubular structures in the physiology of reproduction requires further examination.

The hard parts of the corallum are included in and nourished by soft tissues.[4] This is invariably the case in every species up to a certain period of growth. In some it is true during all the stages of their development, whilst in many species only the upper part of the corallum is in contact with the soft tissues after a certain height has been attained.

Thus, in the *Caryophyllia clavus* the outside of the corallum is covered by soft tissues from its narrow base to its calicular margin and the inside also. The wall, the costæ, the septa, the pali, and the columella are covered by a membrane which sends processes into their dense structure. The nutrition, growth, and in some instances the absorption of the hard tissues, are carried on by means of the membrane and those processes, and so long as the hard and soft parts are in contact, the first cannot be said to be independent of the latter.

In corals where the growth is accompanied by the formation of dissepiments in the interloculi, the whole of the interior of the corallum below the dissepiments nearest the calice, is not in contact with the soft parts; it has ceased to be nourished by them, and it is to all intents and purposes dead. Moreover, the external membrane does not descend for any considerable distance below the calicular margin, and the lower parts of the costæ and wall are as dead as the lower parts of the interior of the corallum. This is the case in most of the large and luxuriantly growing compound corals, and only a few lines on their surface may be living, the rest is dead. Each portion of the *endotheca*, as it springs from the septa or wall, is formed by the fine membrane and is included in it; as growth proceeds the curved, straight, horizontal, or vertical dissepiment is lined on each surface

[1] Plate II, fig. 17. [2] Plate II, fig. 11. [3] Plate II, fig. 2.
[4] Plate I, fig. 17, diagram.

by the soft tissue, but as the dissepiment closes off the space beneath it the inferior layer of membrane is absorbed, and finally is no longer to be noticed. This is the case with the *exothecal* structures also; the exothecal layers, the cœnenchymal cells, and the perithecal cells, are formed by the membranes, and as the cells become closed the included membrane is absorbed. All the granular and spiniform ornamentation of the sclerenchyma is also formed in the soft tissues, and the more or less dense *epitheca* results from the development of a tissue from the base of the corallum.

This last is called the foot-secretion, and covers the results of the growth of the membrane which develops the wall and costæ.

The deposit of earthy and inorganic matter in living corals is not, then, a simple concretionary process, but is essentially a vital one; it follows certain laws, and its extent and amount depend on the nutrition of the individual. When the influence of the soft tissue is no longer felt the hard parts become harder and denser and are subject to various changes in their mineral condition.

In those corals whose calices are not separate, but are continuous and running into series, the tentacules, as a rule, are small, numerous, and are often partly hidden by a ridge of membrane.[1] There are several mouths to the elongated and tortuous calices.

The microscopic structure of the soft tissues of the *Sclerodermic Zoantharia* has been ably studied by many observers, and the following extract from the description of the soft parts[2] of *Cladocora cæspitosa* by the late M. Jules Haime contains information sufficiently exact for the present purpose.

" The surface of the corallum is more or less convex. When extended the polypes touch each other with the extremity of their tentacules, and when they are seen from above there is no interval between them. The tentaculiferous disc is never more than two or three millimètres above the calicular margin of the polyperites, and the lateral and inferior continuation of the disc only descends one or two millimètres below the margin. When a polyperite is cut longitudinally it will be readily observed that the soft tissues are not prolonged much deeper internally in the visceral chamber, so that in the adult coral, which is usually several centimètres long, only about five or six millimètres of its upper part are covered by the soft tissues. This limited portion is bounded inferiorly by the uppermost of the series of horizontal dissepiments. All the rest of the corallum appears to be dead, and is ordinarily covered with Serpulæ and Nullipores.

"When the tentacules are fully extended, the diameter of the circle formed by their extremities is about one and a half times as large as that of the calice. The margin of the calice is usually visible on account of the transparency of the soft parts covering it.

"The tentaculiferous disc is horizontal, but towards the middle of it there is a slight

[1] Plate II, figs. 14, 16, 17.

[2] 'Hist. Nat. des Corall.,' vol. ii, page 589 *et seq.* See description of Plate II.

concave track, and the mouth projects in the form of a more or less oblong truncated cone. There are from sixteen to eighteen internal folds, faintly shown, however, on the rim around the mouth.

"The tentacules are of the same number as the septa whose summits they envelope, and there are always from thirty-two to thirty-six. They are evidently equal in size and in length. Their length is nearly equal to that of the diameter of the corallite. They are elongated, swelling a little above their insertion, and then becoming very slender as far as their free extremity, which is terminated by a small knob-shaped enlargement.

"The polypes can contract to various extents. Several very characteristic movements may be noticed, however. A slight agitation of the surrounding water or the contact with small particles, suffices to cause a shortening of some or all of the tentacules, although the disc does not alter its shape or position.

"When the exciting cause acts more decidedly and continuously, the shortening of the tentacules increases, the disc retreats, and the protractile mouth elongates in advance of the calice. This state of things is very usual in disturbed or decomposing water. If the animal itself is shaken or is touched, it retracts its disc into the calicular fossa, and nothing is to be seen of the soft parts but some small elevations corresponding with the tentacules. Finally, a violent shock or a prolonged irritation produces so complete a retractation that the tentacules disappear completely, and the white colour of the septa is seen. The calice looks as if it were dried, and there is only a light brown tissue in the interseptal loculi. In this last case the water which usually distends the tissues has been gradually expelled, and they are so reduced in volume that they are readily withdrawn into the interseptal and columellar spaces.

"The disc and the tentacules are of a transparent brown colour, and when the sun shines, a brilliant green tint may be seen within the tentacules. This coloration evidently depends in some instances upon the light. But it is necessary to remark, that the primary and secondary tentacules and those of the third cycle which are flanked by quaternary are those which show this green tint in their insides. The peculiarities of these tentacules coincide with the presence of pali, which are situated beneath and within them.

"When the mouth opens, as it often does when the polype is semi-retracted, the papillæ of the columella are visible. The stomach is very short, and is almost reduced to a rim, which is confounded with the lips.

"The tentacules are not smooth, but are covered with a multitude of small wart-shaped prominences, of a transparent white colour; they are equal in size, and measure a tenth of a millimètre in width. The terminal bulb presents a narrow central canal, which communicates both with the tentacular cavity and with the external medium. The three layers of tissue which constitute the tentacules have the same general characters as in the *Actiniæ*, but the four layers of the tegumentary covering are not to be detected.

"1. The first envelope is quite transparent, and is composed principally of nematocysts of three dimensions, those of medium size being the commonest; also, of very simple cells, either irregular in shape, or oblong or pyriform; and of small rounded and transparent globules, which form the innermost layer.

"There are no cells in the external tegument which produce the colour of the polype. The white warts which project on the surface are made up of a mass of large, transparent and elongated vesicles.

"The nematocysts, which form the most important part of the integument of the tentacules, are slender and cylindrical, one of their extremities being smaller than the other.[1] They contain a thread regularly rolled up as a spiral, and which near the large end terminates in a straight and central portion. The thread when unrolled is about two tenths of a millimètre in length. The nematocysts are perpendicular to the tentacular surface, and their large end is the most external; the internal thread makes its exit by this extremity.

"The terminal bulb[2] of the tentacules is almost entirely composed of these filiferous capsules; there are two other kinds in it unlike those just described, some larger and stouter, and others much narrower and more slender. The first are elliptical, slightly attenuated at one of their ends, and they contain a thread rolled into a slack spiral. This thread shoots out from the small end of the cell. The remaining nematocysts do not appear to have a proper cell-wall; they are cylindrical, slightly smaller at both ends, and very slender; they are formed by a filament very closely rolled into a dense spiral, which unrolls itself like the wires used in some elastic clothing.[3]

"The structure of the skin is the same over the whole surface of the polype. The nematocysts of the second size are the most common. A certain number of those of the largest size are found in the stomacho-buccal rim. The cilia are very distinct at this spot, and around the disc also, although they are very delicate; they are rare and feeble on other parts of the polype; they are very indistinct on the tentacules, and are wanting on the bulb.

"2. The middle or muscular layer is formed by transverse and vertical fibres which are excessively slender and sparely distributed. Very thin oblique muscular fibres may be seen at the bases of the tentacules.

"3. The internal membrane is formed by a layer of transparent cells tolerably adherent to each other, and by a layer of colour-bearing globules which are spherical or slightly oval in shape.

"It is these cells which give the colour to the polype; they are filled with irregular-shaped grains, of a bright brown colour; they themselves are secreted in certain transparent vesicles, and present the greatest resemblance both in shape, colour, and structure to the globules which float free in the tentacular cavities of young sea-anemones. It is probable they have a corresponding function in their early age. Near the top of the

[1] Plate II, fig. 1. [2] Plate II, fig. 3. [3] Plate II, fig. 6.

tentacules these colour-bearing cells are arranged in small irregular groups, but elsewhere they become more numerous.

"The internal membrane lines the interseptal loculi, where its presence is rendered evident by its colour; it is stopped inferiorly by the last sclerenchymatous dissepiments. The mesenteric folds formed by this membrane present a few colour-cells. The folds[1] give attachment to the simple "boyaux pelotonnés" which float in the large interseptal loculi along the smaller septa, and which often show themselves on the columella when the mouth is half open and the polype is slightly contracted. Their walls are almost entirely composed of nematocysts of the largest size, and their surface is furnished with large and strong cilia; they are frequently affected by peristaltic movements, and they are attached to the tentaculiferous disc by strong muscular fibres."

IV.—Reproduction and Multiplication.

Ovular Reproduction; Gemmation; Fissiparous and Serial Growth; Reproduction.

The *mesenteric folds* and the twisted *tubular processes,* whose ends are free in the visceral cavity, appear to be the organs which develop the male and female elements.

It would appear that all corals are not bisexual, but the majority are so. Spermatozoa were asserted to exist in the *tubular processes*, but their description tallied with that of the thread-processes of *nematocysts.* Milne-Edwards dispelled this illusion, and the true male elements have been discovered. The presence of ova in the *mesenteric folds* and in the *tubular processes* has been noticed and in the latter position by Michelotti and Duchassaing in large compound corals.[2] The ova are matured in the folds and processes, and then escape into the visceral cavity, and are expelled through the stomach and mouth. They have some power of active locomotion, and select favorable localities for their resting-place. The young polypes have faint traces of the future sclerenchyma, and grow rapidly when once fixed, provided they are well nourished.

As growth proceeds, the structure of the wall determines the shape of the corallum; and its *simple or compound character is regulated by the particular methods of the multiplication of the individual.* Some corals are always simple or solitary, others for a considerable period, and some for a very short time. The kind of *gemmation* or *budding* determines the massive, dendroid, encrusting, &c., nature of corals.

It appears to be very rare for buds to fall from the parent corallum and to form independent individuals.

By gemmation is meant the development of corallites from the tissues of a parent corallum. A very small patch of the membrane in immediate contact with the sclerenchyma of the parent appears to pucker, and septa are rapidly formed within the enlargement which occurs; tentacules have already appeared, and the small bud proceeds as if it

[1] Plate II, fig. 2. [2] Op. cit. [3] Op. cit.

were an independent organism as regards its growth, but its membranes are continuous with those of the parent. In many corals the base of the bud and the visceral cavity of the parent are at first continuous; but in others the membrane reflected over the septa, the margin of the wall, the external surface of the wall or of the base, produces the gemmation.

The gemmation may take place, then, on any part of a coral. It may occur within the calice, on the calicular margin, on any part of the wall between the calice and the base, and it may happen at the base. The direction of the line of growth of the bud has much to do with the future shape of the corallum, and the power of growth of the parent corallite after the development of the bud also.

The parent corallite may not grow after the production of a bud from its external wall; the bud becomes a perfect corallite, and gives origin to a bud in its turn. This repetition may go on, and a corallum results, formed by an ascending series of simple corallites; or the parent corallite may elongate after giving off a succession of whorls of buds which do not in their turn always develop others. The space between the whorls and the individual buds becomes filled up with exotheca and cœnenchyma. A dendroid corallum results, as in the genera *Madrepora*[1] and *Stylophora.*

Again, straight cylindrical corallites give off one or two buds, and all continue to grow, passing upwards, the calices keeping on one level, and the corallites being parallel. This determines the massive corals of many *Astræidæ.*

A corallum with geometrical calices whose walls are soldered together buds within the calices;[2] the parent calice and the bud grow, and the coral both expands laterally and increases in height. This produces a very common form of compound coral.

Certain corals never raise themselves far from the foreign substance they rest upon; the base gives off a bud, which, stolon-like, gives forth others, and all turn upwards slightly.

From these considerations it is evident that there is a necessary division of the gemmation into *calicular, basilar,* and *lateral.*

Calicular gemmation takes place from the interseptal loculi near the columellary space, and either midway between it and the wall, or just within the calicular margin. One or more buds may grow at once, and the budding may or may not be fatal to the parent. A pseudo-calicular gemmation is occasionally seen in simple corals which are only oviparous. It is produced by one of the young polypes settling on the parent accidentally, and growing to its detriment.[3]

The true calicular gemmation is well seen in the simple forms of the genus *Cyathophyllum,*[4] in a new genus from the Lias (*Lepidophyllum*), and in the genera *Stauria,*[5] *Isastræa,*[6] &c.

[1] Plate IV, fig. 18.
[2] Plate IV, fig. 11.
[3] Plate IV, figs. 8, 10.
[4] Plate IV, fig. 10.
[5] Plate III, fig. 15.
[6] Plate IV, fig. 11.

Gemmation from the wall—the lateral form—may occur at the top so as to affect the calicular margin, and at any place between this and the base. The gemmation may be solitary, alternate, whorled, numerous, or irregular; and the parent may or may not grow after the development of the buds.[1]

The genera *Cladocora, Solenastræa, Oculina, Lophohelia, Madrepora, Heliastræa, Stylocœnia, Stylina, Astrocœnia, Stephanocœnia,* &c., furnish examples of lateral and marginal gemmation.

The basilar gemmation is especially to be observed in the genera *Rhizangia, Astrangia, Phyllangia,* and other *Astrangiaceæ.*

Fissiparous growth.—Many corals increase in dimension and become cæspitose, gyrate, laminar, or massive, by a repetition of a fissiparous process in the calice or calices. The general nature of this method of calicular division and subsequent growth may be seen in Plate IV, figs. 12, 13. The calice is fairly bisected through the columella or columellary space by the growth of two or more opposite septa, and the wall appears to curve inwards, whilst the parts on either side grow independently and separate with varying rapidity. The process may be more or less speedily repeated in the new calices, and as they separate and grow upwards they may or may not be enveloped in cœnenchyma.

Very differently shaped corals thus result.

The genus *Dichocœnia* offers examples of massive corals where there is fissiparous growth and much cœnenchyma. The genus *Favia* has its fissiparous individuals in close contact, and the species of *Thecosmilia* yield long, dendroid, and cæspitose forms.

Serial growth.[2]—Corals of the genus *Diploria, Latimæandra, Rhipidogyra, Pectinia, Teleiophyllia, Thysanus, Manicina,* &c., have either faint traces of calices running laterally into each other, or else the septa follow each other in a longer or shorter series, which is sometimes straight, at others twisted. The occurrence of cœnenchyma, and the particular manner in which the "series" may be joined laterally, determine the shape of the corallum. In the *Latimæandræ* the faint traces of calices may be seen. In *Diploria* and *Mæandrina*[3] the septa are in series, and form a massive coral; whilst in the *Teleiophylliæ* and *Thysani*,[4] where there is a long series, the corallum is simple and pedicillate.

Gemmation occurs both in fissiparous and serial corallites.

V.—PHYSIOLOGY.

The ovules of corals are projected from the visceral cavity through the pyloric constriction, the stomach, and the mouth, by the contraction of the tissues of the disc; and the cilia of the cavities assist the transit. Cilia cover the small ovule and move it onwards

[1] Plate IV, fig. 16, 17. [2] Plate IV, fig. 14. [3] Plate IV, fig. 15.
[4] Plate IV, fig. 14.

with the assistance of the currents in the water; when it comes in contact with a hard substance, or rests, out of a current, on soft ground, the base adheres, and the minute tentacular disc is gradually developed, and finally expands. The young polypes are carried here and there; they exercise no volition, and only those which find a fit base upon which to rest live on to maturity. Either the young corallum adheres fixedly through life, or is so buried in mud or sand as to be immovable.

The locomotion of corals, therefore, is confined to the early period of their existence, is more or less passive, and the organs concerned in it are the cilia. The cilia vary in length, and their movement is vigorous; their activity is increased by light, warmth, and a highly aërated pure sea-water.

The adhesion to the foreign substance occurs by means of the outer membrane: if the base of the future corallum is to be small and pedunculate, the membranes at the base grasp some irregularity of the surface of the stone or shell, as the case may be, or envelop the body should it be small. As the hard parts are developed by the inner membranes, they pass around or envelope the substance, and fix the coral permanently. Occasionally, specimens are found with erosions at the base, as if they had suffered a violent rupture from the supporting substance and had continued to exist.

When a broad and flat base occurs, either the membranes and the subsequently developed sclerenchyma fill up the irregularities on the surface of the substance upon which the polype has rested, or are attached to it by a secretion of the epitheca. When corals rest on soft mud or sand, and become immersed, the tentacular disc appears just above the surface, and the body of the coral is very generally found covered by the epithecal membrane and its badly organised calcareous secretion. It is especially these corals that have large lateral growths, large costæ and processes; and they may be broad at the base, or quite the reverse.

The epitheca acts as an anchor and as a sheathing to the coral.

It has already been noticed, that the skeleton of the coral—its sclerenchyma—is developed and nourished by the inner membrane; and the retreat of this membrane, as well as the apparent death of all the hard parts below its level, have been explained. It will be found that the inner membrane permeates the hard tissues, that these are developed as granules in its intercellular spaces, and that, as the granules become hard, close, and solid, the nourishing influence of the membrane gradually ceases. In perforate corals the membrane is always in contact with the reticulate sclerenchyma, and the interiors of adjacent corallites are constantly in mutual relation.

Considering the weight of many individual corals, and the tenuity of the soft parts, this development of sclerenchyma is very wonderful. It must be remembered, that in many large compound corals only the few upper lines of the corallites are really nourished by the soft parts; all the rest has been gradually developed and left by them.

The density of the sclerenchyma differs more in species than in individuals, and size has nothing to do with it. As a rule, very quickly growing corals are less dense than others, and the tissues in contact with the membranes are the least resisting. The calcareous and other salts which form the sclerenchyma are derived from the matters assimilated by the coral during its digestive and respiratory processes; their deposition is a vital and not a mechanical process, and its amount is regulated by those conditions which affect the general nutrition of the individual.

The following analyses of recent corals are selected from those made by Silliman:[1]

	Porites.	Madrepora.	Pocillopora.	Mæandrina.	Astræa.	(Heliastræa?)
Carbonate of lime	95·84	94·807	94·583	93·559	96·471	91·782
Phosphates and Fluorides	2·05	0·745	1·050	0·910	0·802	2·100
Organic matter	2·11	4·448	4·397	5·536	2·727	6·118

The fluorides, phosphates, &c., yielded the following results (per cent. of their precipitate) in three examinations.

	1.	2.	3.
Silica	22·00	12·5	8·70
Lime	13·03	7·5	16·74
Magnesia	7·66	4·2	45·19
Fluoride of calcium	7·83	26·34	0·71
Fluoride of magnesium	12·48	26·62	2·34
Phosphate of magnesia	2·70	8·0	0·34
Alumina and Iron	16·00	14·84	25·97
Oxide of iron	18·30		

Silliman arrived at the following conclusions respecting the proportions of the phosphates, fluorides, and other salts:—"Fluorine is present in much larger proportion than phosphoric acid. The silica exists in the coral in its soluble modification, and probably is united to the lime. The free magnesia existed as carbonate, and was thrown down as caustic magnesia by the lime-water."

The dead and living tissues are liable to be perforated by parasitic borers; and the surface of the coral below the soft tissues is often covered with Bryozoa, Serpulæ, &c.

The inner membrane develops the buds, and it has an absorbing as well as a depositing power.

Food is obtained by living corals through the agency of the tentacules, the spiral threads, the cilia of the disc, and the lips. It consists of Animalcula, small Crustacea, the ova of Mollusca, and the spores of Algæ and smaller marine plants. Myriads of organisms may be seen in every small glass of water taken from the tropical seas, and the growth and nutrition of the coral-polypes can be readily accounted for.

[1] B. Silliman in Dana's 'Structure and Classification of Zoophytes,' Appendix, p. 124 *et seq.*

The *nematocysts* of the tentacules[1] and of the general surface are the destroying weapons; their missiles paralyse and slay, whilst the spiral threads envelope and kill as well. The spiral threads are observed in the corals with "serial" calices especially,[2] and the tentacules are not well developed in those species. The threads appear at the calicular margin, and have openings through which they pass to and fro from the visceral cavity. They are sometimes noticed in simple corals with well-developed tentacules.[3]

Anything destroyed by the nematocysts of the tentacules, or killed by the spiral threads, either falls on to the disc, or is passed on to the mouth directly and without the agency of the cilia. The cilia are especially useful in passing small bodies towards the lips; and these, when protruded, are moved in all directions seeking food.

Once within range of the lips, the food is grasped by their sphincter and passed into the stomach.

The movement of the tentacules and of the lips is produced by the contraction of the second or muscular tissue. All the tissues are very excitable, and contractions are readily produced by irritation; but the muscles act with a remarkable coordination, considering the absence of the organs of vision and of all nervous structures.

The stimulus of light acts very decidedly, so does that of heat, and direct contact produces that series of changes which has been described by M. J. Haime.

The stomach dissolves more or less of what goes into it, and passes the solution into the visceral cavity through the pylorus, whilst the fæces are returned and rejected. No acid reaction has been obtained from the stomachal membrane. Much water passes through the stomach and into the visceral cavity.

The visceral cavity receives the primarily assimilated food and the water which passes through the stomach; all this is brought in contact with the irrigatory system—with the tissues lining the interloculi covering the septa, &c., with the mesenteric folds and the tubular processes, as well as with the inferior surface of the disc and the bases of the tentacules. Finally, this watery medium kept in agitation by the cilia of the visceral membranes is now and then expelled through the tentacular orifices. A process of absorption goes on, and the results of secondary assimilation appear to be the deposit of the sclerenchyma and the nutrition of the soft tissues.

Doubtless, the external tissues with their nematocysts have a power of retaining and more or less absorbing nourishment without the process of digestion.

The respiration of corals appears to be carried on by the tentacules, the membrane lining the intermesenteric spaces—the irrigatory system, and by the general surface.

[1] Plate II, figs. 1, 3, 5, 6, 7, 8. [2] Plate II, fig. 17. [3] Plate II, fig. 11.

Well-aërated water of a certain temperature and containing minute organisms is absolutely necessary for the nutrition and respiration of corals; and mud and sediment held in suspension by brackish water, or by water very slightly saline, are very noxious.

Corals soon die when exposed to such adverse influences; and it is probable that the contractions which are noticed on some simple forms are due to periods when nourishment was scarce and the sea-water impure.

Corals are often phosphorescent; and this is very constantly observed when they have been removed from the sea and allowed to drain away on stones.

There are no special structures in the mesenteric folds which account for the process of absorption, and the method of the development of the male and female elements of generation in them is not satisfactorily determined. The tubular processes allow the ova to escape, and the ciliary motion of the visceral cavity tends to their ejection. The generation of corals is said to require a temperature of not less than 75°; but it must be remembered that very temperate seas have their corals, and that the coast of Norway and of Scotland abounds with them.

Without entering into the question of the geographical and bathymetrical distribution of corals, it may be safely determined that the perforate corals are the most rapid growers, and have the largest amount of soft tissues; they are usually found where the sea is the best aërated and full of organisms, just as some of the most solid of the aporose corals are to be found in calm water and at great depth.

It is the comprehension of the stomach, pylorus, mesenteric folds, and tubular processes within one cavity that distinguishes true *Madreporaria* from the *hydroid Acalephs.* The tabulate corals have been classified amongst these last, but upon insufficient data. Whenever the polype of a tabulate coral is proved to have its digestive and reproductive organs in separate cavities, then the views of Agassiz will be justified, but not till then; the tabulæ are not necessarily calicular bases, for they may often be separated from the continuous septa and columellæ.

VI.—Classification.

In examining a fossil coral, attention must be first of all paid to the structure of its wall and septa. It must be determined whether the first is aporose,[1] or, on the contrary, perforate,[2] and whether the septa are assignable to systems of cycles which follow the disposition of the rugosa or not. Should there be a tubulate structure of the wall and a rudimentary condition of the septa, it should be noted. Finally, the existence of horizontal tabulæ in the endotheca[3] must be ascertained.

[1] Plate I, figs. 1, 2, 3, 4, 14, 15.
[2] Plate III, figs. 3, 4; Plate IV, fig. 18.
[3] Plate III, figs. 9, 10, 11, 16.

There is a vast difference between the economy of a coral with imperforate and a coral with porose walls, and a method of diagnosis arises from it. The aporose and perforate sections are at once natural and easily distinguished.

The horizontal tabulæ may be found in perforate as well as in aporose corals, but the absence of vesicular endotheca and of the usual endothecal arrangements may be so marked that a section can be very fairly marked off. Nevertheless, the gradation of dissepiments into horizontal tabulæ[1] is witnessed in many *Rugosa*, and is not feebly marked even in some corals of the section *Aporosa*.

The tubulate wall and defective septa offer materials for a doubtful section, for they are very closely matched by some aporose forms.

The *Rugosa*[2] are so peculiar in their septal arrangement that, as a rule, they are distinguished at once; but their diagnosis will be carefully elaborated in a future page.

When the *section* of a coral has been determined, the existence or deficiency of endothecal structures becomes diagnostic. The existence of endotheca refers very definitely to the nutrition and growth of the species, and is readily discoverable.

The method of multiplication, the existence of fissiparous or serial calices, and the independence or the soldered condition of the corallites, must be then noticed.

The existence of pali and the nature of the septal arrangement must be made out, and the absence or presence of a columella determined. The nature of the columella, the shape of the calices, the size and ornamentation of the septa and costæ, must be examined, and the plain or incised condition of the septal margin decided. The existence of exotheca, cœnenchyma, peritheca, and epitheca is to be discovered, and the peculiarities of the structures noticed. The height and breadth, and the habit of the coral should be estimated. There are, then, many data for the foundation of a classification; and the following tables have been drawn up of that of the genera which are most likely to be found in the British Secondary and Tertiary rocks.[3]

[1] Plate IV, fig. 2. [2] Plate III, figs. 15, 18, 19, 20.

[3] The tables have been selected from the 'Hist. Nat. des Coral.,' Milne-Edwards and Jules Haime, and have been altered where requisite.

DIAGNOSTIC CLASSIFICATION.

ZOANTHARIA SCLERODERMA (MADREPORARIA), whose visceral chamber is	open throughout, or more or less divided transversely by endothecal dissepiments. The septal structures	well developed; the sclerenchyma	compact .	Section—*Aporosa*.
			perforate .	,, *Perforata*.
		rudimentary		,, *Tubulosa*.
	subdivided into stages by tabulæ. Septal structures	rudimentary, and of the hexameral type . .		,, *Tabulata*.
		well developed, and of the quaternary type .		,, *Rugosa*.

SECTION I. 1. *Madreporaria aporosa*, having the interseptal loculi	perfectly free and open. Septa	formed by two laminæ			Family—1. *Turbinolidæ*.
		formed by three laminæ			2. *Dasmidæ*.
	more or less subdivided by	dissepiments. The visceral cavity	becoming obliterated from below upwards . .		3. *Oculinidæ*.
			not becoming so, but being simply partitioned off by endotheca.	A distinct cœnenchyma surrounding the corallites . . .	4. *Stylophorinæ*
				No cœnenchyma .	5. *Astræidæ*.
		Synapticulæ			6. *Fungidæ*.

FAMILY 1. *Turbinolidæ*	with pali . . .	Sub-Family—1. *Caryophyllinæ*.
	without pali	2. *Turbinolinæ*.

Sub-Family. *Caryophyllinæ*, having	pali before one cycle. The wall	more or less elevated and naked. The columella	with its surface chicoracée; pali	broad			Genus—*Caryophyllia.*
				narrow and tall			„ *Bathycyathus.*
			with a papillary surfaｓe				„ *Brachycyathus.*
		horizontal and covered with epitheca. Columella	lamellar				„ *Discocyathus.*
			fascicular and papillary				„ *Cyclocyathus.*
	pali before more than one cycle. The columella	fascicular. The pali	wanting before the septa of the last cycle. The corallite	fixed or subpedicillate. The wall	naked. The base	narrow	„ *Trochocyathus.*
						broad	„ *Paracyathus.*
					with epitheca		„ *Thecocyathus.*
			existing before all the septa. The corallite free and discoid				„ *Leptocyathus.*

Sub-Family. *Turbinolinæ*, having the wall	naked or with a partial epitheca. The columella	styliform		*Turbinolia.*
		lamellar		*Sphenotrochus.*
		wanting		*Smilotrochus.*
	with a pellicular epitheca. Columella	never essential, wanting, or partly parietal		*Flabellum.*

Family. *Dasmidæ.*	Corallum simple, pedicillate, without epitheca; costæ large; septa with large grains	*Dasmia.*

Family. *Oculinidæ*, with unequal septa.	Pali before several cycles of septa. The columella	papillary	*Oculina.*
		styliform	*Synhelia.*
	No pali. Columella well developed and spongy		*Diplohelia.*

FAMILY. *Astræidæ*, with the septa	smooth or very slightly granular on the free margin	*Eusmilinæ* (Sub-family).	Corallites simple	(Division)	*Trochosmiliaceæ*
			Corallites aggregated	,,	*Stylinaceæ*.
	more or less dentate, serrate, and ragged	*Astræinæ* (Sub-family), with the corallites	simple	,,	*Lithophyllaceæ simplices*.
			or fissiparous, forming a cæspitose bush	,,	— *cæspitosæ*.
			or a linear series; more or less confluent	,,	— *mæandroides*.
			aggregated; multiplication by budding; corallum massive; calices rarely in series	,,	*Astræaceæ*.
			,, ; multiplication by fissiparity	,,	*Faviaceæ*.
			corallites free by their walls; multiplication by budding	,,	*Cladacoraceæ*.
			,, ; reproduction by basilar stolons	,,	*Astrangiaceæ*.

DIVISION. *Trochosmiliaceæ*, having the epitheca	rudimentary or wanting. Endothecal dissepiments	very abundant. The costæ	simple, and no columella	Genus—*Trochosmilia*.
		few in number. The columella	spongy	*Parasmilia*.
			none	*Cœlosmilia*.
	membraniform and complete. The columella		lamellar	*Peplosmilia*.
			styliform	*Axosmilia*.

Stylinaceæ, with the corallites united by their costæ. The columella	styliform.	Calicular margins free and circular	*Stylina*.
	rudimentary or none.	Calicular margins free and circular	*Cryptocœnia*.
		Calicular margins hidden by the septo-costal rays	*Convexastræa*.

Lithophyllaceæ simplices, with a strong and membranous epitheca; no columella . . . . *Montlivaltia*.

Lithophyllaceæ cæspitosæ, having their corallites	free laterally, or united in series whose walls are never completely soldered. Epitheca	rudimentary or none. Costæ	simply granular and distinct. The columella	rudimentary or none; mural "collerettes" exist . . .	*Calamophyllia.*
				well developed, "no collerettes" .	*Rhabdophyllia.*
		well developed and	confounded with the wall. The septa	numerous, with subspiniform teeth	*Thecosmilia.*
				few, with feebly developed teeth .	*Cladophyllia.*
			calicular gemmation very decided; corallites short . .		*Lepidophyllia.*
Lithophyllaceæ mæandroides, with the	corallites massive; the series soldered by their walls, which form simple elevations.	Series radiating			*Stelloria.*
Faviaceæ, having the septa non-confluent.	Corallites united by costæ and exotheca.				*Favia.*
Astræaceæ, having the calicular margins	free. Multiplication usually by extra-calicular gemmation. The costæ	rudimentary. The corallites united by exotheca			*Solenastræa.*
	united. Corallites increasing by superior and marginal and intra-calicular gemmation. Septa having their edges	oblique internally, and little or not at all confluent externally; teeth of septa	subequal, the calices	limited . .	*Isastræa.*
				forming series .	*Latimæandra.*
		more or less horizontal, and being confluent externally. Endotheca	well developed; walls compact . . .		*Plerastræa.*
			rudimentary; septo-costal rays confluent .		*Thamnastræa.*
		Columella styliform.	Projections on calicular interspaces . .		*Stylocœnia.*
			No projections		*Astrocœnia.*
	united; corallites free inferiorly; an epitheca				*Elyastræa.*
Cladocoraceæ, without pali					*Goniacora.*

Astrangiaceæ, with a complete epitheca . . . . *Cryptangia*.

SECTION. *Fungidæ*, having the wall or common plateau
- porose, and usually echinulate . . . . Sub-family—*Funginæ*.
- neither porose nor echinulate . . . . ,, *Lophoserinæ*.

Funginæ, with septa
- formed by laminæ nearly continuous; the wall distinct and granular . . . . . . Genus—*Micrabacia*.
- subporose; wall indistinct; corallites
  - simple . . . . . . ,, *Anabacia*.
  - compound . . . . . . ,, *Genabacia*.

Lophoserinæ, having the corallum compound; the corallites
- rather tall; "collines" independent of the common plateau; septa numerous; epitheca complete . . *Comoseris*.
- very short; calices not separated by collines, but nearly radiate; columella papillary . . *Protoseris*.

Section of *Madreporaria perforata*,
- wall well developed and simply porous; principal septa lamellar . . . . . . Family—*Madreporides*.
- whole corallum formed by reticulate sclerenchyma; septa as trabicular processes . . . . ,, *Poritides*.

Madreporides whose corallum has
- not an independent cœnenchyma . . . . Sub-Family—*Eupsamminæ*.
- an abundant cœnenchyma. The six primary septa
  - unequally developed, two larger than the others . . . . . ,, *Madreporinæ*.
  - equally developed . . . . ,, *Turbinarinæ*.

Group				Genus
Eupsamminæ, having the corallum	simple,	with base largely fixed		Genus—*Balanophyllia.*
		discoid, with a horizontal wall		*Stephanophyllia.*
	compound; epitheca rudimentary; columella	distinct and well developed generation	by budding	*Dendrophyllia.*
			by fissiparity	*Lobopsammia.*
		rudimentary, or none		*Stereopsammia.*

Madreporinæ = 1 genus. Cœnenchyma abundant; two of the septa opposite and united . . . . *Madrepora.*

Group		Genus
Turbinarinæ, with the corallum fixed, in form	foliaceous and encrusting	*Astræopora.*
	arborescent	*Dendracis.*
Poritidæ, without pali, the septa	not confluent	*Litharæa.*
	confluent	*Microsolena.*
with pali		*Porites.*

Amongst the Corals with tabulæ, the genus *Axopora*[1] may be distinguished by its reticulate sclerenchyma, its defective septa, and its large fasciculate columella.

Amongst the rugose Corals, the genus *Holocystis*[2] may be recognised by its lamellar endotheca, well developed septa and costæ, and its styliform columella.

---

[1] *Axoporæ* in the Brockenhurst and Bracklesham beds.

[2] *Holocystis* in the Lower Greensand.

VII. CORALS FROM THE TERTIARY[1] FORMATIONS.

I. *Corals from Brockenhurst and Roydon.*

The fossiliferous bed at Brockenhurst in Hampshire was discovered during the formation of a railway; it was diligently examined, and it has produced some most interesting mollusca and corals.

The molluscan fauna has much in common with those of the beds in Germany about Magdeburg, Bernburg, Aschersleben, Egeln, Helmstädt, and Latdorf,[2] and with those of the strata at Tongres, near Liége. Moreover, some of its most characteristic species are found in the Middle Headon beds at Colwell Bay and at Whitecliff Bay, in the Isle of Wight.[3]

The Brockenhurst bed lies immediately upon a freshwater formation,[4] the fossils of which are specifically identical with those of the freshwater beds of the Lower Headon; and it is covered by unfossiliferous sands.

The fossils from Roydon probably came from a well.

Corals are not found in the Middle Headon beds, but they abound at Brockenhurst; and it may therefore be admitted that the strata at the latter locality are the purely marine and oceanic representatives of the former.[5]

The specimens of fossil corals from Brockenhurst are tolerably perfect; they are generally covered with a red argillaceous sand; and they often contain selenite and sulphide

[1] It is necessary in using the terms "Tertiary," "Eocene," &c., to remember that there has been a constant and gradual development of "species" from the first appearance of life on the globe to the present day, and that the terms are only useful as parts of a scientific nomenclature. There is only an arbitrary distinction to be made between any of the successive formations and systems. Hence I have felt very disinclined to term the Brockenhurst beds Lower Oligocene, although they are clearly the equivalents of the German beds so called by Beyrich, and of the Tongrien Inférieur of Dumont.

[2] Beyrich, 'Ueber den Zusammenhang der Norddeutschen Tertiärbildungen, zur Erläuterung einer geologische Uebersichtskarte; Abhandl. der K. Akad. der Wissenschaften zu Berlin,' 1855.

Rœmer, in Dunker's 'Palæontographica,' 1862, 1864; Reuss, "Zur Fauna des Deutschen Oberoligocäns," 'K. Akad. der Wiss.,' Nov. 1864.

[3] Von Koenen, "Oligocene Deposits," 'Quart. Journ. Geol. Soc.,' Dec., 2nd 1863. (Mr. F. Edwards' researches formed the basis of this paper.) F. Finch, Dr. Sc., has assured me of the truth of this statement from the results of his personal observation.

[4] Von Koenen, op. cit.

[5] 'Mem. Geol. Survey Great Britain,' on the "Tertiary Fluvio-marine Formation of the Isle of Wight," by Edward Forbes, edited by R. Godwin-Austen, F.R.S., and others, 1856. 'Mem. Geol. Surv.,' "The Geol. of Isle of Wight; Explan. Sheet 10," by H. W. Bristow, F.R.S., 1862. These publications contain admirable and exhaustive descriptions of the Headon series.

of iron. Many have been rolled; and in all the original carbonate of lime of the sclerenchyma has been but slightly altered.

Family—*ASTRÆIDÆ.*

Sub-family—ASTRÆINÆ.

Tribe—ASTRÆACEÆ.

Genus—Solenastræa.[1]

The generic characters of the *Solenastrææ* are as follows:[2]

The corallum is usually massive, convex, cellular, and light: the corallites are long, and are united by a well-developed exotheca, and not by the costæ, which are never large enough to come in contact with those of neighbouring corallites. The costæ are always more or less rudimentary. The calicular margins are free and circular: the columella is spongy, and usually but feebly developed. The septa are very thin, and are formed by well-developed laminæ: their margin is dentate, and the lowest teeth are the largest. The endothecal dissepiments are simple, numerous, and close. The gemmation is extra-calicinal.

The species already recorded have been separated into those with rudimentary and those with distinct columellæ,[3] but they are all well and easily distinguished from the six forms about to be described.

1. Solenastræa cellulosa, *Duncan.* Pl. V, figs. 1—7.

The corallum is rather short, and appears to increase in breadth: its upper surface is irregular, and covers more space than the lower.

The corallites are inclined, distant, parallel, and are connected by a cellular exotheca which here and there forms a denser connecting tissue.

The calices are unequal in size and irregular in outline;[4] they project considerably above

[1] 'Compt. Rend. de l'Acad. des Sc.,' vol. xxvii, p. 494, 1848. Milne-Edwards and J. Haime.

[2] 'Hist. Nat. des Corall.,' vol. ii, p. 495, Milne-Edwards et J. Haime.

[3] Edwards and Haime, op. cit., page 497.

[4] Plate V, fig. 7.

the upper surface of the cœnenchyma which separates them;[1] their margins are sharp, the fossa is shallow, and the columella rudimentary.

The septa are thin and well marked; they form six systems, and there are four cycles. The primary, secondary, and tertiary septa are very much alike, and extend well towards the centre of the calice. The septa of the fourth and fifth orders are short, and do not extend far from the wall.[2]

The columella is rudimentary, and consists of a few processes derived from the inner margins of the septa. It is made to appear larger than it really is, by the frequent development of the endotheca near the inner septal margins. In some calices it appears as if one of the larger septa crossed over the columellary space, and became connected to the opposite one.

The endotheca is greatly developed; the dissepiments incline so much that transverse sections of corallites or worn calices show numerous transverse bars between the septa:[3] really these bars are but sections of oblique dissepiments.

The costæ, when covered by the exotheca, are rudimentary, and exist as faint unequal ridges which are slightly moniliform; but where they are above the cœnenchyma, and close to the calicular margin, they are feebly developed, but distinct, unequal, alternately large and small, and bluntly dentate.[4]

The exotheca is abundant, and consists of small square cells, rectangular cells, and of a tissue in which the cells form a dense cœnenchyma.[5] It passes from corallite to corallite, and is marked on its upper or free surface by faint ovoid and rather flat elevations, which are close in some places but distant in others.[6] The upper layer of the exotheca often grows up the sides of the corallites to the calices.

Height of corallum $\frac{6}{10}$th inch. Diameter of corallites $\frac{2}{10}$th inch (in the largest).

Locality.—Brockenhurst. In the Museum of Practical Geology, London.

2. Solenastræa Koeneni, *Duncan*. Pl. V, figs. 8, 9.

The corallum is short, gibbous, and irregular.

The corallites are rather but unequally distant from one another. The calices are hardly exsert, and are very shallow and open. There are no costæ visible, and a continuation of the cœnenchyma upwards reaches the calicular margin.[7]

The septa are in six systems and there are four cycles. The septa are thin, delicate, wide apart, unequal, and occasionally not quite straight.[8]

[1] Plate V, fig. 1. [2] Plate V, figs. 3, 4. [3] Plate V, fig. 3.
[4] Plate V, figs. 5, 6. [5] Plate V, figs. 2, 5. [6] Plate V, fig. 7.
[7] Plate V, fig. 8. [8] Plate V, fig. 9.

The columella hardly exists, and it is formed by a few offshoots from the inner margins of the septa.

The endotheca is scanty.

The wall is thin above, but thick low down in the corallites.

The exotheca is well developed; its cells are small, and its upper or free surface is but faintly marked.

Height of corallum $\frac{1}{8}$ inch. Diameter of corallites $\frac{2}{10}$th inch.

Locality.—Brockenhurst. In the Museum of Practical Geology, London.

3. Solenastræa Reussi, *Duncan*. Plate V, figs. 10—16.

The corallum is tall, with an irregular upper surface. The corallites are subturbinate, with wide calices and narrow bases; they are irregular in their distances, but are connected more by bands or layers of dense exotheca than by a cellular cœnenchyma, but both structures exist.[1] The calices are very slightly exsert, and irregular in shape and distance. The fossa is shallow, and the margin is thin. The columella is very rudimentary. The septa are very distinct, unequal, not always straight, thin; and the highest orders are rudimentary, but exist as small projections. There are six systems and five cycles.[2] The laminæ are marked with granules in a series of slanting rows.[3]

The endotheca is very scanty and highly inclined.[4]

The wall is not very stout.

The costæ where uncovered by exotheca are distant, very slightly prominent, straight, unequal, and very bluntly dentate.

The exotheca forms layers which curve around the corallites, and connect them together at certain heights only, the intermediate parts being uncovered by exotheca; the uppermost layer is more or less granular, and reaches to the calicular margin.[5] The layers are formed by elongated and very thick cells, and they rarely are square and thin. The gemmation is extra-calicular, but several buds spring from the same wall, very close to each other.[6]

Height of individual corallites $\frac{6}{10}$th inch.

Diameter of the calices $\frac{3}{10}$th inch.

Locality.—Brockenhurst. In the collection of Frederick Edwards, Esq., F.G.S.

[1] Plate V, figs. 10, 11, 14.
[2] Plate V, fig. 16.
[3] Plate V, fig. 15.
[4] Plate V, fig. 15.
[5] Plate V, fig. 12.
[6] Plate V, fig. 10.

4. SOLENASTRÆA GEMMANS, *Duncan*. Plate VI, figs. 1—7.

The corallum is tall, its base is small, and the calicular surface is very irregular.

The corallites are very unequal, they are sometimes crowded and for the most part are separated by cœnenchyma; they are not very exsert, as a rule, but many pass up above the level of the common cœnenchyma and exhibit their wall marked with small costæ.

The exotheca is dense, and resembles layers of membranous epitheca more than a cellular exotheca. It is found here and there only, so that much of the wall of many corallites is free. The exotheca spreads across from corallite to corallite in wavy horizontal layers, and the costæ are hidden by it. But where the exotheca is wanting the costæ vary greatly in their size and development.[1]

The calices are irregular in shape, size, and distance; the fossa is shallow, and the columella is rudimentary. The calicular margin is rather blunt. The septa are long, delicate, very ragged on their sides, from their connection with the endotheca, and but slightly granular.[2] There are four cycles and six systems; the primary and secondary septa extend well inwards, and their ends, which are occasionally enlarged, are connected by ragged and irregular processes; the tertiary are smaller; and the septa of the fourth and fifth orders are almost rudimentary. Sections of corallites show the wall to be moderately thick.

The costæ are unequal, and are either plain, short and rounded, short and moniliform, short and bluntly dentate, or even almost vesicular. They are rudimentary when covered by the exotheca.

The endotheca is very abundant and highly inclined.[3] The gemmation is peculiar, and causes the species to resemble in its growth some of the *Cladacoraceæ:* the bud separates widely from the parent, and then passes upwards and soon gives forth a bud which takes the same course.

Height of corallum several inches. Diameter of corallites $\frac{2}{10}$th inch.

Locality.—Brockenhurst. In the collection of Frederick Edwards, Esq., F.G.S.

5. SOLENASTRÆA BEYRICHI, *Duncan*. Plate VI, figs. 8—13.

The corallum is massive, short, and has a very irregular calicular surface. The corallites are short, and widen out rapidly from a comparatively small base.

The calices are large, very irregular in shape, generally close, and they are separated by the cœnenchymal exotheca; the fossa is shallow, the columella is rudimentary, the wall at the margin is stout, and the septa are thin, often wavy, and rugged laterally.

[1] Plate VI, figs. 2, 3, 4. [2] Plate VI, fig. 7. [3] Plate VI, fig. 7.

There are six systems of septa, and four complete cycles; moreover, in the largest corallites there are many rudimentary septa of the fifth cycle. The septa are unequal in the sectional view; often larger (the primary) at the inner end than midway; and may extend across the columellary space. A little below the calice the wall is very thick, and the endotheca is most abundant and very inclined. The costæ exist above the level of the common cœnenchyma; they are alternately large and small, but always short, ill developed, and faintly dentate.

The exotheca is greatly developed; its cells are irregular in shape, not elongate, but more or less square in outline;[1] it covers up the corallites, leaving them free to a small extent only. The upper surface of this exothecal cœnenchyma is faintly granular.

Height of corallum 1 inch. Great diameter of calices $\frac{3}{10}$th to $\frac{4}{10}$th inch.

Locality.—Brockenhurst. In the collection of Frederick Edwards, Esq., F.G.S.

6. Solenastræa granulata, *Duncan*. Plate VI, figs. 14—18.

The corallum is short, and its upper surface presents much granular cœnenchyma[2] between the calicular ends of the corallites. The corallites are small and distant; and in well-preserved specimens are seen to project somewhat above the common exothecal cœnenchyma, but in worn fossils they are but slightly elevated, and present a very thick wall. The calice is circular in outline, its fossa is shallow, its margin is thin; and the columella is rudimentary.

There are six systems of septa and four cycles of them; they are unequal, the primary being much the largest,[3] and all except those of the fourth and fifth orders have a paliform elevation near the columellary space.

The septa are rugged laterally, from their connection with much endotheca, which is highly inclined.

The costæ are seen above the surface of the cœnenchyma as short ridges alternately large and small, and they appear to emerge into the large granules on the free surface of the cœnenchyma; where the corallites are not covered by exotheca below the free surface, the costæ are also visible.

The exotheca is cellular and banded.[4] The occurrence of the bands admits of much corallite wall being costulated.

The free surface of the exotheca is dense and covered with large granules.

Height of corallum $\frac{1}{2}$ inch. Diameter of calices $\frac{2}{10}$th inch.

Locality.—Brockenhurst and Roydon. In the collection of Frederick Edwards, Esq., F.G.S., and in the Museum of Practical Geology, London.

All these species present the most important generic characteristic of the *Solenastrææ*, and they are all very closely allied. The principal specific distinctions are in the amount

[1] Plate VI, fig. 9. [2] Plate VI, fig. 14. [3] Plate VI, fig. 17.
[4] Plate VI, figs. 15, 16.

and structure of the exotheca, in the method of gemmation, and in the septal development. These distinctions render the division of the Brockenhurst *Solenastrææ* into six species absolutely necessary. This increase in the number of the species proves that the genus must have been a large one; and the resemblance of the specific forms to varieties (from the really slight structural distinctions) is what is generally noticed in the case of large genera.

These new species belong to a division of the genus which is not represented elsewhere; it is characterised by the high septal number, the deficient columella, and the amount of inclined endotheca.

The recent *Solenastrææ* are found in the Red Sea, the Caribbean Sea, and in the Indian and Pacific Oceans. The horizontal endotheca and low septal number distinguish all these species from those of Brockenhurst.

The fossil species of the genus are *Solenastræa Verhelsti*, Ed. and H., *Solenastræa Turonensis*, Michelin sp., and *Solenastræa composita*, Reuss sp.

Solenastræa Verhelsti is an Eocene form from Ghent; and its rudimentary third cycle of septa, very close corallites, and its paucity of slightly oblique and subconvex endotheca, distinguish it at once and very decidedly.

Solenastræa Turonensis has very long and close corallites with three cycles of septa and a well-developed columella; its very scanty and very feebly inclined endotheca, and the wide-apart exothecal dissepiments, separate it from the form from Ghent, as well as from those from Brockenhurst. MM. Milne-Edwards and J. Haime determine that this species and *Solenastræa composita* are identical. The Touraine form is of course from the Upper Miocene.

The following is a scheme of the classification of the Tertiary *Solenastrææ*:

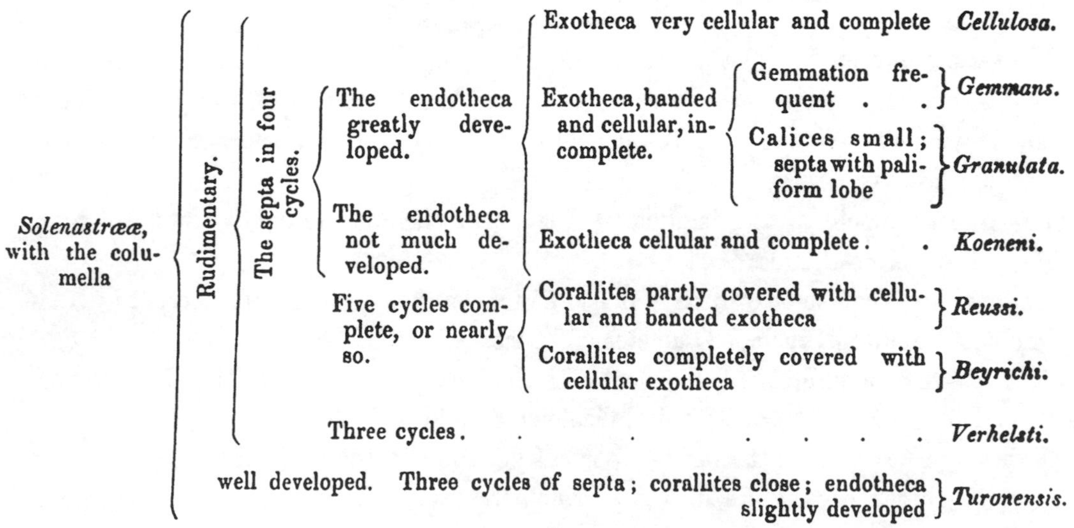

SECTION—*MADREPORARIA PERFORATA.*

FAMILY—*Madreporidæ.*

SUB-FAMILY—EUPSAMMINÆ.

Genus—BALANOPHYLLIA.

BALANOPHYLLIA GRANULATA, *Duncan.* Plate VII, figs. 1—5.

The corallum is short, has a very large and encrusting base, and is constricted immediately below the calice. There is no epitheca, and the costæ are large and very distinct.[1]

The calice is oval in outline, is compressed, and is marked by very small and equal costæ externally; it has a small columella and very numerous septa.

The septa are delicate, wavy, and granular; there are six systems of them and five complete cycles, with half of a sixth.

Very large, equal, rather wavy, flat, and rounded costæ are seen at the edge of the base; they bifurcate inferiorly[2] here and there, and are profusely granular, as well as connected by many cross bars.

As the costæ approach the constriction they diminish in size, become thinner, more numerous, and less granular, until, close to the calicular margin, they are almost linear. All are connected by the cross bars. The granules often are large enough to stand up well in relief.

Height of corallum $\frac{1}{2}$ inch. Diameter of base $1\frac{1}{3}$rd inch. Diameter (greatest) of calice $\frac{6}{10}$th inch.

Locality.—Brockenhurst. In the collection of Frederick Edwards, Esq., F.G.S.

The genus *Balanophyllia* (Wood) has received much attention since MM. Edwards and Haime's 'Monograph on the British Fossil Corals' was written. These authors have described in the 'Histoire Naturelle des Coralliaires' (vol. 3) some new species.

Since that work was completed Reuss has described three species from the "lower marine sand" of Weinheim: and F. Roemer and Philippi have each discovered a new species in the fossiliferous beds of Latdorf. Moreover, the South Australian Tertiary beds contain species.

The *Balanophyllia calyculus*, Wood; *B. verrucaria*, Pallas, sp.; *B. cylindrica*, Michelotti, sp.; *B. Italica*, Michelin, sp.; *B. tenuistriata*, Ed. and H.; *B. desmophyllum*, Lons-

[1] Plate VII, figs. 1, 2.

[2] Plate VII, figs. 2, 3.

dale sp.; *B. Bairdiana*, Ed. and H.; *B. geniculata*, D'Archiac; *B. Cumingii*, Ed. and H., and *B. subcylindrica*, Philippi sp., may be arranged together to form a subgenus characterised by forms with broad adherent bases. The following species will fall into another subgenus whose forms have the base more or less pedicillate:—*Balanophyllia prælonga*, Michelotti, sp.; *B. Gravesii*, Michelin, sp.; *B. sinuata*, Reuss; *B. inæquidens*, Reuss; *B. fascicularis*, Reuss, and *B. Australiensis*, Duncan.

The new species from Brockenhurst, *B. granulata*, must be received into the first subgenus. The absence of epitheca, the profusely granular costæ, and the existence of part of the sixth cycle of septa, distinguish *B. granulata* from all the species already described.

There is nothing in the species *B. granulata* to connect it with any geological horizon; for the *Balanophylliæ* without epitheca range from the Eocene to the present day. The species *B. granulata* has only a generic alliance with those described by Reuss, Roemer, and Philippi.

Genus—Lobopsammia.

Lobopsammia cariosa, *Goldfuss*, sp. Plate VII, figs. 6—10.

The corallum has a wide base, above which it is slightly constricted. It rises in the form of a short cylindrical trunk, terminated by several gibbous processes, which support calices and project outwards.

The under surface of the base has a concavity[1] which is lined and surrounded for a short distance by a dense epitheca; the costæ radiate around the margin of the epitheca, and ascend the outside surface of the corallum, pursuing very irregular and wavy courses, being thin, rounded, equal, and joined laterally by numerous cross bars of exotheca.

The costæ, which are very faintly granular, have this same peculiarity[2] on the upper surface of the corallum between the gibbous calices.

The calices are irregular in shape, and so speedily commence to elongate prior to dividing fissiparously, that simple ones are rarely seen. They are, nevertheless, in the figure of eight, and are situated on the ends of the gibbous projections; their margins are irregular, the fossa is shallow, and the columella is very feebly developed.

The septa are very numerous, and form at least five cycles in six systems; they are unequal, stout, and often bifurcate near the columella.

Height of corallum about one inch; diameter of trunk $\frac{6}{10}$ths inch; greatest diameter of calices $\frac{4}{10}$ths inch.

Locality. Brockenhurst, Acy, Auvert, and Valmondois.

[1] Plate VII, fig. 10. [2] Plate VII, fig. 7.

In the Museum of Practical Geology, London, and in the collection of Frederick Edwards, Esq., F.G.S.

Lobopsammia cariosa is a common fossil at Brockenhurst, and the specimens differ in the stoutness of the corallum and distinctness of the costæ. There is a so-called species, *L. dilatata*, Roemer,[1] from Latdorf;[2] but it is not worthy of more than the title of a variety of our widely diffused form. The same may be determined with respect to *L. Parisiensis*, Michelin, sp.

Section—*MADREPORARIA PERFORATA.*

Family—PORITIDES.

Sub-family—PORITINÆ.

Genus—Litharæa.

Litharæa Brockenhursti, *Duncan.* Plate VII, figs 17, 18.

The corallum is massive, irregular in shape, and has an uneven upper surface. The corallites are close, and are very rarely separated by much reticulate cellular structure; they are rather short, and vary in their diameter in different parts of the corallum. The walls are well marked.

The calices are shallow, close, and generally quadrangular. The margins are formed by trabecular tissue, and the septa are irregular, unequal, wavy, and are often enlarged at the inner end; their laminæ are much perforated; they are in six systems, and there are three cycles, the primary being the largest; the others are often very small. The laminæ are faintly dentate laterally.

The columella is slightly developed, and appears to be formed by processes from the septal ends. Diameter of the calices $\frac{4}{30}$ths inch.

Locality. Brockenhurst. In the collection of Frederick Edwards, Esq., F.G.S.

The scanty cœnenchyma, the shallow and quadrangular calices, the three cycles of unusually perforate septa, the ill-developed columella, and the shape of the corallum, distinguish this species from *Litharæa Websteri* and the *Litharææ* of the French Tertiaries.

The genus ranges from the Maestricht Chalk to the Faluns at Dax.

[1] Roemer in Dunker's 'Palæontographica,' 1862—1864.

[2] In the Lower Oligocene.

Section—*MADREPORARIA TABULATA.*

Family—MILLEPORIDÆ.

Genus—Axopora.

Axopora Michelini, *Duncan.* Plate VII, Figs. 11—15.

The corallum is large, very irregular in shape, and marked by inequalities of the surface. The cœnenchyma is abundant, very finely reticulate, and is dotted by numerous and very small calices, which are not very deep, and often irregular in shape; they are not separated by ridges. The columella is formed by longitudinal fibres, and projects but slightly at the bottom of the calice; it is slender, very long, and often wavy.

There are no septa.

The tabulæ are horizontal, not numerous, very small, and do not go through the columella, and divide the corallite off perfectly.

A variety of this species is in the form of a flat cake, and its corallites are very long and thin.[1]

Locality. Brockenhurst. In the collection of Frederick Edwards, Esq., F.G.S.

Axopora is a very remarkable genus, for its corallites have no septa, but a great columella and tabulæ. The tabulæ do not pass through the fasciculate columella, and yet they cut off all the space below them from that nearer the calice.

The species are not numerous; they were probably rapid growers, and the structures entering into their composition are so simple that it is very difficult to determine specific distinctions.

The *Holoræa Parisiensis*, which is synonymous with *Alveolites Parisiensis*, Michelin, and which was described by MM. Milne-Edwards and Jules Haime, in the first part of their Monograph, has been determined by them to be an *Axopora*. The *Axopora Michelini* is a very large and fine form, and is closely allied to *Axopora Solanderi*, Defrance, sp., and less so to *A. Fisheri*, Dunc., but it differs very decidedly from *A. Parisiensis.*

[1] Plate VII, figs. 13, 15.

Section—*MADREPORARIA PERFORATA.*

Family—MADREPORIDÆ.

Genus—Madrepora.

1. Madrepora Solanderi, *Defrance.* Plate VIII, figs. 12—14.

The corallum is arborescent; the branches are subcylindrical.

The calices are sunken in the very porous cœnenchyma, and they are large and wide apart.

This is the description given by MM. Milne-Edwards and J. Haime,[1] and the following is from Michelin:[2]

M. ramosa, porosa; ramis subcylindricis, sæpe compressis, raro coalescentibus, granulosis; stellis universis, rotundis; lamellis 12 fragilissimis, 6 maximis, aliis parvulis.

The Brockenhurst specimen shows the granulated cœnenchyma and the septa; but it proves that the calices, like all others of the genus, were more or less prominent before being worn.

Localities. Brockenhurst. Mary près Meaux (Seine et Marne), Auvert, Graux, and Valmondois. In the collection of Frederick Edwards, Esq., F.G.S.

2. Madrepora Roemeri, *Duncan.* Plate VIII, figs. 8—11.

The corallum is partly foliaceous and partly ramose, but the branches coalesce.

The calices are very distant and, in unworn portions of the corallum, are on the top of conical and very costulated projections. The calicular margin and the conical base produce a "tubuliform calice."

The costæ are projecting, wavy, rounded, and are lost in a very granular and almost echinulate cœnenchyma.

The septa are stout, and twelve in number.

Locality. Brockenhurst. In the Museum of Practical Geology, London.

3. Madrepora anglica, *Duncan.* Pl. VIII, figs. 1—7.

The corallum is in the shape of a stout trunk, with numerous aborted branches which give it a very gibbous appearance.

[1] Op. cit., vol. iii, p. 162. [2] 'Icon. Zooph.,' p. 165.

The calices are either scattered irregularly over the papillate cœnenchyma or are aggregated in sets; a parent corallite being surrounded by its buds. The calices are small but slightly projecting, tubuliform and finely costulated, the costæ being lost in the irregular, porose, and papillate common tissue. Some are not costulated, but are sunken in the cœnenchyma, and all are circular in outline with thickish walls.

The septa are as is usual in the genus; and the opposite primary septa frequently join by their inner ends. There are six large and six small septa.

The cœnenchyma is highly cellular, and its free surface is almost aciculate with sharp papillæ. *Locality.* Brockenhurst. In the Museum of Practical Geology, London.

These species of the genus *Madrepora* are all new to the British coral-fauna. *M. Solanderi* is an indifferent species, for there may have been any amount of ornamentation on the cœnenchyma, and the calices may have been very prominent and costulate, but nearly every detail has been worn off the specimens. Many well-characterised species, were they worn and rolled, would present the appearance of the typical specimen of *M. Solanderi.*

Madrepora Roemeri is well characterised by its form, its distant tubuliform calices with costulated external surfaces, and by its very granular and echinulate cœnenchyma. The species most closely allied to *M. Roemeri* is *M. granulosa,* Edwards and Haime, a recent form from the Ile de Bourbon.

The *Madrepora Anglica* is a well-marked species, and is allied to *M. crassa,* Edwards and Haime, a recent form whose locality is unknown.

The genus *Madrepora* comprehends at least ninety-two species, of which only eight are fossil. The Paris Basin and the Turin Miocene have hitherto been the localities whence the fossil species have been collected; and now the Brockenhurst beds must be admitted amongst the strata whose remains indicate the former existence of coral-reefs exposed to a furious surf and the wash of a great ocean.

The Brockenhurst *Madreporæ* do not resemble, except generically, the species from Turin.

The recent species are found all over the Pacific, the Indian Ocean, the Caribbean Sea, and one species has retained its position in the White Sea, near Archangel (*M. borealis,* Edwards and Haime).

As yet the very fossiliferous Tertiary strata of the islands of the West Indies have not yielded any fossil *Madrepora.*

REMARKS ON THE CORAL-FAUNA OF BROCKENHURST.

The coral-fauna of Brockenhurst and Roydon consists of thirteen species:—*Solenastræa cellulosa, S. Koeneni, S. Reussi, S. gemmans, S. Beyrichi, S. granulata, Balanophyllia granulata, Lobopsammia cariosa, Axopora Michelini, Lithuræa Brockenhursti, Madrepora Anglica, M. Roemeri, M. Solanderi.*

Two of the species, viz., *Lobopsammia cariosa*, Goldf., sp., and *Madrepora Solanderi*, Defrance, sp., are found in the Eocene beds of the Paris Basin; they have not, however, been noticed either in the London Clay or in the Bracklesham and Barton beds in England.

The *Madrepora Solanderi* is a species of very doubtful value, and the reasons for this assertion have been already given.

The *Lobopsammia cariosa* is found under the name of *L. dilatata*, Roemer, at Latdorf.

The *Litharæa* and the *Axopora* from Brockenhurst have no very close specific alliance with the forms of the genus found in the London Clay and the Bracklesham beds.

The Nummulitic coral-fauna[1] of Italy, Sinde, &c., has no species in common with that of Brockenhurst; and the researches of Reuss and Roemer in the coral-faunæ of the Tertiary series termed Lower, Middle, and Upper Oligocene, have not produced any results which enable me to correlate any one of those series with the coraliferous beds at Brockenhurst.

The Miocene coral-fauna has no specific relationship with that under consideration.

It becomes evident from these considerations that the new coral-fauna has very slight resemblances and affinities with those already described.

The Brockenhurst corals are, therefore, very remarkable; the absence of simple forms and the presence of species of *Madrepora*, *Axopora*,[2] and *Solenastræa* indicate the former existence of a vigorous polype-growth, and of all the physical conditions now observed near and about coral-reefs. The great size of the trunk of *Madrepora Anglica* is especially significant. It may be still true that this coral-fauna was a local one, for at the present day the distinction between reef-, barrier-, and simple coast-corals is sufficiently determinable.

The coral-fauna of the so-called Lower Oligocene beds of Germany is associated with the mollusca which characterise the Brockenhurst beds and their equivalents in the Headon series of the Isle of Wight.[3] It is distinct from the coral-fauna of Brockenhurst, although the correlation of the strata can be established from the study of the Mollusca; hence the probabilities of the Latdorf coral-fauna being that of a coast-line, and of the Brockenhurst being that of an oceanic and reef area, are great.

The coral-fauna of Brockenhurst is more recent than that of Barton and evidently flourished under very different physical conditions. It is older than the Falunian and Crag-faunæ.

[1] The coral-fauna of the London Clay, and of the Bracklesham and Barton beds, and of the Paris Basin, is contained to a certain extent in the great Nummulitic coral-fauna of Southern Europe and India; but there were clearly two coral-provinces during the early Tertiary period, just as there are at the present day—the West Indian and the Pacific.

[2] *Axopora* is represented in existing reefs by many tabulate corals.

[3] Von Koenen, "Die Fauna der Unter-Oligocänen Tertiär-Schichten von Helmstädt," 'Zeitschrift der Deut. geol. Gesell.,' Band xvii, 1865.

VIII. Corals from the Eocene of the Isle of Wight and from the London Clay.

Order—*ZOANTHARIA.*

Family—TURBINOLIDÆ.

Tribe—Turbinolinæ.

Genus—Turbinolia.

1. Turbinolia affinis, *Duncan.* Plate IX, figs. 1, 2, 3.

The corallum is slightly truncated inferiorly, and it is conical low down, but cylindro-conical above; it is symmetrical and very small.

The costæ are well developed and obtuse; the largest are swollen out inferiorly, and all are moderately prominent; not very thick, but very distinct.

The intercostal spaces are wide on account of the costæ being separated by a portion of the wall, which is very visible at the bottom of the spaces.

There are no dimpled markings on this portion of the wall.

There are very decided markings on the sides of the costæ produced by rudimentary exotheca.

The wall is thin.

The calice is circular in outline.

The septa are thin, delicate, unequal, rather ragged, granular, and slightly enlarged near the columella. There are three perfect cycles of septa and six systems.

The numbers of the septa and costæ are the same.

The columella is not very projecting above the base of the calicular fossa, and is rather elongated and ovoid.

The height of the corallum is $\frac{2}{10}$ths inch, and the diameter of the calice nearly $\frac{1}{10}$th inch.

This species is more closely allied to the rare *Turbinolia firma*, Edwards and Haime, than to any of the other members of the genus. The broad intercostal spaces and the markings on the sides of the costæ in the new species distinguish it from *Turbinolia firma*.

Locality. High Cliff, Isle of Wight. In the collection of Frederick Edwards, Esq., F.G.S.

2. Turbinolia exarata, *Duncan*. Plate IX, figs. 4, 5, 6, 7.

The corallum is conical inferiorly and cylindrical superiorly, so as to be rather subturbinate. Its base is small and narrow, although the costæ are very projecting there.

The costæ are greatly developed; they are subequal, very prominent, and thin; their free margin is rather sharp, and not much narrower than their base.

The largest costæ are very prominent inferiorly, and the tertiary arise at the distance of about one quarter of the whole height of the corallum from the base.

The costæ are very wide apart, and the base or bottom of the intercostal spaces is wide, very visible, and it is not marked by any dimpling.

The sides of the costæ are strongly marked with a rudimentary exotheca, which is attached to the wall close to the base of the costæ (fig. 7).

The wall is very thin.

The calice is circular in outline, very deep, and its margin is rendered very distinct by the well-developed costæ.

The septa are slender, thin, and unequal; they form three perfect cycles, and there are six systems.

The septa and costæ correspond.

The columella is very small, cylindrical, pointed, and in the typical specimen there are two papillæ on its free surface.

Height $\frac{4}{10}$ths inch. Diameter of the calice $\frac{2}{10}$ths inch.

This very interesting species resembles the *Turbinolia Prestwichi*, Edwards and Haime, in some points; but it has no vestige of a fourth cycle of costæ; moreover, the new species has not the truncated base of *Turbinolia Prestwichi*, and its third cycle of costæ arise high up.

The width of the intercostal furrows and the absence of well-marked dimpling are very distinctive peculiarities of *Turbinolia exarata*.

Locality.—The species is found at Brook, Hampshire (New Forest). In the collection of Frederick Edwards, Esq., F.G.S.

3. Turbinolia Forbesi, *Duncan*. Plate IX, figs. 8, 9, 10, 11.

The corallum is very small, conico-cylindrical, and has rather a sharp base.

The costæ are very stout, obtuse, and slightly prominent; the largest are often wavy in their upward course, and all are separated by wide intercostal furrows or spaces. There is a well-marked but very small costa situated high up in the corallum and in each intercostal space.

There are large and distinct exothecal markings on the sides of the costæ; but the existence of dimples on the wall at the bottom of the intercostal spaces is too doubtful to be safely asserted.

The calice is unsymmetrical, from its peculiar septal arrangement; its marginal wall is very thin, and the fossa is deep. The columella is angular in its transverse outline, and is often very prominent.

The septa are unequal, straight, and delicate. There are no septa corresponding with the rudimentary costæ; their arrangement gives the idea of there being two systems of three cycles, the septa of the third cycle being deficient; but there are really six systems.

In four systems there are three cycles of septa, and the rudimentary costæ are of the fourth and fifth orders; and in the remaining systems there are two cycles of septa with the rudimentary costæ of the third order.

Height of corallum $\frac{3}{20}$ths inch. Diameter of the calice $\frac{1}{20}$th inch.

The cyclical arrangement and the rudimentary costæ distinguish this species from all the others.

Locality. High Cliff, Isle of Wight. In the collection of Frederick Edwards, Esq., F.G.S.

The genus *Turbinolia*, thus enriched by the discovery of three new species, was so elaborately described by MM. Milne-Edwards and J. Haime, that it only remains to place these species in their proper position in the genus.

The following scheme will point out their correct affinities:

TURBINOLIÆ.

TURBINOLIÆ with four cycles of septa; the fourth more or less incomplete.	*Turbinolia costata.*	1.
,, ,, ,, ,, ,, ,,	— *dispar.*	2.
,, three cycles of septa	— **exarata.*[1]	3.
	— **Dixoni.*	4.
	— **firma.*	5.
	— **affinis.*	6.
	— *Pharetra.*	7.
	— *laminifera.*	8.
	— *Nystana.*	9.
	— *attenuata.*	10.
	— *pygmæa.*	11.
,, three cycles of septa, with costæ of a fourth cycle .	— **sulcata.*	12.
	— **Bowerbanki.*	13.
	— **Fredericiana.*	14.
	— **Prestwichi.*	15.
,, three incomplete cycles of septa . . .	— **minor.*	16.
	— **Forbesi.*	17.
	— **humilis.*	18.

---

[1] The species marked with an asterisk are British.

Turbinolia attenuata, Keferst.
— *laminifera*, Keferst.
— *pygmæa*, Roemer.
These species require further examination; they were discovered in the "Unter-Oligocän" of Germany, are very minute forms, and are probably the young of other species.

FAMILY—*CARYOPHYLLIACEÆ.*

Tribe—TROCHOCYATHACEÆ.

Genus—TROCHOCYATHUS.

1. TROCHOCYATHUS AUSTENI, *Duncan.* Pl. IX, figs. 15—17.

The corallum is rather tall, slightly curved and compressed; it is rounded at the base, and its sides are marked with slightly prominent but not spined or crested costæ.

The calice is elliptical, much compressed, and slightly angular at its extremities; its long axis is on a lower plane than the short axis, and its margins are raised into several angular processes, on account of the primary and secondary septa being less exsert than the tertiary.

The fossa is moderately deep; and the columella is long, and not very visible. The septa are thin, rather close, and very subspinose laterally.

The septa are in six systems, and there are four perfect cycles. The septa are unequal, and are not very exsert: the primary and secondary septa are on a lower level than the others, and correspond to the largest and most prominent costæ.

There are small pali before the primary, secondary, and tertiary septa.

The costæ are distinct from the base, and granular; the primary and secondary are the largest, and all are broader than the septa. Height of corallum, $\frac{9}{10}$ths inch. Great diameter of calice, $\frac{5}{10}$ths inch. Small diameter of calice, $\frac{3}{10}$ths inch.

This species belongs to the striated *Trochocyathi*;[1] and its tall and curved form, with its four cycles of septa, bring it in close relation with *Trochocyathus elongatus*, Edwards and Haime.[2] The angular calicular margin is wanting in this last species, whose corallum is moreover slightly twisted.

It is very evident that the new species is the representative of *Trochocyathus elongatus. Trochocyathus elongatus* is found at Quartier-du-Vit, near Castellane (Basses Alpes), in an Eocene formation, and *Trochocyathus Austeni* was discovered at Bracklesham.

In the collection of Frederick Edwards, Esq., F.G.S.

2. TROCHOCYATHUS INSIGNIS, *Duncan.* Pl. X, figs. 1—4.

The corallum is tall, compressed, slightly curved inferiorly, and it has a large calice and a sharp base.

The calice is ovoid, and its axes are on the same plane.

[1] 'Hist. Nat. des Corall.,' vol. ii, p. 27. [2] 'Ann. des Sc. Nat.,' 3rd ser., vol. ix, p. 305, 1848.

The septa are small, thin, wavy, unequal, and have very long and sharp lateral spines. The septa are in six systems, but the four cycles are incomplete. The four cycles are complete in two systems, but are incomplete in one of the halves of each of the other systems. There are therefore eight septa in two systems and six in the rest.

The columella is small and situated deeply.

The pali are small, and are situated before all the septa, except those of the last cycle.

The costæ are subequal, broad, very slightly rounded, and barely prominent; they are generally marked by three rows of granules, and at the calicular margin they become conical, and ornamented with a prominent and wavy ridge-like process, which passes downwards, becoming soon lost in a faint fissure, which may be seen on most of the costæ low down.

Height, $\frac{9}{10}$ths inch. Great diameter of calice, $\frac{1}{2}$ inch. Small diameter of calice, between $\frac{3}{10}$ths and $\frac{4}{10}$ths inch.

This species is readily distinguished from all other striated *Trochocyathi* by its shape, septal arrangement, small pali, and the curious ornamentation of the costæ.

Locality. Whetstone (London Clay).

In the collection of N. T. Wetherell, Esq., F.G.S.

These are the only *Trochocyathi* which are known in the London Clay, and it is very doubtful if *Trochocyathus sinuosus*, Brongniart, sp., was ever found there.[1]

Genus—Paracyathus.

1. Paracyathus cylindricus, *Duncan.* Plate IX, figs. 18—21.

The corallum is cylindrical, straight, tall, and has a flat base, whose diameter is nearly equal to that of the corallum. There is a constriction just above the base, the wall is often marked with growth-rings, and in some corallites the calice is slightly expanded.

The calice is circular in outline, its fossa is shallow, and the columella very small.

The septa are slightly exsert, and in some calices more so than in others; they are delicate, are marked with large granules laterally (fig. 21), and have an irregular upper margin.

There are six systems of septa, and three perfect cycles; moreover, in one half of four or more systems a septum of the fourth cycle is developed. The septal number is therefore very irregular, and there are from twenty-eight to thirty septa in the calice. The

[1] See 'Corals of the London Clay,' MM. Edwards and J. Haime, page 22.

pali are small and lobular, and appear to be placed before all the septa except those of the fourth cycle.

The costæ are distinct from the base upwards, are subequal, slightly prominent, and granular. The intercostal grooves are very distinct. Near the calicular margin the costæ are often found projecting outwards and becoming exsert.

Height of the corallum $\frac{1}{4}$—$\frac{1}{3}$ inch. Diameter of the calice $\frac{2}{10}$ths inch.

Locality, Bramshaw, New Forest. In the collection of Frederick Edwards, Esq., F.G.S.

2. Paracyathus Haimei, *Duncan*. Plate IX, figs. 12—14.

The corallum is short and broad, and its base is nearly as broad as the calice.

The wall is thin.

The calice is irregularly elliptical, and its long axis is on a lower plane than the short axis. The margin is sharp and irregular, the fossa is not deep, and the columella does not occupy very much space.

The septa are slender, crowded, unequal, granular, and slightly exsert. There are six systems, and the arrangement of the cycles is very irregular. There are two systems in which the septa of five cycles are complete, two in which they are incomplete, and two presenting septa of four cycles only. The primary septa are readily distinguished, and all the septa are long and often flexuous. The tertiary septa join the secondary in some systems.

The pali are present before all the septa, except those of the last cycle.

The columella is spongy.

The costæ are thin, sharp, laminate, and project; they are often slightly flexuous, and their free margin is moniliform. The intercostal spaces are wide and deep.

There are traces both of exotheca and of endotheca.

Height of corallum $\frac{2}{10}$ths inch. Great diameter of calice $\frac{8}{10}$ths inch.

Locality, Barton. In the collection of Frederick Edwards, Esq., F.G.S.

These *Paracyathi* are closely allied to the species already described from the London Clay, by MM. Milne-Edwards and J. Haime.

P. Haimei differs, however from its nearest ally, *P. crassus*, in its septal arrangement, in the sharpness and ornamentation of the costæ, and in the size of the intercostal spaces.

P. cylindricus has some resemblance to some varieties of *P. caryophyllus*, but the septal arrangement, the small columella, and the very small pali, distinguish it.

Family.—OCULINIDÆ.

Tribe.—Oculinaceæ.

Genus.—Oculina.

1. Oculina incrustans, *Duncan.* Plate IX, figs. 22—24.

The corallum is small and encrusting. There is much cœnenchyma, but it is not granular on the surface; it is marked near the calices by very faint costal ridges.

The calices are arranged without order, and are situated upon more or less prominent eminences; they are usually circular in outline, but there are indications of fissiparity. The calicular margin is sharp, the fossa is shallow from the presence of a large and prominent columella, and the spaces bounded by the columella, the margin, and the primary septa are deep.

The primary and secondary septa are long and nearly equal; they reach the columella and appear to be extended over its upper surface, but this appearance is really produced by the pali.

There are four cycles of septa, and six systems; but the septa of the fourth and fifth orders are very small. All the septa are delicate, rather narrow, and very unequal, except in the case of the primary and secondary.

The pali are before all the septa, except those of the last cycle; they are small and indistinct.

The columella is bulky, projected, rounded, and probably was papillated.

The costæ are very faintly marked, are not straight, and can hardly be said to exist.

Height of calicular projections $\frac{3}{20}$ths inch. Diameter of calice $\frac{2}{10}$ths inch.

Locality. Bracklesham. In the Sharpe Collection of the Geological Society.

The deficiency of granular cœnenchyma, the existence of additional septa, the bulky columella and the thin pali, distinguish this species from *O. conferta.*

2. Oculina Wetherelli, *Duncan.* Plate X, figs. 5—7.

The corallum is short, has a very broad base for its size, is constricted above the base, and expands into a calice. It increases by gemmation just below the calicular margin; many buds are aborted.

The surface is very finely granular under high magnifying powers, but smooth to the naked eye.

The calice is nearly circular in outline, and has a moderately thick wall and a deep fossa.

Its septa are delicate, unequal, thin, and belong to four cycles, there being six systems. The primary are the longest, and there are small pali before all except the septa of the fourth and fifth orders.

The columella is small, blunt, and delicately papillose.

There are no costæ.

Height of corallum $\frac{1}{10}$th—$\frac{3}{10}$ths inch. Diameter of calice $\frac{1}{10}$th—$\frac{2}{10}$ths inch.

Locality. Ballad's Lane, Finchley (London Clay). In the collection of N. T. Wetherell, Esq., F.G.S.

This species is closely allied to *O. conferta* and *O. incrustans*, and but remotely to *O. Halensis*.[1] The gemmation, the small columella and pali, and the septal arrangement, distinguish the new species.

SECTION—*MADREPORARIA PERFORATA.*

FAMILY—MADREPORIDÆ.

Sub-family—EUPSAMMINÆ.

Genus—DENDROPHYLLIA.

DENDROPHYLLIA ELEGANS, *Duncan*. Plate X, figs. 15—19.

The corallum has a broad encrusting base which gradually tapers into a tall, slender, and straight stem, terminated by a calice. Gemmation occurs close below the calicular margin on the outside wall, and the branches are in whorls, are long, and do not coalesce.

The calices are either circular in outline or compressed; they are deep, have a very irregular cellular margin, and a very regular septal arrangement; they vary in size, and are peculiarised by long, thin, and delicate septa, and large interseptal loculi.

There are six systems of septa, and four complete cycles; all the septa are well developed, laminar, and project very decidedly from the wall. The primary and secondary are straight and project well inwards; and processes from them develop the columella. The tertiary septa are small, but well produced; and the septa of the fourth and fifth orders meet externally to the tertiary septa and proceed to the columella. The laminæ are sharply granular, but not irregularly so, and their perforations are decided.

The columella is formed by processes from the ends of the septa, and is small.

The costæ are close, rounded above, and wider and more flattened below. The upper

[1] Fossil Corals from Sinde: 'Ann. and Mag. Nat. Hist.,' April, 1864.

costæ are granular either in series of one or of two rows, whilst the lower present many irregular rows. The cross bars of the exotheca are numerous.

Height of corallum 2 inches. Diameter of calice $\frac{3}{10}$ths inch.

Locality.—Bracklesham. In the Dixon Collection in the British Museum.

This species, very closely allied to *D. dendrophylloides*, is distinguished from it by the habit of growth, by there being multi-granulose costæ, and by the development of the higher orders of the septa.

Section—*MADREPORARIA PERFORATA.*

Family—MADREPORIDÆ.

Sub-family—Turbinarinæ.

Genus—Dendracis.

Dendracis—generic characters.[1]

Corallum arborescent; cœnenchyma very dense, granulated on the surface; calices submammiform; no columella; septa few in number and barely exsert.

Dendracis Lonsdalei, *Duncan.* Plate X, figs. 11—14.

The corallum consists of stems branching laterally, both the stems and branches being nearly cylindrical.

The cœnenchyma is very abundant, is covered with blunt conical dentations, and the calices are rare, but slightly elevated, and very small. The calices seem to be defects in the cœnenchyma rather than independent structures. They are wide apart, circular, shallow, and have no columella. The septa are twelve in number, very large at the margin, and every other one has a thin continuation which passes inwards. The central space is deep. There are no costæ (fig. 12).

The transverse section of a stem shows its cellular nature, and that it consists of superimposed cœnenchymal cells (fig. 13).

Diameter of stems $\frac{2}{10}$ths inch. Diameter of calices $\frac{1}{40}$th inch.

Locality. Bracklesham. In the Dixon Collection in the British Museum.

The wide apart and rare calices, and the strongly echino-dentate cœnenchyma, distinguish this species from the *Dendracis Gervillii*, Defrance, sp.

The new species is attached to the under part of the base of Lonsdale's typical specimen of *Porites panicea* described in Dixon's 'Geology of Sussex' (pl. i, fig. 7).

[1] 'Hist. Nat. des Corall.,' vol. iii, p. 169.

Section—*MADREPORARIA PERFORATA.*

Family—PORITIDÆ.

Sub-family—Poritinæ.

Genus—Porites.

Porites panicea, *Lonsdale.*[1] Plate X, figs. 8—10.

The corallum is flat and encrusting, and its upper surface is irregular.

The calices are small, circular, and either crowded or rather distant. In the first instance, the outer margins of the septa are in close contact, and in the second there is more or less granular cœnenchyma between the calices.

The calices vary in the depth of their fossæ, but the septa are always thick externally and thin internally; they are granular superiorly and laterally. There are six large and six small septa; the largest are connected by pali with a solid columella. All are rather exsert.

The longitudinal section shows the corallites to be deep, to have some endotheca, to be very porose, and to be united by a cœnenchyma of very distinct cells. The amount of this cœnenchyma varies according to the approximation of the corallites.

Height of corallum $\frac{3}{4}$ths inch. Diameter of calices $\frac{1}{20}$th inch.

Locality. Bracklesham. In the Dixon Collection in the British Museum.

There can be no doubt about this coral possessing a granular cœnenchyma, a columella, and pali. It is not the *Astræa panacea* of Michelin,[2] which is really an *Astræopora,*[3] having neither columella nor pali. The *Porites panicea* has more lamellate septa and a more decided cœnenchyma than the other species of the genus, and it unites the genera *Astræopora, Porites,* and *Litharæa.* The species has no resemblance to the *Porites incrustans,* Defrance, from the Miocene of Turin, nor has it close alliances with any of the recent forms.

[1] 'Dixon, 'Geol. and Foss. of Sussex,' pl. i, fig. 7.

[2] Michelin 'Iconogr.,' pl. 44, fig. 11.

[3] Pictet, 'Paléont.,' vol. iv.

Section—*MADREPORARIA TABULATA.*

Family—MILLEPORIDA.

Genus—Axopora.

Axopora Fisheri, *Duncan*. Plate X, figs. 20—22.

The corallum is large; it has an oval encrusting base, and a gibbous and tumid upper surface and sides.

The cœnenchyma is coarsely reticulate even for an *Axopora*, and is very abundant.

The calices are larger than usual in the genus, are very distinct, rather distant, and are separated by irregular elevations of the cœnenchyma.

The columella is large, is very simple and prominent, and is rounded and rather sharp. The tabulæ are very wide apart.

Height of the corallum 1½ inch.

Locality. Bracklesham. Collected by the Rev. Osmond Fisher, F.G.S.

The coarse cœnenchyma and the size of the calices, with the nature of the encrusting base, distinguish this species from those already described.

IX.—LIST OF BRITISH TERTIARY CORALS FROM THE CRAG, BROCKENHURST BEDS, AND THE EOCENE OF THE ISLE OF WIGHT AND THE LONDON CLAY.

I.—No new species have been discovered in the British Crag since the publication of the Monograph of the British Fossil Corals by MM. Milne-Edwards and Jules Haime. Those noticed and described in that monograph are as follows:

1. *Sphenotrochus intermedius*, Münster, sp.
2. *Flabellum Woodi*, Edwards and Haime.[1]
3. *Cryptangia Woodi*, Edwards and Haime.
4. *Balanophyllia calyculus*, Searles Wood.

---

[1] The species should be called *Flabellum semilunatum*, Wood, but doubtless Mr. Searles Wood will be satisfied with the distinction MM. Milne-Edwards and Jules Haime conferred on him.

II.—The following species have been described from the Brockenhurst beds:

1. *Solenastræa cellulosa*, sp. nov.
2. — *Koeneni*, ,,
3. — *Reussi*, ,,
4. — *gemmans*, ,,
5. — *Beyrichi*, ,,
6. — *granulata*, ,,
7. *Balanophyllia granulata*, ,,
8. *Lobopsammia cariosa*, Goldfuss, sp.
9. *Axopora Michelini*, sp. nov.
10. *Litharæa Brockenhursti*, ,,
11. *Madrepora Anglica*, ,,
12. — *Roemeri*, ,,
13. — *Solanderi*, Defrance.

III.—The following list includes all the species from the London Clay, the Bracklesham beds, and the Barton beds:

1. *Turbinolia sulcata*, Lamarck.
2. — *Dixoni*, Edwards and Haime.
3. — *Bowerbanki*, ,,
4. — *Fredericiana*, ,,
5. — *humilis*, ,,
6. — *minor*, ,,
7. — *firma*, ,,
8. — *Prestwichi*, ,,
9. — *affinis*, sp. nov.
10. — *exarata*, ,,
11. — *Forbesi*, ,,
12. *Leptocyathus elegans*, Edwards and Haime.
13. *Trochocyathus sinuosus*, Brongniart, sp.
14. — *Austeni*, sp. nov.
15. — *insignis* ,,
16. *Paracyathus crassus*, Edwards and Haime.
17. — *caryophyllus*, Lamarck, sp.
18. — *brevis*, Edwards and Haime.

19. *Paracyathus Haimei*, sp. nov.
20. — *cylindricus*, „
21. *Dasmia Sowerbi*, Edwards and Haime.
22. *Oculina conferta*, „
23. „ *incrustans*, sp. nov.
24. „ *Wetherelli*, „
25. *Diplohelia papillosa*, Edwards and Haime.
26. *Stylocœnia emarciata*, Lamarck, sp.
27. — *monticularia*, Schweigger, sp.
28. *Astrocœnia pulchella*, Edwards and Haime.
29. *Stephanophyllia discoides*, „
30. *Balanophyllia desmophyllum*, „
31. *Dendrophyllia elegans*, sp. nov.
32. — *dendrophylloides*, Lonsdale, sp.
33. *Stereopsammia humilis*, Edwards and Haime.
34. *Dendracis Lonsdalei*, sp. nov.
35. *Porites panicea*, Lonsdale.
36. *Litharæa Websteri*, Bowerbank, sp.
37. *Axopora Fisheri*, sp. nov.
38. — *Parisiensis*, Michelin, sp.

Number of Species.[1]

Crag . . .	4.
Brockenhurst . . .	13.
London Clay, Bracklesham, Barton . . .	38.
Total Tertiary Species . .	55.

[1] The corals from Lenham and the ferruginous sands of the North Downs are only found as indeterminable casts.

A MONOGRAPH

OF THE

BRITISH FOSSIL CORALS.

SECOND SERIES.

BY

P. MARTIN DUNCAN, M.B.LOND., F.R.S.,

FELLOW OF, AND SECRETARY TO, THE GEOLOGICAL SOCIETY.

Being a Supplement to the

'Monograph of the British Fossil Corals,' by MM. MILNE-EDWARDS *and* JULES HAIME.

PART II.

CORALS FROM THE WHITE CHALK, THE UPPER GREENSAND, THE RED CHALK OF HUNSTANTON, THE UPPER GREENSAND OF HALDON, THE GAULT, AND THE LOWER GREENSAND.

Pages 1—46; Plates I—XV.

LONDON:

PRINTED FOR THE PALÆONTOGRAPHICAL SOCIETY.

1869—1870.

PRINTED BY

ADLARD AND SON, BARTHOLOMEW CLOSE.

CONTENTS OF SUPPLEMENT TO THE CRETACEOUS CORALS.

		PAGE
I.	Introduction	1
II.	Corals from the White Chalk; Description of Species	2
III.	List of New Species	15
IV.	List of Species	16
V.	Corals from the Upper Greensand; Description of	18
VI.	List of Species	23
VII.	Corals from the Red Chalk of Hunstanton; Description of	23
VIII.	Corals from the Upper Greensand of Haldon	27
IX.	Corals from the Gault; Description of New, and Notes on Old Species	31
X.	List of Species from the Gault	38
XI.	Corals from the Lower Greensand; Description of New, and Notices of Old Species	39
XII.	List of New Species	43
XIII.	List of Species from the Lower Greensand	43
XIV.	List of Species from the Cretaceous Formations	44

A MONOGRAPH

OF THE

BRITISH FOSSIL CORALS.

(SECOND SERIES.)

PART II.—No. 1.

CORALS FROM THE CRETACEOUS FORMATIONS.

INTRODUCTION.

NOTWITHSTANDING several years have elapsed since MM. Milne-Edwards and Jules Haime wrote their description of the Corals of the British Crctaccous series, and vast additions have been lately made to the faunæ of the Chalk, Upper Greensand, Gault, and Lower Greensand, very few new *Madreporaria* have been discovered in these Upper Secondary deposits.

A few species which had been described by Mr. Lonsdale before MM. Milne-Edwards and Jules Haime wrote their Monograph for the Palæontographical Society, but which those authors did not consider sufficiently distinguished, appear, from the study of new specimens and the examination of the original types, to be worthy of re-publication. These species, with some others known in Continental Cretaceous deposits, but not hitherto noticed in Great Britain, and several new species, are described and illustrated in this Part.

Some important varieties of the species described by MM. Milne-Edwards and Jules Haime and by Mr. Lonsdale have been studied and described, and some illustrations of the specific forms themselves have been added in consequence of the reception of fine specimens.

Corals from the Upper and Lower White Chalk.[1]

MM. Milne-Edwards and Jules Haime noticed and described nine species from these formations. One of these species had been previously described by Mantell and another by Reuss, so that seven species were added to our British fauna through the industry of the great French Zoophytologists.

During the last few months I have thoroughly examined the specimens offered to me and those which had been studied by Milne-Edwards and Jules Haime, Lonsdale, and Mantell. I can add ten new species to the list of the Corals from the White Chalk, and five good varieties of formerly known species. It is necessary, also, to admit a species of Mr. Lonsdale's, and to suppress one of MM. Milne-Edwards and Jules Haime's.

Section—*APOROSA.*

Family—TURBINOLIDÆ.

Division—Caryophyllaceæ.

Genus—Caryophyllia.

MM. Milne-Edwards and Jules Haime adopted for a Coral from the Upper Chalk the name *Cyathina lævigata.* They published this name in their "Monog. des Turbinolides" ('Ann. des Sciences Nat.,' 3me série, vol. ix, p. 290, 1848), and in their 'Monograph of the Corals of the Upper Chalk' (Pal. Soc., 1850). Lonsdale named the same coral *Monocarya centralis,* Dixon (' Geol. of Sussex,' 1850), and probably *Monocarya cultrata* also.

In 1850 D'Orbigny ('Prodr. de Paléont.,' t. ii, p. 275, 1850) gave the Coral the specific name *cylindracea,* it having become evident that Reuss was the primary discoverer of the species in 1846. In his 'Kreideformation,' p. 61, pl. xiv, figs. 23—30, Reuss gave the name *Anthophyllum cylindraceum.* The genus of the Coral is evidently *Caryophyllia* in the sense adopted by Charles Stokes in 1828.

[1] The following authors have written upon this subject:

Parkinson, 'Organic Remains of a Former World,' &c., 1811.
Mantell, 'Geol. of Sussex,' 1822; and 'Trans. Geol. Soc.,' 2nd series, vol. iii, 1829.
Fleming, 'British Animals,' 1828.
Phillips, 'Illust. Geol. York,' part i, 1829.
S. Woodward, 'Syn. Table of Brit. Org. Remains,' 1830.
R. C. Taylor, in 'Mag. Nat. Hist.,' vol. iii, p. 271, 1830.
MM. Milne-Edwards and Jules Haime, op. cit.
Lonsdale, in Dixon's 'Geol. Sussex,' 1850.

MM. Milne-Edwards and Jules Haime, having all this information before them, very properly admit the generic and specific names to be *Caryophyllia cylindracea,* Reuss, sp. ('Hist. Nat. des Corall.,' vol. ii, p. 18).

This species is very polymorphic, and the pali of some specimens are very like the outer terminations of the columellary structures in some *Parasmiliæ*. Very frequently it is hardly possible to determine in *Caryophyllia cylindracea* which are pali and which the ends of the columellary fasciculi. Moreover, in some specimens the base is small and the costæ reach low down, whilst in others the base is normal and large, the costæ being abnormal from their length.

There is a new species of this genus in the Dunstable Chalk and another in the Chalk of Sussex. There are thus three species of *Caryophyllia* in the Upper Chalk of England:

1. *Caryophyllia cylindracea,* Reuss, sp.
2. „ *Lonsdalei,* Duncan.
3. „ *Tennanti* „

1. CARYOPHYLLIA CYLINDRACEA, *Reuss,* sp. Pl. I, figs. 7—12.

In the British Museum, Dixon Collection.

2. CARYOPHYLLIA LONSDALEI, *Duncan.* Pl. I, figs. 1—3.

The corallum has a large and encrusting base, and the stem is cylindro-conical and straight. There is a slight curve near the base.

The calice is circular, small, not very open, and moderately deep.

The columella is small, and is terminated by rod-shaped processes.

The septa are slightly exsert, the primary especially. There are three complete cycles, and the septa of the higher orders of the fourth cycle are not developed in every system. The primary, secondary, and tertiary septa are very alike. They have a wavy inner edge, and are granular.

The pali are situated before the tertiary septa, and are knob-shaped and rather flat from side to side.

The costæ are nearly equal at the calicular margin, and pass downwards as flat, band-like prominences, separated by shallow intercostal grooves. They are continued to the base, but are hidden midway by an epithecal growth.

Height of the corallum, $\frac{5}{8}$ths inch. Breadth of the calice, $\frac{1}{3}$rd inch.

Locality. Dunstable. In the Collection of the Rev. T. Wiltshire, F.G.S.

This species is readily distinguished by its costæ, and is more closely allied to *C. cylindracea* than to any other form.

3. Caryophyllia Tennanti, *Duncan*. Pl. I, figs. 4—6.

The corallum has a large base, a curved cylindrical stem, and an inclined elliptical calice. It is short in relation to its broad base.

The calice is open and shallow.

The columella is small, and terminates in twelve knob-shaped endings to the fasciculi.

The septa are unequal, and there are five incomplete cycles.

The laminæ are marked with curved lines of granules, are wavy and unequal.

The pali are higher than the columellary processes, are wavy, flattened, and curved.

The costæ are sub-equal in the upper third, but are not seen below.

Height, $1\frac{1}{3}$rd inch. Length of calice, $\frac{5}{8}$ths inch.

Locality. Sussex; Upper Chalk. In the Collection of Professor Tennant, F.G.S.

This species connects the Cretaceous *Caryophylliæ* with those of the Tertiary and Recent systems.

Family—TURBINOLIDÆ.

Division—Turbinoliaceæ.

Genus nov.—Onchotrochus.

The corallum is simple, tall, slender, rather hook-shaped or clavate, and presents evidences of irregular growth.

There is no endotheca.

The costæ are rudimentary, and there is no columella.

The septa are few in number.

The epitheca is pellicular and striated.

The genus is somewhat allied to *Smilotrochus, Stylotrochus,* and very distantly to *Flabellum.*

Onchotrochus serpentinus, *Duncan*. Pl. VI, figs. 1—4.

The corallum is tubulate, curved superiorly, and straight and tapering inferiorly. A sudden diminution in the diameter of the upper part of the corallum exists.

The costæ are quite rudimentary.

The epitheca is marked with fine transverse striations.

The septa are continuous with what appear to be rudimentary intercostal spaces.

The laminæ are twelve in number; they project into the circular calice, but are not exsert. A section proves that they are very stout, even low down in the corallum.

Length of the corallum, 1 inch. Diameter of the calice, $\frac{1}{8}$th inch.

Locality. Charlton, Kent. In the Collection of the Rev. T. Wiltshire, F.G.S.

This species is mimetic of *Parasmilia serpentina*, Ed. and H., from the same geological horizon, just as *Trochosmilia cylindrica* is mimetic of *Parasmilia cylindrica*. The *Stylotrochi* of the Cambridge Upper Greensand are closely allied to this species, which is found in the Grey Chalk and Lower White Chalk.

Family—ASTRÆIDÆ.

Genus—Trochosmilia.

Sub-genus—Cœlosmilia.

It is a great question whether *Cœlosmilia* can stand as a genus. It is impossible to separate its species from those of *Trochosmilia* by an external examination, and sections prove that there is no columella and a very scanty endotheca. Still there is an endotheca, and the visceral cavity of the Coral was not open from top to bottom, as in the *Turbinolidæ*. It is true that there is a facies common to the *Cœlosmiliæ*, and that they are a natural group; but, in fact, they do not differ from a *Trochosmilia* with scanty endotheca. On studying the genus *Trochosmilia* it will be noticed that many of its species have never been described with reference to their endotheca. Many were determined from one or two specimens, and sections of the majority have not been made. Now, *Trochosmilia sulcata*, Ed. and H., has very little endotheca; it is a species from the Gault, and the *Cœlosmiliæ* are all from the Cretaceous, Eocene, and recent Coral-faunæ. In placing *Cœlosmilia* as a sub-genus, but included in *Trochosmilia*, it must be admitted that the classification becomes simpler and more natural. Since MM. Milne-Edwards and Jules Haime published their 'Hist. Nat. des Coralliaires,' some new species of *Cœlosmilia* have been published or described.

The following species have been described:

1. *Cœlosmilia poculum*, Ed. & H., recent.
2. „ *Faujasi*, „ White Chalk, Ciply.
3. „ *punctata*, „ „
4. „ *laxa*, „ Norwich Chalk.
5. „ *Edwardsi*, D'Orb., Sezanne.
6. „ *Atlantica*, Martin, sp., Timber Creek, New Jersey.
7. „ *excavata*, Hagenow, sp., Chalk of Rugen.
8. „ *radicata*, Quenstedt, Nattheim.

The new species are—

9.	*Cœlosmilia*	*elliptica,*	Reuss,	Castel Gomberto.
10.	„	*Javana,*	Duncan, MS.,	Java.
11.	„	*cornucopiæ*	„	Trimmingham Chalk.
12.	„	*Wiltshiri*	„	Norwich Chalk.
13.	„	*Woodwardi*	„	White Chalk, England.
14.	„	*granulata,*	„	„
15.	„	*cylindrica,*	„	„

The species *cornucopiæ*, *Wiltshiri*, *Woodwardi*, *granulata,* and *cylindrica* are new to British palæontology, and are very characteristic of the Upper Chalk.

There are in the Upper Chalk three well-marked varieties of *Cœlosmilia laxa,* Ed. & H.

An analysis of the species produces the following results.

1. The species *Atlantica*, *punctata*, *Edwardsi*, *excavata*, and *radicata*, either pertain to other species or are really indeterminable.

2. The species whose septal arrangement shows more cycles than four or which have some septa of the fifth cycle are—

Cœlosmilia poculum.
„ *Faujasi.*
„ *Javana.*
„ *cornucopiæ.*
„ *Wiltshiri.*
„ *Woodwardi.*
„ *elliptica.*

3. The species whose septal arrangement shows three cycles or four cycles, or some septa of the fourth cycle, are—

Cœlosmilia granulata.
„ *cylindrica.*
„ *laxa.*

4. The species with large bases and with more than four cycles are—

Cœlosmilia poculum.
„ *elliptica.*

5. The species with a large base and with more than three cycles of septa, but not more than four, is—

Cœlosmilia cylindrica.

TROCHOSMILLÆ. *Sub-genus*—Cœlosmilia (having scanty endotheca).

With wide bases; the costæ	hardly prominent, replaced inferiorly by granules; five cycles; corallum straight	*Trochosmilia* (*C.*) *poculum*, Ed. & H., sp.
	cristiform superiorly, with intermediate costæ; five cycles; corallum curved	,, (*C.*) *elliptica*, Reuss, sp.
	very distinct, flat, wide; intercostal spaces linear; four cycles; corallum cylindrical	,, (*C.*) *cylindrica*, Duncan.
With pedicel or a small trace of a former attachment; five cycles; the costæ . .	throughout indistinct, plane and subequal	,, (*C.*) *Faujasi*, Ed. & H., sp.
	alternately large and small; subcristiform above, crossed by ornamentation	,, (*C.*) *Javana*, Duncan.
	subcristiform throughout; subequal above	,, (*C.*) *cornucopiæ*, ,,
	very distant, distinct and subcristiform, smaller much ornamented .	,, (*C.*) *Wiltshiri*, ,,
	long, the principal cristiform, many smaller between them; corallum long and cornuted	,, (*C.*) *Woodwardi*, ,,
Four cycles (or part of); the costæ	well marked, distant, very granular; intercostal spaces very granular and strongly marked; corallum curved	,, (*C.*) *granulata*, ,,
	distant, distinct, and cross-marked in intercostal spaces . .	,, (*C.*) *laxa*, Ed. & H., sp.

Genus—Trochosmilia.

Sub-genus—Cœlosmilia.

1. Trochosmilia (Cœlosmilia) laxa, *Ed. & H.* Pl. III, figs. 11—17; Pl. IV, figs. 9—12.

In examining good specimens of this species I found the fourth cycle of septa to be present. Its laminæ are small, but decidedly visible. Consequently the calice as drawn by MM. Milne-Edwards and Jules Haime ('Monog. Brit. Foss. Corals,' P. I, Pl. VIII, fig. 4 *c*) is incorrect. The following description will apply to three varieties of the species.

Variety 1.—The corallum is conico-cylindrical and straight.

The costæ are intensely granular inferiorly, and two large costæ are separated by three smaller. Near the calice the larger costæ have a wavy cristiform ridge upon them, the intermediate costæ being very granular, with chevron patterns, or they may be moniliform. At the calicular margin the costæ are nearly flat and granular. The fourth cycle of septa is distinct.

Variety 2.—Inferiorly in structure as variety 1. Superiorly the principal costæ are very cristiform, and well marked with a secondary ridge. The chevron markings of the intermediate costæ are very distinct.

Variety 3.—Costæ inferiorly wavy and sparely granular. Superiorly the costæ are subcristiform and plain, the continuity of the crests being defective. The intermediate costæ are broken and moniliform, and here and there chevroned.

Localities.—Norwich Chalk; Wiltshire Chalk. In the British Museum and in the Salisbury Museum.

2. Trochosmilia (Cœlosmilia) cornucopiæ, *Duncan.* Pl. III, figs. 6—10.

The corallum is strongly curved in the plane of the smaller axis, and it is compressed superiorly, and is finely pedunculate. The growth rings and swellings are moderately developed.

The costæ are subequal above, and cristate and unequal inferiorly.

The septa are numerous and very unequal. There are five cycles of septa and six systems. The primary septa are very exsert, and the secondary are less so. The septa of the fifth cycle are very small.

The calice is elliptical, and the fossa very deep, the larger septa joining those opposite at its bottom.

There are traces of epitheca.

Height, 1 inch. Breadth of calice $\frac{3}{8}$ths inch; length of calice, 1 inch. Depth of fossa $\frac{3}{8}$ths inch.

Locality. Trimmingham; Upper Chalk. In the Collection of the Rev. T. Wiltshire, F.G.S.

3. TROCHOSMILIA (CŒLOSMILIA) WILTSHIRI, *Duncan*. Pl. III, figs. 1—5.

The corallum is tall, curved, finely pedicillate, and is not compressed.

The growth-rings are distinct.

The costæ are very distinct and unequal, and they reach from base to calice. The smaller intermediate costæ are ornamented with chevrons and horizontal lines. The larger costæ have a secondary crest upon their free surface.

The septa are unequal, slender, and not crowded.

The calice is circular.

There are five cycles of septa, but the fifth is incomplete in some systems. The primary septa are large, slightly exsert, and extend far inwards.

The calicular margin is very thin, and the fossa is deep.

Height, 1$\frac{2}{3}$rds inch. Diameter of the calice, $\frac{2}{3}$rds inch.

Locality. Norwich; Upper Chalk. In the Collection of the Rev. T. Wiltshire, F.G.S.

4. TROCHOSMILIA (CŒLOSMILIA) WOODWARDI, *Duncan*. Pl. IV, figs. 5—8.

The corallum is tall, cornute, slightly pedicillate, and narrow.

The growth-markings are distinct.

The costæ are distinct from base to calice. Two large subcristiform and very distinct costæ bound three intermediate small and more or less moniliform costæ. Sets of these costæ occur around the corallum.

The septa are crowded, wavy, and unequal. Many unite laterally, and the largest reach far into the axial space.

The calice is circular, and the wall is very thin.

Height, 2 inches. Breadth of the calice, $\frac{3}{8}$ths inch.

Locality. Chalk of South of England. In the British Museum (Dixon Collection).

5. Trochosmilia (Cœlosmilia) granulata, *Duncan.* Pl. IV, figs. 1—4; Pl. VI, fig. 9.

The corallum is tall and slightly curved, and it has a long pedicel, with a very distinct base.

The corallum is slightly compressed, and bulges here and there.

The costæ are well marked, distant, subequal, and intensely granular. The larger costæ are more distinct inferiorly and midway than close to the calicular margin; they are cristiform in some places, notched by chevron-shaped ornamentation in others, and occasionally sharply pointed or absent. The spaces between the larger costæ are wide, faintly convex, and are marked longitudinally by small costæ, and transversely by wavy or chevroned ornamentation.

The whole external surface of the corallum is very granular.

The calicular wall is very thin, and the calice is elliptical.

There are three perfect cycles of septa, and some orders of the fourth cycle in some of the systems. The septa are wide apart, slightly exsert, unequal, and slender. They do not reach far inwards at once, but dip downwards with a gentle curve.

In a section the inner margin of the larger septa is wavy.

The endotheca is scanty.

Height, $1\frac{2}{3}$rds inch. Length of calice, $\frac{5}{6}$ths inch; breadth, $\frac{2}{3}$rds inch.

Locality. Norwich, and Chalk of south of England. In the British Museum (Dixon Collection).

6. Trochosmilia (Cœlosmilia) cylindrica, *Duncan.* Pl. V, figs. 1—3.

The corallum is tall, cylindrical, and very slightly bent. The calicular opening is smaller in diameter than the rest of the corallum.

The costæ are nearly equal, broad, slightly rounded, and are separated by shallow, narrow, and undulating intercostal grooves. The costæ are profusely ornamented with transverse ridges, straight, curved, or angular, and with large granules.

The calicular edge is very thin, and the broad convex costæ are continuous with slender, unequal septa.

There are four cycles of septa. The primary are exsert, and the laminæ of the higher orders are very small.

There is no columella, the larger septa are united by a few short attachments from their inner margins.

The endotheca is scanty.

Height, several inches. Breadth of the calice, $\frac{5}{6}$ths inch.

Locality. Norwich, Upper Chalk. In the Collection of the Rev. T. Wiltshire, F.G.S.

The sub-genus *Cœlosmilia* is represented in the British Chalk by one species formerly known, by three varieties of it, and by five new species.

1. *Trochosmilia* (*Cœlosmilia*) *laxa*, Ed. & H.
 „ „ „ Varieties 1, 2, 3, Duncan.
2. „ „ „ *cornucopiæ*, Duncan.
3. „ „ „ *Wiltshiri*, „
4. „ „ „ *Woodwardi*, „
5. „ „ „ *granulata*, „
6. „ „ „ *cylindrica*, „

These *Trochosmiliæ*, with a slight amount of endotheca—what there is of it is generally low down—are very characteristic of the Upper Chalk, and their presence suggests that the Upper Chalk of Norwich and Trimmingham is, from the evidence of its Corals, as well as from the proofs already asserted from its Mollusca, on a higher horizon than the Upper Chalk, usually so called, in the south-east district. The Coral evidence brings the Norfolk Chalk closer in relation with the Faxoe, Rugen, and Ciply deposits.

The affinity between *Trochosmilia* (*C.*) *cornucopiæ* and *Cœlosmilia excavata*, Hagenow, sp. (a doubtful form, but well drawn by Quenstedt), is evident. It is from Rugen. *Trochosmilia Wiltshiri* and *T. Faujasi* from Ciply are closely allied.

The depth of the space between the calicular margin and the top of the upper dissepiment in these species indicates that the corals had great mesenteric, ovarian, perigastric, and water systems. They were probably very rapid growers. The wall is merged into the costal system, which is strengthened by a most unusual cross-bar and cristiform ornamentation; and this development, which is almost epithecal, is complementary to the defective endotheca.

Family—ASTRÆIDÆ.

Division—Trochosmiliaceæ.

Genus—Parasmilia.

MM. Milne-Edwards and Jules Haime described five species of this genus from the Upper Chalk, viz.—

1. *Parasmilia centralis*, Mantell, sp.
2. „ *Mantelli*, Ed. and H.
3. „ *cylindrica*, „
4. „ *Fittoni*, „
5. „ *serpentina*, „

Parasmilia cylindrica and *Parasmilia serpentina* are readily distinguished by their external shape; but, owing to the polymorphic character of *Parasmilia centralis*, it is by no means easy to separate it from *Parasmilia Mantelli* and *Parasmilia Fittoni*.

Parasmilia Mantelli, Ed. and H., was determined from one specimen alone, and it is clearly united to *Parasmilia centralis* by *Parasmilia Gravesana*, Ed. and H., of the White Chalk of Châlons-sur-Marne and Beauvais (Oise). This species has been found in England. Having found many specimens of *Parasmilia centralis* with costæ like those of *P. Mantelli* in some parts of the corallum, and found normal costæ in others, I consider *P. Mantelli* a variety of *P. Gravesana*, and that this last species is a variety and good sub-species of *P. centralis*. *Parasmilia Fittoni*, Ed. and H., has a large columella and a definite structural distinction in its tertiary costæ from *P. centralis*.

The following is a list of the British *Parasmiliæ*:

1. *Parasmilia centralis*, Mantell, sp.
   ,, ,, variety *Mantelli*.
   ,, ,, sub-species *Gravesana*, Ed. and H.
2. ,, *Fittoni*, Ed. and H.
3. ,, *cylindrica*, ,,
4. ,, *serpentina*, ,,
5. ,, *monilis*, Duncan.
6. ,, *granulata*, ,,

1. Parasmilia centralis, *Mantell*, sp.; sub-species *Gravesana*, Ed. and H. Pl. VI, figs. 14—17; pl. V, figs. 8, 9.

MM. Milne-Edwards and Jules Haime notice that *P. Gravesana* is "très voisine de la *P. centralis;* elle s'en distingue seulement par ses côtes."—'Hist. Nat. des Coral.,' vol. iii, p. 173. Pl. V, figs. 10—15.

In the British Museum.

2. Parasmilia monilis, *Duncan*. Pl. V, figs. 4—7.

The corallum is long, much curved, and distorted. It is more or less cylindrical above, and contracted here and there. Inferiorly it is pedunculate, the peduncle being small, curved, and long.

The costæ are nearly equal on the peduncle; and there they are rather subcristiform, a secondary crest being found on each costa. In the intercostal spaces there is either a faint ridge, or a moniliform series of granules. On the body of the Coral the principal costæ are sharp, wavy, granular, and keeled. They have several smaller and less prominent granular costæ between them, and in the intercostal space there is a series of moniliform granules.

The calice is often smaller than the body, and the wall is very thin.

The septa are small, and there are four cycles, the last cycle being rudimentary.

The columella is small.

The height varies from $\frac{1}{4}$ inch to 2 inches, and the diameter from $\frac{1}{2}$ to $\frac{2}{3}$rds inch.

Locality. Gravesend. In the Collection of the Rev. T. Wiltshire, F.G.S.

3. Parasmilia granulata, *Duncan*. Pl. VI, figs. 5—8.

The corallum is tall, nearly straight, finely pedunculate, and cylindro-conical.

The calice is very large, widely open, deep, and has a thin margin.

The columella is well developed.

The septa are barely exsert, reach but slightly inwards, and pass downwards at once. They are very unequal, and alternately large and small, and there are four complete cycles and part of the fifth.

The costæ are subequal near the calice, and the broadest are continuous with the smallest septa. On the body the costæ are subcristiform and in sets of four. On the pedicel they are very granular and very distinct.

Height, $1\frac{1}{3}$rd inch. Breadth of calice, $\frac{1}{2}$ inch. In the British Museum (Dixon Collection).

This species was included by Lonsdale in his genus *Monocarya*, and was termed *M. centralis*. *Parasmilia* has the priority as a genus, and the species is evidently not *P. centralis*.

The position of the genus *Parasmilia* is somewhat like that of *Cœlosmilia*, but MM. Milne-Edwards and Jules Haime have created the genus *Cylicosmilia* for *Parasmiliæ* with abundant endotheca. Now, in careful sections (pl. VI, figs. 12, 13) I find that *P. centralis* and its varieties have endothecal dissepiments reaching close to the calicular fossa. The genus must, therefore, absorb *Cylicosmilia*; and *C. Altavillensis*, Defrance, sp., of the Eocene of Hauteville, must become *Parasmilia Altavillensis*, Defrance, sp.

Reuss has described an Eocene *Parasmilia* from Monte Grumi which is closely allied to the *Parasmilia centralis* series.

Order—*ZOANTHARIA APOROSA.*

Family—OCULINIDÆ.

Genus—Diblasus, *Lonsdale.*

This genus was established by Lonsdale in Dixon's 'Geol. of Sussex,' 1850, pp. 248—254, pl. xviii, figs. 14—28), and was described by the learned zoophytologist with all that critical acumen which characterises him. MM. Milne-Edwards and Jules Haime, whilst they acknowledge the genus to be "voisin des *Synhelia*" ('Hist. Nat. des Corall.,' pl. 2, p. 115), do not give it a place in their classification. I have, therefore, carefully studied and drawn the specimens from the Dixon Collection in the British Museum, and have great pleasure in doing justice to Mr. Lonsdale by inserting his genus with slight alterations, to meet the terminology of the day.

Genus—Diblasus, *Lonsdale* (amended).

The corallum is encrusting, and very irregular in shape.

The calices are wide apart, and projecting.

The intercalicular tissue is costulate.

The septa are unequal.

There are no pali.

The columella is formed by the junction of the larger septa, and does not exist as a separate structure.

Gemmation marginal and intercalicinal.

The genus is clearly not closely allied to *Synhelia,* for it has no palular or true columellary structures. It approaches the genus *Astrohelia,* which is a transition genus, bringing the *Oculinidæ* in relation with the *Astræinæ* through the *Cladangiæ* (Milne-Edwards and Jules Haime, 'Hist. Nat. des Corall.,' vol. ii, p. 111).

1. Diblasus Gravensis, *Lonsdale.* Pl. II, figs. 1—11.

The corallum is very irregular in shape and size.

The calices project, and are irregular in their projection and size.

The costæ are granular, equal, subequal, and unequal in different parts of the same corallum.

There are three cycles of septa, and sometimes some of the fourth cycle are seen.

Some primary septa nearly reach those opposite to them, and form a rudimentary columella. They are dentate, crowded, and are granular laterally.

Diameter of usual-sized calices, $\frac{1}{8}$th inch.

Locality. Gravesend Chalk. In the British Museum (Dixon Collection).

The condition in which the specimens of this species are found is very remarkable. The inside of nearly every calice has been worn away, so that the mural edges of the septa are all that remain. The perfect calices appear to have shrunk from the surrounding cœnenchyma, and in many places the costæ have been worn off.

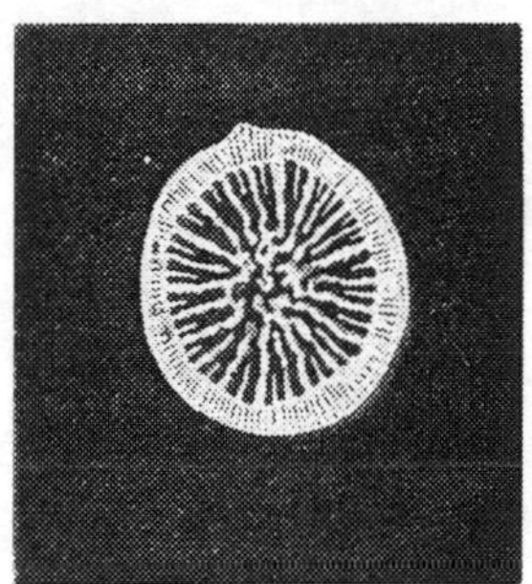

A large Calice magnified.

There are several specimens of Corals from the Lower Chalk, which cannot, however, be identified, on account of their fragmentary condition. Most probably, *Onchotrochus serpentinus,* Duncan, is a Lower as well as an Upper Chalk form.

The following is a complete list of the Fossil Corals from the Upper and Lower White Chalk of Great Britain :

III.—List of New Species.

1. *Caryophyllia Lonsdalei,* Duncan.
2. „ *Tennanti,* „
3. *Onchotrochus serpentinus,* „
4. *Trochosmilia* (*Cœlosmilia*) *cornucopiæ,* Duncan.
5. „ „ *Wiltshiri,* „
6. „ „ *Woodwardi,* „
7. „ „ *granulata,* „
8. „ „ *cylindracea,* „
9. *Parasmilia monilis,* Duncan.
10. „ *granulata,* „
11. *Diblasus Gravensis,* Lonsdale.

IV.—List of the Corals from the Upper and Lower White Chalk.

1. *Caryophyllia cylindracea*, Reuss, sp.[1]
2. „ *Lonsdalei*, Duncan.
3. *Caryophyllia Tennanti*, Duncan.
4. *Onchotrochus serpentinus*, „[2]
5. *Trochosmilia laxa*, Ed. and H., sp., and varieties 1, 2, 3.[3]
6. „ *cornucopiæ*, Duncan.
7. *Trochosmilia Wiltshiri*, „
8. „ *Woodwardi*, Duncan.
9. „ *granulata*, „
10. „ *cylindracea*, „
11. *Parasmilia centralis*, Mantell, sp., varieties 1, 2.[4]
12. „ *cylindrica*, Ed. and H.
13. „ *Fittoni*,[5] „
14. „ *serpentina*, „
15. „ *monilis*, Duncan.
16. „ *granulata* „
17. *Diblasus Gravensis*, Lonsdale.
18. *Synhelia Sharpeana*, Ed. and H. }[6]
19. *Stephanophyllia Bowerbanki*, Ed. and H. }

The list of species presents a remarkable assemblage of forms. The *Caryophylliæ* are represented in existing seas, from low spring-tide level to 80 or 200 fathoms. The West Indian, the Mediterranean, the south-west and the north-east British seas, are favourite localities. With one exception, the *Caryophyllia Smithi*, they are always deep water forms; and this Coral is evidently a littoral variety of *C. borealis*. The *Oculinidæ* of the present day are usually found under the same conditions as the *Caryophylliæ*, and doubtless the *Parasmiliæ* and *Trochosmiliæ* were dwellers in from 10 to 200 fathoms.[7]

There are no forms which indicate shallow waters, or anything like a reef. The Coral fauna was a deep-sea one.

[1] Synonym, *Cyathina lævigata*.

[2] Lower Chalk.

[3] Varieties or sub-species not hitherto described.

[4] Varieties or sub-species not hitherto described.

[5] See the remarks upon the propriety of absorbing *P. Mantelli*. M. de Fromentel has described *Caryophyllia decemeris* from Southfleet. Much experience in these species inclines me to believe that the decemeral arrangement is a monstrosity. There has only been one specimen of this species found.

[6] Lower Chalk.

[7] Dr. W. Carpenter, F.R.S., dredged up living *Oculinidæ* from the great depth of 530 fathoms, in the autumn of 1868.

NOTE.—CORALS IN FLINTS.

THE flints of the Upper Chalk often contain Corals. Usually the destructive silicification has produced such loss of structures as to render the specific and often the generic diagnosis impossible. No new species have been distinguished in the flints.

The flint pebbles of the Woolwich series and the basement bed of the London Clay were derived from the Upper Chalk principally. In breaking up a series of the pebbles Mr. J. Flower, F.G.S., discovered several Corals. A cast of a Trochosmilian (Cœlosmilia, sp. —?) is represented below.

Cast of a Coral from a pebble.

Several young simple Corals were noticed by Mr. Flower, but their structures are very badly preserved.

Section of simple Corals in flint.

The most interesting fossil of the series is a perforate Coral, with a most delicate lace-like structure of its cœnenchyma. Within this Coral is an aporose form, probably a *Caryophyllia.*

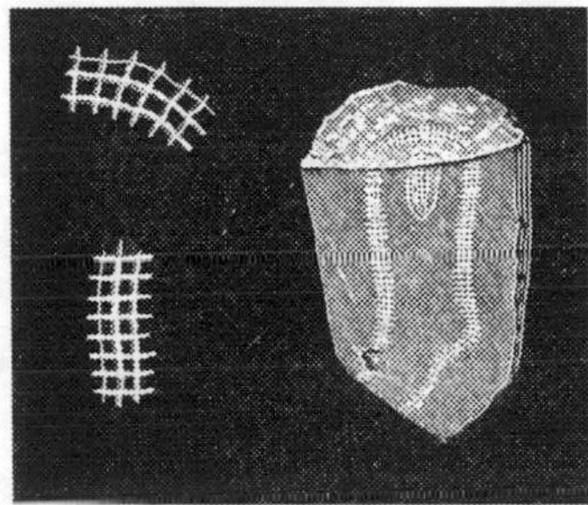

Sections of Coral in flint.

The perforate structure resembles that of the Alveoporæ.

The only example of an aporose Coral which is invariably surrounded by another structure is in Cryptangia, a genus whose species are always immersed in Celleporæ. It is possible that this Caryophyllia of the Chalk was, like *Cryptangia parasita,* always immersed in a mass of cellular Alveopora.

V.—Corals from the Upper Greensand.[1]

The scanty Coral-fauna of the Upper Greensand was described by MM. Milne-Edwards and Jules Haime; and although some years have elapsed since the publication of the first part of the 'British Fossil Corals,' Pal. Soc., and the beds have been well searched, very few additions can be made to the list of the *Madreporaria*. The following is the list of the published species (1850):

1. *Peplosmilia Austeni*, Ed. and H.
2. *Trochosmilia tuberosa*, „
3. *Parastræa stricta*, „
4. *Micrabacia coronula*, Goldfuss, sp.

In their 'Hist. Nat. des Corall.,' vol. ii, MM. Milne-Edwards and Jules Haime make some alterations in the synonyms of the genera, and add a species to the list. They do not give any further information respecting some doubtful species noticed by Mr. Godwin-Austen and Prof. Morris. Their amended list is as follows:

1. *Peplosmilia Austeni*, Ed. and H.
2. *Smilotrochus tuberosus*, „
3. „ *Austeni*, „
4. *Favia stricta*, Ed. and H.
5. *Micrabacia coronula*, Goldfuss, sp.

Family—TURBINOLIDÆ.

Division—Turbinoliaceæ.

Genus—Smilotrochus.

Trochosmilia tuberosa, Ed. and H., has no endotheca, and therefore is of necessity included amongst the *Turbinolidæ*. The genus *Smilotrochus* was determined in order to receive the species.

[1] The following authors have written on this subject:
W. Smith, 'Strata Identified by Organic Fossils,' 1816.
Godwin-Austen, 'Trans. Geol. Soc.,' 2nd series, vol. vi, p. 452, 1842.
Morris, 'Cat. of British Fossils,' p. 46, 1843.
MM. Milne-Edwards and Jules Haime, op. cit.

Genus—Smilotrochus, *Ed. and H.*

The corallum is simple, straight, cuneiform, free, and without a trace of former adhesion. There is no columella.

The wall is naked and costulate.

There is no epitheca.

The simple costæ are distinct from the base to the calice.

This is the simplest form of *Aporose Zoantharia,* and its structures only comprise a wall, septa, and costæ. *Flabellum* has an epitheca in addition, and *Stylotrochus* of De Fromentel is a *Smilotrochus* with a styliform columella, the septa uniting also by their thickened internal margins. *Onchotrochus,* nobis, has a pellicular epitheca, no columella: but, like *Stylotrochus,* the septa are united internally.

1. Smilotrochus tuberosus, *Ed. and H.*

Trochosmilia tuberosa, *Ed. and H.*
Turbinolia compressa (?), *Morris.*

This species, with five cycles of septa, was described in the 'Monograph of the Brit. Foss. Corals, Upper Greensand,' Milne-Edwards and Jules Haime.

2. Smilotrochus Austeni, *Ed. and H.* Pl. VII, fig. 12.

This species is thus described in the 'Hist. Nat. des Corall.,' vol. ii, p. 71:

The corallum is regularly cuneiform, very much compressed below, and slightly elongate.

The calice is elliptical; the summit of the larger axis is rounded.

Forty-eight costæ, subequal, straight, fine, and granular.

Height of the corallum, about ⅓rd inch.

Locality. Farringdon.

MM. Milne-Edwards and Jules Haime do not mention where the specimen is deposited.

3. Smilotrochus elongatus, *Duncan.* Pl. VII, figs. 1—6.

The corallum is tall, straight, and nearly cylindrical.

The columellary space is large.

The septa are fine and unequal, especially in length. There are four cycles of septa.

Height, about an inch.

Locality. Upper Greensand of Cambridgeshire. In the Collection of James Carter, Esq.

4. Smilotrochus angulatus, *Duncan.* Pl. VII, figs. 7, 8.

The corallum is conical, hexagonal, and slightly curved at its very fine inferior extremity. It is broad superiorly, has six prominent angles, and is slightly compressed.

The septa are fine, unequal, and each plane between the angles has a system of four cycles.

The columellary space is large.

Height, $\frac{3}{4}$ths to 1 inch. Breadth, $\frac{1}{2}$ inch.

Locality. Upper Greensand, Cambridge. In the Collection of James Carter, Esq.

Genus—Onchotrochus.

Onchotrochus Carteri, *Duncan.* Pl. VIII, figs. 1—14.

In the young corallum there is a flat and rounded expansion at the base, by which it was attached to foreign substances, but this is lost as growth proceeds.

The corallum is either straight or slightly curved, is tall, very slender, cylindro-conical, clavate, and enlarged here and there.

The worn specimens are more or less angular in transverse outline.

The costæ are angular projections, which extend from base to calice; they are subequal, wide apart, and are connected and covered with a fine, striate, pellicular epitheca, which readily disappears.

The growth-markings are very common.

The calice is circular and shallow.

The septa are stout at the walls and wedge-shaped; they are rounded superiorly, and do not extend far inwards. There are twelve septa, and they are subequal. The septa in sections often appear to be equal, and their inner ends are joined, and the axial space is filled up by a deposit of coral structure; but the reverse is the case occasionally, and the irregularity of the septa may often be well seen. The septa are continuous with the costæ.

Height, $\frac{1}{3}$rd—$\frac{2}{3}$rds—1 inch. Diameter of costæ, $\frac{1}{12}$th—$\frac{1}{10}$th inch.

Locality. Cambridge Greensand. In the Collections of James Carter and Rev. T. Wiltshire, F.G.S.

The species has great resemblance to the lower part of *Onchotrochus serpentinus*, nobis. Very careful examination of sections and calices proves that there is no columella, that the inner ends of the septa produce a false one, and that the styloid appearance is due to fossilization.

The discovery of better specimens may, perhaps, lead M. de Fromentel to consider his *Stylotrochus*, which resembles this form, to be of the same genus.

Family—ASTRÆIDÆ.

Division—Stylinaceæ.

Genus—Cyathophora, *Michelin.*

This genus has the usual characters of compound *Astræinæ*, but the dissepiments act as tabulæ, and shut in the calice below, just as in some of the Liassic *Isastrææ*. There is no columella. The curved dissepiments are not noticed, and the family of the genus must remain unsettled, for the minute structure is clearly tabulate. The genus flourished in the Lower and Middle Oolites, and the only Cretaceous species is that under consideration, and which has been described by D'Orbigny from the Craie tuffeau of Martigues.

1. Cyathophora monticularia, *D'Orb.*, sp. Pl. VIII, figs. 15—18.

The septa are rather thick.

There are three cycles, but the third is often deficient in one or two systems.

Locality. Haldon. In the Collection of the Geological Society.

Division—Faviaceæ.

Genus—Favia, *Ehrenberg.*

This genus has absorbed the *Parastræaceæ*, so that the old *Parastræa stricta*, Ed. and H., is now named *Favia stricta*, Ed. and H.

1. Favia minutissima, *Duncan*. Pl. VII, figs. 9—11.

The corallum is encrusting, gibbous, and small.

The calices are very small, close, and with very scanty intercorallite tissue.

There are twelve septa.

The costæ are continuous.

Diameter of the calices, under $\frac{1}{12}$th inch.

Locality. Haldon. In the Collection of the Geological Society.

This is the smallest of the *Faviæ*.

Division—Astræaceæ.

Genus—Thamnastræa.

Thamnastræa superposita, *Michelin*, sp. Pl. VII, figs. 13—17.

MM. Milne-Edwards and Jules Haime thus notice this species ('Hist. Nat. de Corall.,' vol. ii, p. 559):

"M. Michelin's specimen is very young; it is encircled by a strongly folded epitheca, which is formed of two layers.

"No columella is distinguishable.

"The septa are tolerably strong and unequal.

"There are three cycles, with the rudiments of a fourth in one or two systems."

The superposition of the calices is remarkable, and I cannot but place a Coral found in the Irish Upper Greensand by Ralph Tate, Esq., F.G.S., in this species.

Locality. Ireland; Upper Greensand. In the Collection of R. Tate, Esq., F.G.S.

VI.—List of Species from the Upper Greensand.

1. *Onchotrochus Carteri*, Duncan.
2. *Smilotrochus tuberosus*, Ed. and H.
3. „ *Austeni*, „
4. „ *elongatus*, Duncan.
5. „ *angulatus*, „
6. *Peplosmilia Austeni*, Ed. and H.
7. *Cyathophora monticularia*, D'Orbigny.
8. *Favia stricta*, Ed. and H.
9. „ *minutissima*, Duncan.
10. *Thamnastræa superposita*, Michelin.
11. *Micrabacia coronula*, Goldfuss, sp.

VII.—Corals from the Red Chalk of Hunstanton, Norfolk.

The Red Chalk of Hunstanton contains several forms of *Madreporaria.* The small fauna has this peculiarity—its species belong to the group of the *Fungidæ* without exception. The specimens are small, usually much worn at the calicular end, and are readily distinguished by their mammiliform appearance and white colour. There are no compound *Fungidæ* in the Red Rock, but only such small simple forms as would now characterise the presence of physical conditions unfavorable for Coral life. The recent simple *Fungidæ* are found at all depths. Vast numbers of fossil specimens are to be collected in the Lower Chalk of Gosau, a few exist in the Upper Greensand and in the Neocomian formations. In the existing Coral-fauna no simple *Fungidæ* are found in the West Indian Seas, whilst the Red Sea, Pacific, and Indian Oceans, abound with them. It is probable that peculiar conditions are necessary for their development.[1]

List of the Species of Corals in the Red Chalk of Hunstanton.

1. *Micrabacia coronula*, Goldfuss, sp.
   „ „ variety, major.
2. *Cyclolites polymorpha*, Goldfuss, sp.
3. *Podoseris mammiliformis*, Duncan.
4. „ *elongata*, Duncan.

[1] For a notice of the geology of the Red Chalk, see Rev. T. Wiltshire's communication to the Geol. Soc., Feb., 1869.

Family—FUNGIDÆ.

Sub-Family—Funginæ.

Genus—Micrabacia.

There are specimens of a small form of *Micrabacia coronula,* Goldf., sp., and of a large variety, in the Red Rock (pl. IX, fig. 1). The species is well known in the Upper Greensand of England and in the Chalk of Essen. There is another species, which is hardly distinguishable from *M. coronula* in the Neocomian of Caussols (Var.).

The variety of the species found in the Red Rock rather resembles the Neocomian species in its diameter and flatness. The genus had a very short vertical range, and was represented in later times by the *Stephanophylliæ.*

Sub-Family—LOPHOSERINÆ.

Genus—Cyclolites.

This genus almost characterises the geological horizon of the Craie tuffeau; Gosau, Ile d'Aix, les Martigues, Vaucluse, Corbières, Uchaux, &c., having deposits in which numerous species have been found. A few species are found in the White Chalk, and in the Eocene and Miocene deposits. There are some doubtful Neocomian species, and the genus is extinct.

Cyclolites polymorpha, *Goldfuss,* sp. Pl. IX, fig. 18.

The corallum is very irregular in shape, generally sub-elliptical, and not very tall.

The highest point of the calice is not central, and the central fossula is very variable in its place.

The septa are very numerous, thin, close, flexuous, crenulate, and occur in series of fours.

The solitary specimen of this form is small, but the fossula and the septa are tolerably distinct.

Locality. Hunstanton. In the Collection of the Rev. T. Wiltshire, F.G.S.

SUB-FAMILY—LOPHOSERINÆ.

Genus nov.—PODOSERIS, *Duncan.*

The corallum has a large concave base, by which it is attached to foreign bodies.

The epitheca commences at the basal margin, and is stout and reaches the calicular margin.

The height of the corallum varies.

The calice is generally smaller than the base, and is convex.

The septa are numerous and unequal, the largest reaching a rudimentary columella.

The central fossula is circular and small.

The costæ are seen when the epitheca is worn; they are distinct, connected by synapticulæ, and are straight.

The genus has been created to admit *Micrabaciæ* with adherent bases and more or less of a peduncle.

1. PODOSERIS MAMMILIFORMIS, *Duncan.* Pl. IX, figs. 2—15.

The corallum is short, straight, and broad. The base is concave, and is either larger than the calice or there is a constriction immediately above it, and it is slightly smaller than the calice.

The calice is round, convex, depressed in the centre, and is bounded by the epitheca.

The laminæ are stout, unequal, curved superiorly, and often join.

There are five cycles in six systems, the last cycle being very rudimentary.

The synapticulæ are numerous.

The costæ are straight and subequal, and are smaller than the septa.

The ornamentation of the septo-costal apparatus varies, and there may be an almost moniliform series of enlargements on the septa, or they may be plain.

The columella is formed principally by the ends of the longest septa.

The height of the corallum appears to be determined by the growth of the body between the base and the calice.

Height of the corallum, $\frac{1}{4}$ inch. Breadth at the calicular margin $\frac{1}{3}$rd inch.
" " " $\frac{1}{6}$th " " " " $\frac{1}{4}$ "
" " " $\frac{1}{2}$ " " " " $\frac{11}{30}$ths "

Monstrosities are often found amongst specimens of this species.

Locality. Hunstanton. In the Collection of the Rev. T. Wiltshire, F.G.S.

2. Podoseris elongata, *Duncan*. Pl. IX, figs. 16, 17.

The corallum is tall, with a broad, circular, and slightly concave base, a long, conico-cylindrical stem; and a small calice, much narrower than the base.

The epitheca is in bands.

The costæ are alternately large and very small, somewhat distant, wavy, and united by synapticulæ, many of which are oblique.

The septa frequently unite by their axial ends. There appear to be five cycles of septa.

The base of the corallum has a cellular tissue, probably from the fossilization of some body to which it was adherent.

Height $\frac{5}{8}$ inch. Breadth of base $\frac{1}{2}$ inch. Calice $\frac{1}{4}$ inch.

The shape of this species is most unusual.

Locality. Hunstanton. In the Collection of the Rev. T. Wiltshire, F.G.S.

A MONOGRAPH

OF THE

BRITISH FOSSIL CORALS.

(SECOND SERIES.)

PART II.—No. 2.

I.—CORALS FROM THE UPPER GREENSAND OF HALDON.

SOME time after the 'Supplement to the Monograph of the Fossil Corals of the Upper Greensand' was published several very interesting specimens of fossil Corals were submitted to examination from the deposit at Haldon, in Devonshire.[1] It was necessary to describe them, for they had not been previously noticed, and this could not be done before the Corals from the Red Chalk were published. The Corals from Haldon should have been described amongst those of the Upper Greensand. It is, of course, evident that the list of Upper Greensand species (p. 23) is incomplete.

MADREPORARIA APOROSA.

FAMILY—ASTRÆIDÆ.

Sub-family—EUSMILINÆ.

Genus—PLACOSMILIA.

1. PLACOSMILIA CUNEIFORMIS, *Ed. and H.* Pl. X, figs. 1—5.

The corallum is much compressed, and deltoid in shape.

The costæ are delicate, close, slightly prominent, and subequal.

[1] Mr. Vicary, of Exeter, had collected the fossils himself, and pointed out to me their siliceous condition of fossilization.

The calicinal fossa is very narrow, long, and shallow.

The septa are close, alternately thick and thin. They number (in full-sized calices) 176.

The columella is lamelliform and indistinct.

Locality. Haldon. In the Collection of William Vicary, Esq., F.G.S., Exeter.

The specimen figured in Pl. X is a young corallum, and has only five cycles of septa. Its granular costæ and the peculiar striation of its septa are very characteristic.

The height of the specimen is $\frac{1}{2}$ inch, and the length of the calice is rather more. The breadth is $\frac{3}{10}$ths inch.

The *Placosmiliæ* hitherto described are from the Craie tuffeau and the Hippurite Chalk of Soulage and Bains de Rennes (Corbières), Les Martigues, Uchaux, Obourg near Mons, and Gosau.

2. Placosmilia Parkinsoni, *Ed. and H.* Pl. X, figs. 6 and 7.

Placosmilia consobrina, *Reuss.*

The corallum is tall, compressed, conical, and slightly curved.

The costæ are fine and separated by decided intercostal spaces.

The calice is subelliptical in shape.

The fossa is narrow and shallow.

The columella is feebly developed.

There are five cycles of septa, and the laminæ are very unequal.

Locality. Haldon. In the Collection of William Vicary, Esq., F.G.S., Exeter.

The specimen from Haldon is somewhat rolled and worn. The height is $\frac{3}{10}$ths inch. The breadth of the calice is $\frac{3}{10}$ths inch, and its length is $\frac{6}{10}$ths inch.

Placosmilia Parkinsoni has been found at Gosau, in the Corbières, and at Uchaux.

3. Placosmilia magnifica, *Duncan.* Pl. X, figs. 11—13.

The corallum is compressed, short, very elongate, and the calicular margin is curved and rounded.

The calice is very long, curved, rounded at each end, compressed, very open, and shallow.

The septa are unequal, distant, large, and curved; they correspond to costæ of the same size. There are five cycles of septa.

The columella is lamellar, very much developed, thick, continuous, long, and slightly prominent in the calicular fossa.

The costæ are unequal and distant.

The exotheca is inclined and very strongly developed.

Height of the corallum, $1\frac{1}{4}$ to $1\frac{1}{2}$ inch. Length of the calice, $2\frac{1}{2}$ to $3\frac{1}{10}$th inches. Breadth of the calice, $\frac{9}{10}$ths to $1\frac{1}{10}$th inch.

Locality. Haldon. In the Collection of William Vicary, Esq., F.G.S., Exeter.

This fine species is strongly Placosmilian, and might be taken as the type of the genus.

Genus—Peplosmilia.

Peplosmilia depressa, *E. de Fromentel.* Pl. X, figs. 8—10.

The corallum is not very tall, and shows traces of epitheca.

The calice is shallow and round.

The septa are well developed and thin. There are more than four cycles, and probably a fifth exists in full-grown individuals.

The columella is very thin and narrow.

Height, $\frac{1}{2}$ inch. Breadth of calice, $\frac{6}{10}$ths inch.

Locality. Haldon. In the Collection of William Vicary, Esq., F.G.S., Exeter.

M. de Fromentel, Pal. Franç., Terr. Crét.,' pl. 46, fig. 1, 1863, and page 241, states that his specimens came from the Upper Greensand of Mans.

The specimen from Haldon is fragmentary, and its columella is defective, but it is so like M. de Fromentel's delineation of *Peplosmilia depressa* that there is no doubt about its being of that species.

Division—Astræaceæ.

Genus—Astrocœnia.

Astrocœnia decaphylla, *Ed. and H.* Pl. XI, figs. 1—6.

This species, described by MM. Milne-Edwards and Jules Haime ('Ann. des Sci. nat.,' 3me série, t. x, p. 298, 1849) was subsequently named *Astræa reticulata* by D'Orbigny (1850), and was noticed as *Astrocœnia magnifica* by Reuss in his great work on the Corals of Gosau ('Denkschr. der Wien Akad. der Wissensch.,' t. vii, p. 94, pl. 8, figs. 4—6, 1854).

Reuss's admirable delineation of the species enables the British form to be recognised at once, and it even possesses the curious transverse arrangement of the walls of some calices which renders the comprehension of Reuss's sixth figure rather difficult.

The *Astrocœniæ* have been fully considered in the 'Monograph of the Liassic Corals' (Pal. Soc., 1867).

Astrocœnia decaphylla is a rather variable species, on account of the preponderance or deficiency, as the case may be, of cœnenchyma. The size of the costæ is limited by the cœnenchyma, and when this is very deficient they are almost rudimentary.

There are ten principal and ten secondary septa; the secondary are the smallest, and do not reach the styliform columella like the primary. They are slightly spined towards their inner margin. The costæ are small. The columella is well developed, and is essential and styliform. The shape of the calices varies; in some places they are circular, and in others polygonal.

Locality. Haldon. In the Collection of William Vicary, Esq., F.G.S., Exeter.

The British specimens are not to be distinguished from those of the Hippurite Chalk of Gosau, or of the Craie tuffeau of Corbières.

Astrocœnia decaphylla was a very persistent form. It resembles in some of its peculiar structures the Astrocœnias of the Lias, and a specimen from the Miocene coralliferous strata of Jamaica[1] cannot be distinguished from the form from Gosau.

Genus—Isastræa.

Isastræa Haldonensis, *Duncan*. Pl. XI, figs. 7 and 8.

The corallum is hemispherical.

The calices are large, irregular in size, very deep, and rather quadrangular.

The wall is thin.

The septa are crowded, small, long, and there are five cycles of them in the largest calices.

There is a disposition to serial growth in some calices.

Diameter of the largest calices, nearly $\frac{1}{2}$ inch.

Locality. Haldon. In the Collection of William Vicary, Esq., F.G.S., Exeter.

The depth and size of the calices, their thin walls, and the numerous septa, distinguish this species, whose closest allies are *Isastræa lamellosissima*, Michelin, sp., from the Craie tuffeau of Uchaux, *Isastræa Haidingeri*, Ed. and H., from the same formation at Piesting, in the Eastern Alps, and *Isastræa tenuistriata*, M'Coy, sp., of the Inferior Oolite.

[1] Duncan, "West Indian Corals," 'Quart. Journ. Geol. Soc.,' Nov., 1863, vol. xix, page 440.

LIST OF UPPER GREENSAND CORALS FROM HALDON.

1. *Placosmilia cuneiformis*, Ed. and H.
2. „ *Parkinsoni*, „
3. „ *magnifica*, Duncan.
4. *Peplosmilia depressa*, E. de Fromentel.
5. *Astrocœnia decaphylla*, Ed. and H.
6. *Isastræa Haldonensis*, Duncan.

Peplosmilia Austeni, Ed. and H., and *Favia stricta*, Ed. and H., are also found at Haldon. They have been already noticed as Upper Greensand forms.

II.—CORALS FROM THE GAULT.

Only six well-marked species of Corals were known to MM. Milne-Edwards and Jules Haime as having been found in the Gault. They were all simple or solitary forms, and such as one would expect to find in moderately deep water. It is evident that the area occupied by the English Gault was not the Coral tract of the period. The resemblance of the Coral-faunas of the Gault and the London Clay is somewhat remarkable, and probably the physical conditions of the area during the deposition of the strata were not very dissimilar.

The following pages contain the descriptions of some species which were not known to MM. Milne-Edwards and Jules Haime, and some notices of the most important forms they described.

MADREPORARIA APOROSA.

FAMILY—TURBINOLIDÆ.

Sub-family—CARYOPHYLLINÆ.

Division—CARYOPHYLLIACEÆ.

Genus—CARYOPHYLLIA.

MM. Milne-Edwards and Jules Haime have changed the generic term *Cyathina* into that of its predecessor *Caryophyllia;* consequently *Cyathina Bowerbanki*, Ed. and H., is now called *Caryophyllia Bowerbanki*, Ed. and H. ('Hist. Nat. des Corall.,' vol. ii, p. 18).

A very interesting variety of this species is in the Rev. T. Wiltshire's Collection, and has its costæ running obliquely to the long axis of the corallum. They are profusely granulated (Pl. XII, figs. 8, 9).

Division—Trochocyathaceæ.

Genus—Trochocyathus.

1. Trochocyathus Harveyanus, *Ed. and H.*

This species was described by MM. Milne-Edwards and Jules Haime in their 'Monograph of the British Fossil Corals,' Part I, p. 65. They associated it with two species, which are, as they suggest, indistinguishable, viz. *Trochocyathus Koenigi* and *Trochocyathus Warburtoni*. The first of these species is the *Turbinolia Koenigi* of Mantell.

An examination of a series of specimens attributed to *Trochocyathus Harveyanus*, Ed. and H., and the consideration of the value of the *Trochocyathi* just mentioned, have led me to recognise five forms of *Trochocyathi breves*, all closely allied and well represented by the original type of *Trochocyathus Harveyanus*, Ed. and H. When placed in a series with this *Trochocyathus* at the head, there is a gradation of structure which prevents a strictly specific distinction being made between the consecutive forms; but when the first and the last forms are compared alone, no one would hesitate to assert that there is a specific distinction between them. All the forms are simple, short, and almost hemispherical; all have four cycles of septa, and the same proportion of pali. These are the primary and most essential peculiarities of the genus.

The costæ differ in their size, prominence, ornamentation, and relation to the septa in some of the forms; and the exsert nature of the septa, their granulation, and the size of the corallum, also differ. The structural differences are seen in many examples, and are therefore more or less persistent; nevertheless it is found that, whilst several specimens have the septa springing from intercostal spaces instead of from the ends of the costæ, one or more, having all the other common structural peculiarities, present septa arising from the costal ends. This method of origin can hardly constitute a specific distinction. I propose to retain *Trochocyathus Harveyanus* as the type of a series of forms the sum of whose variations in structure constitutes the species.

Variety 1 (Pl. XIII, figs. 1, 2).—The corallum is nearly double the size of the type; its septa are rather exsert, and are very granular.

The costæ are very prominent, ridged, marked with numerous small pits, and are continuous with the septa.

The epitheca is waved and well developed. The spaces between the larger costæ are more or less augular.

The peduncle is large.

Locality. Gault, Folkestone. In the British Museum.

Variety 2 (Pl. XIII, figs. 3, 4).—The corallum is as large as that of varicty 1, but it is more conical.

The costæ are less pronounced, and the septa, which are more granular than those of variety 1, arise from the intercostal spaces. The costal ends are very elegant in shape, and form a margin of rather sharp curves, side by side.

Locality. Gault, Folkestone. In the British Museum.

Variety 3 (Pl. XII, figs. 1, 3, 4; and Pl. XIII, fig. 13).—The corallum is rather flat, but hemispherical.

The septa are not exsert, and they arise from the costal ends.

The costæ are equal; none are more prominent than others. They are all rather broad, flat, and beautifully ornamented with diverging curved lines. Their free ends are equal and curved.

Locality. Gault, Folkestone. In the Collection of thc Rev. T. Wiltshire, F.G.S.

Variety 4.—The corallum and costæ are like Variety 3, but the septa arise from the intercostal spaces.

Locality. Gault, Folkestone. In the Collection of the Rev. T. Wiltshire, F.G.S.

Variety 5 (Pl. XII, fig. 2).—The corallum is rather more conical inferiorly than in Varieties 3 and 4.

The septa are exsert, and project slightly beyond the costal margin.

The costæ are all rudimentary.

The epitheca is well developed, and reaches up to the septa.

Locality. Gault, Folkestone.

The forms may be distinguished as follows:

With more or less ridged costæ . . .	The type. Variety 1. „ 2.
With nearly equal flat costæ . . .	Variety 3. „ 4.
Costæ rudimentary	Variety 5.
Septa arising from the costal ends . .	The type. Variety 1. „ 3.
Septa arising from the intercostal spaces . .	Variety 2. „ 4.

All the forms have four cycles of septa and pali before the first, second, and third orders.

An ill-developed and monstrous form is shown in Pl. XIV, figs. 1—5.

2. Trochocyathus Wiltshirei, *Duncan.* Pl. XIV, figs. 10—12.

The corallum is straight, conical, and either cylindrical above or compressed. Its base presents the trace of a peduncle for attachment.

The epitheca is scanty and in transverse masses.

The costæ are distinct and subequal.

The calice is very open and rather deep.

The septa are unequal, hardly exsert, and broad at the margin of the calice. There are four cycles of septa, and six systems.

The pali are large, and are placed before all the cycles except the last.

The columella is rudimentary.

Height, $\frac{3}{10}$ths inch. Breadth of calice, $\frac{3}{10}$ths inch.

Locality. Gault, Folkestone. In the Museum of the Royal School of Mines, and in the Collection of the Rev. T. Wiltshire, F.G.S.

This species is closely allied to *Trochocyathus conulus*, Phillips, sp. The compressed calice, the rudimentary columella, and the shape of the corallum, distinguish the new species from *Trochocyathus conulus.*

Genus—Leptocyathus.

1. Leptocyathus gracilis, *Duncan.* Pl. XIII, figs. 5—8.

The corallum is small, flat, and circular in outline.

The costæ are very prominent, and join exsert septa. The primary and secondary costæ are very distinct, and the others less so. All the costæ unite centrally at the base. Many are slightly curved.

The septa are thick externally, very unequal, thin internally, and the largest are more exsert than the others. There are six systems and four cycles of septa.

The pali are small and exist before all the septa.

The columella is very rudimentary.

The calicular fossa is rather wide and shallow.

Height, hardly $\frac{1}{10}$th inch. Breadth, $\frac{3}{10}$ths inch.

Locality. Gault, Folkestone. In the British Museum.

This species is very closely allied to *Leptocyathus elegans*, Ed. and H., of the London

Clay. *Leptocyathus elegans* has not a flat base, and it has very granular septa. Moreover, its costæ are large and small in sets. Nevertheless the alliance is of the closest kind.

Genus—Bathycyathus.

MM. Milne-Edwards and Jules Haime described a species of this genus in their Monograph of the British Fossil Corals,' Part I, pp. 67, 68. Two specimens in the Collection of Rev. T. Wiltshire present all the appearances recognised by those distinguished authors. The costæ are very granular, and not in a simple row. In one specimen the breadth of the base is very great (Pl. XII, figs. 5—7).

Family—TURBINOLIDÆ.

Sub-Family—Turbinolinæ.

Division—Tubinoliaceæ.

Genus—Smilotrochus.

Some species of this genus were described amongst the Corals from the Upper Greensand,[1] and one was noticed as belonging to this geological horizon which should have been included with the Lower Greensand forms.

The Upper Greensand *Smilotrochi* are—

Smilotrochus tuberosus, Ed. and H.
,, *elongatus*, Duncan.
,, *angulatus*, ,,

There are four species of the genus found in the Gault, which are all closely allied. One of them cannot be distinguished from *Smilotrochus elongatus* of the Upper Greensand.

The specimens of this species found in the Upper Greensand are invariably worn and rolled, and are generally in the form of casts; but in the Gault the structural details are well preserved, and even the lateral spines on the septa are distinct.

The Gault forms are shorter and more cylindro-conical and curved than those from the Upper Greensand.

[1] See *antè*, p. 19.

The species of the genus *Smilotrochus* from the Gault are as follows:

1. *Smilotrochus elongatus*, Duncan.
,, *cylindricus*, ,,
,, *granulatus*, ,,
,, *insignis*, ,,

1. Smilotrochus elongatus, *Duncan*. Pl. XII, figs. 10—16; Pl. XIII, figs. 10—12; and Pl. XIV, figs. 13—15.

This species is described at page 19 of the first number of this Part, and is figured in Pl. VII, figs. 1—6.

Locality. Folkestone. In the Collection of the Royal School of Mines.

The lateral spines of the septa are very well marked, and the costæ are equal in size in this species. Its septal number varies, on account of the very late perfection of the fourth cycle of septa.

2. Smilotrochus cylindricus, *Duncan*. Pl. XIV, fig. 16.

The corallum is small, cylindrical, nearly straight, and has a truncated base.

The costæ are equal, very distinct above, and rudimentary below and in the middle. They are marked with a few large granules in one series.

The septa are subequal, very exsert, thin, close, and marked with large granules, few in number. The septa are in six systems, and there are three cycles.

Height, $\frac{3}{10}$ths inch. Greatest breadth, rather less than $\frac{2}{10}$ths inch.

Locality. Gault, Folkestone. In the Collection of the Rev. T. Wiltshire, F.G.S.

3. Smilotrochus granulatus, *Duncan*. Pl. XIV, fig. 17.

The corallum is conico-cylindrical in shape, and has a more or less truncated base.

The costæ are subequal, prominent, very granular, and distinct superiorly.

The septa are subequal, thick, and very granular. The septa are in six systems, and there are three cycles.

Height, $\frac{2}{10}$ths inch. Breadth, $\frac{3}{20}$ths inch.

Locality. Gault, Folkestone. In the Collection of the Rev. T. Wiltshire, F.G.S.

4. SMILOTROCHUS INSIGNIS, *Duncan*. Pl. XII, fig. 17; and Pl. XIV, fig. 18.

The corallum is trochoid, short, and has a wide calice, and a conical and rounded base.

The calice is circular in outline; the fossa is deep and small, and the septa are wide, exsert, curved above, and so marked with one row of granules that their free margin appears to be spined. There are three cycles of septa, and the orders are nearly equal as regards size.

The costæ are large, prominent, broad at their base, and are marked with one row of granules on the free surface.

Height, $\frac{2}{10}$ths inch. Breadth of calice, $\frac{2}{10}$ths inch.

Locality. Gault, Folkestone. In the Collection of the Rev. T. Wiltshire, F.G.S.

An analysis of the genus will be found after the description of the species from the Lower Greensand.

There is a compound or aggregate Madreporarian found in the Gault of Folkestone. It has much endotheca, and resembles worn specimens of the well-known *Holocystis elegans* of the Lower Greensand. The specimens are not sufficiently well preserved for indentification with any genus.

Family—FUNGIDÆ.

Sub-family—FUNGINÆ.

Genus—MICRABACIA.

1. MICRABACIA FITTONI, *Duncan*. Pl. XIV, figs. 6—9.

The corallum is nearly hemispherical in shape. Its base is flat, and extends beyond the origin of the septa in a sharp and uninverted margin. The breadth of the base exceeds the height of the corallum.

The costæ are flat, straight, convex externally at the calicular margin, and equal.

The septa are unequal, much smaller than the costæ. There are four cycles of septa, in six systems.

The synapticulæ between the septa are large.

Height, $\frac{2}{10}$ths. Breadth, nearly $\frac{1}{2}$ inch.

Locality. Gault, Folkestone. In the Collection of the Rev. T. Wiltshire, F.G.S.

The flat base, the flat costæ, and the limitation of the septal number to four cycles, distinguish this species from *Micrabacia coronula*[1] of the Upper Greensand, and from *Micrabacia Beaumontii*[2], Ed. and H., of the Neocomian.

List of New Species from the Gault.

Variety of *Caryophyllia Bowerbanki,* Ed. and H.
Five varieties of *Trochocyathus Harveyanus,* Ed. and H.
Trochocyathus Wiltshirei, Duncan.
Leptocyathus gracilis, „
Smilotrochus elongatus, „
„ *granulatus,* „
„ *insignis,* „
„ *cylindricus,* „
Micrabacia Fittoni, „

III.—List of Species from the Gault.

1. *Caryophyllia Bowerbanki,* Ed. and H., and one variety.
2. *Trochocyathus conulus,* Phillips, sp.
3. „ *Wiltshirei,* Duncan.
4. „ *Harveyanus,* Ed. and H., and five varieties.
5. *Bathycyathus Sowerbyi,* Ed. and H.
6. *Leptocyathus gracilis,* Duncan.
7. *Cyclocyathus Fittoni,* Ed. and H.
8. *Smilotrochus elongatus,* Duncan.[3]
9. „ *granulatus,* „
10. „ *cylindricus,* „
11. „ *insignis,* „
12. *Trochosmilia sulcata,* Ed. and H.
13. *Micrabacia Fittoni,* Duncan.

[1] 'Hist. Nat. des Coral.,' vol. iii, p. 30.
[2] Ibid., p. 30.
[3] Common to the Gault and Upper Greensand.

IV.—Corals from the Lower Greensand.

One species of Coral was described by MM. Milne-Edwards and Jules Haime from the Lower Greensand, in their 'Monograph of the British Fossil Corals.'

Fitton had noticed a compound Coral in the Lower Greensand, and named it *Astræa* in his "Essay on the Strata below the Chalk," 'Geol. Trans.,' 2nd series, vol. iv, p. 352, 1843. In 1847 he called the species *Astræa elegans*, and Lonsdale separated it from the *Astræidæ* under the name *Cyathophora ? elegans* in 1849 (' Quart. Journ. Geol. Soc.,' vol. v, pt. 1, p. 83, pl. iv, figs. 12, 15, 1849).

MM. Milne-Edwards and Jules Haime recognised the quadrate arrangement of the septa of this species, and classified it amongst the *Rugosa*, in the family *Stauridæ*. Their *Holocystis elegans*, Fitton, sp., is a very good species, and specimens are found varying in the size of the corallum and of the calices.

Since the publication of their 'Monograph on the British Fossil Corals,' MM. Milne-Edwards and Jules Haime have named a species from Farringdon *Smilotrochus Austeni* (' Hist. Nat. des Corall.,' vol. ii, p. 71). I have noticed it inadvertently in my description of the Upper Greensand Corals, p. 19, and Pl. VII, fig. 12. In order to complete this part it is introduced here again.

Family—TURBINOLIDÆ.

Division—Turbinoliaceæ.

Genus—Smilotrochus.

1. Smilotrochus Austeni, *Ed. and H.* Pl. VII, fig. 12.

The corallum is regularly cuneiform, very much compressed below, and slightly elongate.

[1] The following authors have written upon the Fossil Corals of the Gault:

MM. Milne-Edwards and Jules Haime, 'Monograph of the British Fossil Corals;' Pal. Soc.

„ „ 'Hist. Nat des Coralliaires.'

Phillips, 'Illust. of Geol. of Yorkshire.'

Mantell's 'Geol. of Sussex,' Lonsdale in.

Fleming, 'British Animals.'

The authors who have written upon the Corals of the Lower Greensand are—

MM. Milne-Edwards and Jules Haime, opp. citt.

Fitton, 'Quart. Journ. Geol. Soc.,' vol. iii, p. 296, 1847.

Lonsdale, 'Quart. Journ. Geol. Soc.,' vol. v, p. 83.

M. de Fromentel has paid especial attention to the French Neocomian Corals; and C. J. Meyer, Esq., F.G.S., has enabled me to study the most interesting species in his collection.

The calice is elliptical; the summit of the larger axis is rounded.

Forty-eight costæ, subequal, straight, fine, and granular.

Height of the corallum, about $\frac{1}{3}$rd inch.

Locality. Farringdon.

MM. Milne-Edwards and Jules Haime do not mention where their specimen is deposited. Mr. Vicary, of Exeter, has a fine specimen of this Coral.

The genus *Smilotrochus* has become of some importance in the palæontology of the Cretaceous rocks. The species are distributed as follows in Great Britain:

Smilotrochus tuberosus, Ed. and H.		
" *elongatus*, Duncan		Upper Greensand.
" *angulatus*, "		
" *elongatus*, "		
" *granulatus*, "		Gault.
" *insignis*, "		
" *cylindricus*, "		
" *Austeni*, Ed. and H.		Lower Greensand.

Smilotrochus elongatus, Duncan, is found in the Gault and Upper Greensand. *Smilotrochus Hagenowi*, Ed. and H., is a fossil from the Maestricht Chalk (Ed. and H., 'Hist. Nat. des Corall.,' vol. ii, p. 71). *Smilotrochus irregularis*, E. de Fromentel, is a small cornute form, with rounded primary costæ and rather an open calice; it is from the Chalk ('Pal. Franç.,' tome viii, livraison 4, Zooph., pl. ix).

Sub-family—Caryophyllinæ.

Division—Caryophylliaceæ.

Genus—Brachycyathus.

1. Brachycyathus Orbignyanus, *Ed. and H.* Pl. XV, figs. 8, 9.

The corallum is very short.

The costæ are indistinct.

The septa are long, very slightly exsert, granulated from below upwards, and there are four cycles in six systems. The primary and secondary septa are equal. The tertiary are a little longer than those of the fourth cycle. All are thin and straight.

The pali are like continuations of the tertiary septa before which they are placed. They are granular.

Height, $\frac{1}{10}$th inch. Breadth, $\frac{6}{10}$ths inch.

Locality. East Shalford, Surrey. Base of the Lower Greensand; found with *Cerithium Neocomiense*, D'Orb.; *Exogyra subplicata*, Tqm.; *Arca Raulini*, Leym.; *Terebratula sella*, Sow. In the Collection of C. J. A. Meyer, Esq., F.G.S.

The specimen upon which the genus was founded was found in the Neocomian formation of the Hautes Alpes, at St. Julien, Beauchène. I have added to the original description, as some portions of the English specimen are better preserved than the type.

Family—ASTRÆIDÆ.

Sub-family—Eusmilinæ.

Division—Trochosmiliaceæ.

Genus—Trochosmilia.

Trochosmilia Meyeri, *Duncan*. Pl. XV, figs. 1—7.

The corallum is small, cylindrical or cylindro-conical. Its base may be wide or very small, and was adherent.

The epitheca is complete.

The costæ are very small, and are occasionally seen where the epitheca is worn.

The calice is rather deep.

The septa are crowded, unequal, spined near the axis, and form six systems. There are four cycles of septa.

The calice is usually circular in outline, but it is occasionally compressed.

The axial space is small.

The endotheca is very scanty.

Height, $\frac{4}{10}$ths inch. Greatest breadth, $\frac{2}{10}$ths inch.

Variety.—The corallum is short, broad, cylindrical, slightly constricted centrally, and has a broad base.

Height, $\frac{3}{20}$ths inch. Breadth, $\frac{8}{20}$ths inch.

Locality. Bargate Stone; upper division of the Lower Greensand. Guildford, Surrey. Found with *Avicula pectinata*, Sow. In the Collection of C. J. A. Meyer, Esq., F.G.S.

These small *Trochosmiliæ* are common in the Bargate Stone, where they were discovered by Mr. Meyer, from whom I have obtained the names of the associated fossils. The presence of epitheca would apparently necessitate these fossils being placed in a new genus, but, after a careful examination of the bearings of the absence or presence of epithecal structures upon the natural classification of simple Corals, I do not think the point sufficiently important to bring about the separation of Mr. Meyer's little Corals from the *Trochosmiliæ*. They form (*i. e.* the type and the variety) a sub-genus of the *Trochosmiliæ*.

Sub-family—Astræinæ.

Division—Astræaceæ.

Genus—Isastræa.

Isastræa Morrisii, *Duncan*. Pl. XV, figs. 10—12.

The corallum is flat and very short. The corallites are unequal, and usually five-sided.

There is no columella.

The wall is thin.

The septa are slender, unequal, and most of them reach far inwards. There are in the perfect calices three cycles of septa in six systems. Usually some of the septa of the third cycle are wanting.

Breadth of a calice, rather more than $\frac{1}{10}$th inch.

Locality. Bargate Stone, Guildford, Surrey; with *Terebratella Fittoni*, Meyer. In the Collection of C. J. A. Meyer, Esq., F.G.S.

This small *Isastræa* is usually found as a cast, and the restored drawing is taken from an impression. The central circular structure is due to fossilization.

The species is closely allied to *Isastræa Guettardana*, Ed. and H., of the Lower Chalk of Uchaux.

Family—FUNGIDÆ.

Sub-family—Lophoserinæ.

Genus—Turbinoseris.

Genus nov.—Turbinoseris. The corallum is simple, more or less turbinate, or constricted midway between the base and calice. The base is either broad and adherent, or small and free.

There is no epitheca, and the costæ are distinct.

There is no columella, and the septa unite literally, and are very numerous.

Turbinoseris De-Fromenteli, *Duncan.* Pl. XV, figs. 13—18.

The corallum is tall, and more or less cylindro-turbinate.

The calice is shallow, and circular in outline.

The septa are very numerous, long, thin, straight, and many unite laterally with longer ones. There are 120 septa, and the cyclical arrangement is confused.

The synapticulæ are well developed.

There is no columella, and the longest septa reach across the axial space.

The costæ are well developed, and often are not continuous with the septal ends.

Height, $1\frac{3}{10}$ths inch. Breadth of calice, $1\frac{2}{10}$ths inch.

Variety.—With a constricted wall and large base.

Locality. Atherfield, in the Lower Greensand. In the Collection of the Royal School of Mines.

The necessity for forming a new genus for this species is obvious. It is the neighbour of *Trochoseris* in the sub-family of the *Lophoserinæ.* This last genus has a columella, and the new one has none.

The species has not been hitherto described, but it has been familiarly known as a *Montlivaltia;* but the synapticulæ between the septa and costæ determine the form to belong to the *Fungidæ.*

V.—List of New Species from the Lower Greensand.

1. *Brachycyathus Orbignyanus,* Ed. and H.
2. *Trochosmilia Meyeri,* Duncan.
3. *Isastræa Morrisii,* ,,
4. *Turbinoseris De-Fromenteli,* ,,

VI.—List of the Species from the Lower Greensand.

1. *Brachycyathus Orbignyanus,* Ed. and H.
2. *Smilotrochus Austeni,* Ed. and H.
3. *Trochosmilia Meyeri,* Duncan.
4. *Isastræa Morrisii,* ,,
5. *Turbinoseris De-Fromonteli,* ,,
6. *Holocystis elegans,* Lonsdale, sp.

VII.—List of the Species from the Cretaceous Formations.

A. *Upper and Lower White Chalk.*

1. *Caryophyllia cylindracea*, Reuss, sp.
2. „ *Lonsdalei*, Duncan.
3. „ *Tennanti*, „
4. *Onchotrochus serpentinus*, „
5. *Trochosmilia laxa*, Ed. and H., sp., and three varieties.
6. „ *cornucopiæ*, Duncan.
7. „ *Wiltshirei*, „
8. „ *Woodwardi*, „
9. „ *granulata*, „
10. „ *cylindrica*, „
11. *Parasmilia centralis*, Mantell, sp., and two varieties.
12. „ *cylindrica*, Ed. and H.
13. „ *Fittoni*, „
14. „ *serpentina*, „
15. „ *monilis*, Duncan.
16. „ *granulata*, „
17. *Diblasus Gravensis*, Lonsdale.
18. *Synhelia Sharpeana*, Ed. and H.
19. *Stephanophyllia Bowerbanki*, Ed. and H.

B. *Upper Greensand.*

20. *Onchotrochus Carteri*, Duncan.
21. *Smilotrochus tuberosus*, Ed. and H.
22. „ *elongatus*, Duncan.
23. „ *angulatus*, „
24. *Peplosmilia Austeni*, Ed. and H.
25. *Cyathophora monticularia*, D'Orbigny.
26. *Favia stricta*, Ed. and H.
27. „ *minutissima*, Duncan.
28. *Thamnastræa superposita*, Michelin.
29. *Micrabacia coronula*, Goldfuss, sp.
30. *Placosmilia cuneiformis*, Ed. and H.
31. „ *Parkinsoni*, „
32. „ *magnifica*, Duncan.
33. *Peplosmilia depressa*, E. de Fromentel.
34. *Astrocœnia decaphylla*, Ed. and H.
35. *Isastræa Haldonensis*, Duncan.

c. *Red Chalk of Hunstanton.*

36. *Cyclolites polymorpha*, Goldfuss, sp.
37. *Podoseris mammiliformis*, Duncan.
38. „ *elongata*, „
39. *Micrabacia coronula*, Goldfuss, sp., and variety.

D. *Gault.*

40. *Carophyllia Bowerbanki*, Ed. and H., and a variety.
41. *Trochocyathus conulus*, Phillips, sp.
42. „ *Wiltshirei*, Duncan.
43. „ *Harveyanus*, Ed. and H., and five varieties.
44. *Bathycyathus Sowerbyi*, Ed. and H.
45. *Leptocyathus gracilis*, Duncan.
46. *Cyclocyathus Fittoni*, Ed. and H.
47. *Smilotrochus elongatus*, Duncan.
48. „ *granulatus*, „
49. „ *insignis*, „
50. „ *cylindricus* „
51. *Trochosmilia sulcata*, Ed. and H.
52. *Micrabacia Fittoni*, Duncan.

E. *Lower Greensand.*

53. *Brachycyathus Orbignyanus*, Ed. and H.
54. *Smilotrochus Austeni*, „
55. *Trochosmilia Meyeri*, Duncan.
56. *Isastræa Morrisii*, „
57. *Turbinoseris De-Fromenteli*, Duncan.
58. *Holocystis elegans*, Lonsdale, sp.

Micrabacia coronula is common to the Upper Greensand and the Red Chalk. *Smilotrochus elongatus* is found in the Gault and in the Upper Greensand.

The number of species of *Madreporaria* in the British Cretaceous formations is therefore fifty-six.

MM. Milne-Edwards and Jules Haime had described twenty-three species before this series was commenced. Of these I have ventured to suppress *Parasmilia Mantelli*, *Trochocyathus Koenigi*, and *Trochocyathus Warburtoni*.

The Coral-fauna of the British area was by no means well developed or rich in genera during the long period during which the Cretaceous sediments were being deposited. The Coral tracts of the early part of the period were on the areas now occupied by the Alpine Neocomian strata, and those of the middle portion of the period were where the Lower Chalk is developed at Gosau, Uchaux, and Martigues.

There are no traces of any Coral reefs or atolls in the British Cretaceous area, and its Corals were of a kind whose representatives for the most part live at a depth of from 5 to 600 fathoms.

A MONOGRAPH

OF THE

BRITISH FOSSIL CORALS.

SECOND SERIES.

BY

P. MARTIN DUNCAN, M.B.Lond., F.R.S., F.G.S.,

PROFESSOR OF GEOLOGY TO, AND HONORARY FELLOW OF, KING'S COLLEGE, LONDON.

Being a Supplement to the
'Monograph of the British Fossil Corals,' by MM. Milne-Edwards *and* Jules Haime.

PART III.

Corals from the Oolitic Strata.

Pages 1—24; Plates I—VII.

LONDON:

PRINTED FOR THE PALÆONTOGRAPHICAL SOCIETY.

1872.

PRINTED BY
ADLARD AND SON, BARTHOLOMEW CLOSE.

CONTENTS OF SUPPLEMENT TO THE OOLITIC CORALS.

	PAGE
Introduction	1
List of Species already described from the Oolitic Strata	4
List of New Species	6
List of all the Species described	7
General Relations of the Oolitic Coral-faunas of Great Britain	9
Description of New Species from the Great Oolite	14
Description of New Species from the Inferior Oolite	16

A MONOGRAPH

OF THE

BRITISH FOSSIL CORALS.

(SECOND SERIES.)

PART III.

1.—INTRODUCTION.

THIS Part concludes the description of the new species of Fossil Corals which have been discovered in the Secondary rocks of Great Britain and Ireland since the appearance of the Monograph by Messrs. Milne-Edwards and Jules Haime, of which this work forms the continuation.

It treats of the Corals from those Jurassic strata which are popularly known as the Oolites; and it will, of course, precede, in the arrangement of the volume, the parts relating to the Liassic Corals, which have already been published.

The following authors have contributed to our knowledge of the Oolitic Corals:— R. Plot, 'Nat. Hist. Oxfordshire,' 1676. J. Walcott, 'Descr. and Fig. of Petref. found near Bath,' 1779. Parkinson, 'Organic Remains,' 1808. W. Smith, 'Strata Identified,' 1816. W. Conybeare and W. Phillips, 'Outlines of the Geol. of Eng. and Wales,' 1822. Fleming, 'British Animals,' 1828. G. Young, 'Geol. Survey of York,' 1828. J. Phillips, 'Geol. of Yorkshire,' 1829. R. C. Taylor, 'Mag. Nat. Hist.,' 1830. S. Woodward, 'Synopt. Table of Org. Rem.,' 1830. E. Bennet, 'Cat. Org. Remains, Wilts,' 1837. Fitton, "Strata below the Chalk," 'Geol. Trans.,' 2nd series, 1843. Morris, 'Cat. of British Fossils,' 1843. M'Coy, 'Ann. Nat. Hist.,' 1848 (several essays). MM. Milne-Edwards and Jules Haime, 'Monog.' (Pal. Soc.), 1851. T. Wright, M.D., F.G.S., 'Cotteswold Club Trans.,' 1866.

An analysis of the work of these authors, with the exception of that of Dr. Wright, is found scattered over the pages of MM. Milne-Edwards and Jules Haime's "Monograph of the Oolitic Corals," Pal. Soc., 1851. No new species of fossil Corals have been described from the Oolitic rocks since that date until very recently. During the last year or two, however, I have added to the species already known five from the Great Oolite, and thirteen from the Inferior Oolite. A careful study of the *Thecosmiliæ* of the Inferior Oolite at Crickley has enabled me to distinguish five very remarkable varieties of *Thecosmilia gregaria*, M'Coy, sp., and to satisfy myself that the relations of the *Thecosmiliæ* of the Lias to the genera *Isastræa*, *Latimæandra*, and others were repeated in the Inferior Oolite. There are specimens of *Thecosmilia gregaria* in Dr. Wright's collection which, had I not had a considerable series to examine from other sources, might have been associated with Reuss's new genus *Heterogyra*, together with *Symphyllia* and *Latimæandra*. The relation of these genera (except *Heterogyra*) to *Montlivaltia* has been noticed in the first Report (Brit. Assoc. Report, Norwich, p. 106 *et seq.*), and there is a clear proof that the same phenomena of evolution may occur consecutively. That is to say, the St. Cassian *Montlivaltiæ* and *Thecosmiliæ* varied and became permanent, compound, and serial Corals of such genera as *Elysastræa*, *Isastræa*, and *Latimæandra;* then the Liassic *Thecosmiliæ* did the same; and now it is evident that a *Montlivaltia* of the Inferior Oolite occasionally took on fissiparous growth, and superadded to others a marginal gemmation and a serial growth, and evolved forms which cannot be distinguished from those of the genera above mentioned and *Symphyllia* and *Heterogyra*. There was evidently an inherent power of variation which declared itself in the same direction during the ages which witnessed the formation of the St. Cassian and the Liassic and the Lower Oolitic deposits; and it is impossible to deny a genetic value to these oft-repeated structural phenomena.

One of the *Thecosmiliæ* from the Inferior Oolite at Crickley, which I have named *Thecosmilia Wrighti*, is very closely related to one of the Lower Liassic species.

It is interesting to find the genus *Cyclolites* represented in the Inferior Oolite by two well-marked species, one of which is like the rest of the forms of the genus in shape, and the other is exceptional in its trochoid form. This last species has, however, all the other characteristics of the genus. The *Cyclolites* are extinct; they flourished in the earlier Cretaceous seas, and lasted during the Miocene. MM. Milne-Edwards and Jules Haime ('Hist. Nat. des Corall.') mention that the genus originated in the Jurassic age, but they produce no evidence to substantiate the assertion.

A form belonging to a new genus of the *Fungidæ* was found by Mr. Mansel at East Coker in the Inferior Oolite In general shape and in the arrangement of the calices the specimen resembles *Dimorphastræa;* but the existence of synapticulæ between the septa and between the costæ necessitates its association with the *Fungidæ*. There is a central calice, and the others are in a circle around it, being separated by long horizontal septo-costal prolongations; the whole is surrounded by an epitheca, and forms a turbinate shape, the free surface being flat and circular. This genus, which I have called *Dimorphoseris*, foreshadows the genera *Cyathoseris* and *Trochoseris* of the Lower Chalk.

Mr. Leckenby discovered the interesting specimens upon which I have founded the genus *Gonioseris*, one of the most extraordinary forms of the *Fungidæ* as yet described.

There are several new species of the genus *Thamnastræa*. *Thamnastræa Browni*, nobis, is remarkable for having in some specimens a long stalk surmounted by a knob-shaped head. The calices are small on the stalk, and very large on the head; so that when the form is examined before it is mature, there is a danger of producing two species instead of one. The stalk often attains the height of three or four inches, In other specimens there is no stalk, and the knob-shaped corallum is sessile.

A large specimen of *Thamnastræa Manseli,* nobis, Inferior Oolite, is pedunculate, short, and very expanded superiorly; the epitheca is well preserved, and the endothecal dissepiments can be seen. This is a very satisfactory species, and I have had it very carefully drawn, so that the suspiciously synapticular endotheca can be proved to be really dissepimental.

A specimen of *Cladophyllia Babeana* is remarkable from the disposition of the Corallites to combine and form serial and fissiparous calices as in *Thecosmilia.* Plate III, figs 1—4.

I am under great obligations to Dr. Holl, F.G.S., Mr. Mansell, F.G.S., Mr. R. Tate, F.G.S., Dr. Wright, F.G.S., Mr. T. C. Brown, Mr. Leckenby, F.G.S., and many other geologists, for the kind loan of specimens.

II. List of Species already described.

MM. Milne-Edwards and Jules Haime described the following Oolitic species[1] in their 'Monograph' (Pal. Soc.), 1851:

Portland Stone.

1. *Isastræa oblonga*, Fleming, *sp.*

Coral Rag.

1. *Stylina tubulifera*, Phillips, *sp.*
2. — *De-la-Bechi*, Ed. & H.
3. *Montlivaltia dispar*, Phillips, *sp.*
4. *Thecosmilia annularis*, Fleming, *sp.*
5. *Rhabdophyllia Edwardsi*, M'Coy, *sp.*
6. *Calamophylla Stokesi*, Ed. & H.
7. *Cladophyllia cæspitosa*, Con. & Phil., *sp.*
8. *Goniocora socialis*, Römer, *sp.*
9. *Isastræa explanata*, Goldfuss, *sp.*
10. — *Greenoughi*, Ed. & H.
11. *Thamnastræa arachnoides*, Parkins, *sp.*
12. — *concinna*, Goldfuss, *sp.*
13. *Comoseris irradians*, Ed. & H.
14. *Protoseris Waltoni*, ,,

Great Oolite.

1. *Stylina conifera*, Ed. & H.
2. — *solida*, M'Coy, *sp.*
3. — *Ploti*, Ed. & H.
4. *Cyathophora Luciensis*, d'Orb., *sp.*
5. — *Pratti*, Ed. & H.
6. *Convexastræa Waltoni*, ,,
7. *Montlivaltia Smithi*, ,,

[1] There are three species common to the Great Oolite and the Inferior Oolite, and one is common to the Coral Rag, the Great and the Inferior Oolite.

8. *Montlivaltia Waterhousei*, Ed. & H.
9. *Calamophyllia radiata*, Lamouroux, *sp.*
10. *Cladophyllia Babeana*, Ed. & H.
11. *Isastræa Conybeari*, „
12. — *limita*, Lamouroux, *sp.*
13. — *explanulata*, M'Coy, *sp.*
14. — *serialis*, Ed. & H.
15. *Clausastræa Pratti*, „
16. *Thamnastræa Lyelli*, „
17. — *mammosa*, „
18. — *scita*, „
19. — *Waltoni*, „
20. *Anabacia orbulites*, Lamouroux, *sp.*
21. *Comoseris vermicularis*, M'Coy, *sp.*
22. *Microsolena regularis*, Ed. & H.
23. — *excelsa*, „

Inferior Oolite.

1. *Discocyathus Eudesi*, Michelin, *sp.*
2. *Trochocyathus Magnevillianus*, Michelin, *sp.*
3. *Axosmilia Wrighti*, Ed. & H.
4. *Montlivaltia trochoides*, „
5. — *tenuilamellosa*, Ed. & H.
6. — *Stutchburyi*, „
7. — *Wrighti*, „
8. — *cupuliformis* „
9. — *De-la-Bechi*, „
10. — *lens*, „
11. — *depressa*, „
12. *Thecosmilia gregaria*, M'Coy, *sp.*
13. *Latimæandra Flemingi*, Ed. & H.
14. — *Davidsoni*, „
15. *Isastræa Richardsoni*, „
16. — *tenuistriata*, M'Coy, *sp*
17. — *Lonsdalei*, Ed. & H.
18. *Thamnastræa Defranciana*, Michelin, *sp.*
19. — *Terquemi*, Ed. & H.
20. — *Mettensis*, „

21. *Thamnastræa fungiformis*, Ed. & H.
22. — *Maccoyi*, ,,
23. *Anabacia hemisphærica*, ,,

Mr. Walton has forwarded me *Zaphrentis ? Waltoni*, Ed. & H., from the Inferior Colite at Dundry, which MM. Milne-Edwards and Jules Haime felt inclined to think was a *remanié* fossil. There is no doubt about the specimen being a *Zaphrentis*, and it is clear that it was derived from an older rock.

III. List of New Species.

Great Oolite.

Thecosmilia obtusa, D'Orb.
Cyathophora insignis, Duncan.
— *tuberosa* ,,
Isastræa gibbosa ,,
Thamnastræa Browni ,,

Inferior Oolite.

Montlivaltia Holli, Duncan.
— *Painswicki*, Duncan.
— *Morrisi* ,,
Thecosmilia Wrighti ,,
Symphyllia Etheridgei, ,,
Thamnastræa Walcotti, ,,
— *Manseli*, ,,
Gonioseris angulata, ,,
— *Leckenbyi*, ,,
Dimorphoseris oolitica, ,,
Cyclolites Lyceti, ,,
— *Beani*, ,,
Podoseris constricta ,,

Including M. d'Orbigny's species there appear to be eighteen new forms which may be added to those formerly described by MM. Milne-Edwards and Jules Haime in their Monograph of the Oolite Corals (Pal. Soc.).

IV. List of all the Species described.

The Oolite fauna may be described as follows :—

	Species.
Portland Stone	1
Coral Rag	14
Great Oolite	28
Inferior Oolite	36
	79

Portland Oolite.

Isastræa oblonga, Fleming, *sp.*

Coral Rag.

Stylina tubulifera, Phillips, *sp.*
— *De-la-Bechi*, Ed. & H.
Montlivaltia dispar, Phillips, *sp.*
Thecosmilia annularis, Fleming, *sp.*
Rhabdophyllia Edwardsi, M'Coy, *sp.*
Calamophyllia Stokesi, Ed. & H.
Cladophyllia cæspitosa, Con. & Phil., *sp*,
Goniocora socialis, Römer, *sp.*
Isastræa explanata, Goldfuss, *sp.*
— *Greenoughi*, Ed. & H.
Thamnastræa arachnoides, Parkinson, *sp.*
— *concinna*, Goldfuss, *sp.*
Comoseris irradians, Ed. & H.
Protoseris Waltoni, „

Great Oolite.

Stylina conifera, Ed. & H.
— *solida*, M'Coy, *sp.*
— *Ploti*, Ed. & H.
Cyathophora Luciensis, d'Orb., *sp.*
— *Pratti*, Ed. & H.

Cyathophora insignis, Duncan.
— *tuberosa* „
Convexastræa Waltoni, Ed. & H.
Montlivaltia Smithi „
— *Waterhousei* „
Thecosmilia obtusa, d'Orb.
Calamophyllia radiata, Lamouroux, *sp.*
Cladophyllia Babeana, d'Orb., *sp.*
Isastræa Conybeari, Ed & H.
— *limitata*, Lamouroux, *sp.*
— *explanata*, M'Coy, *sp.*
— *serialis*, Ed. & H.
— *gibbosa*, Duncan.
Clausastræa Pratti, Ed. & H.
Thamnastræa Lyelli, „
— *mammosa*, „
— *scita*, „
— *Waltoni* „
— *Browni*, Duncan.
Anabacia orbulites, Lamouroux, sp.
Comoseris vermicularis, M'Coy, *sp.*
Microsolena regularis, Ed. & H.
— *excelsa*, „

Inferior Oolite.

Discocyathus Eudesi, Michelin, *sp.*
Trochocyathus Magnevillianus, Michelin, *sp.*
Axosmilia Wrighti, Ed. & H.
Montlivaltia trochoides; „
— *tenuilamellosa*, Ed. & H.
— *Stutchburyi*, „
— *Wrighti*, „
— *cupuliformis*, „
— *De-la-Bechi*, „
— *lens*, „
— *depressa*, „
— *Holli*, Duncan.
— *Painswicki*, „
— *Morrisi*, „

Thecosmilia gregaria,[1] M'Coy, *sp.*
— *Wrighti*, Duncan.
Latimæandra Flemingi, Ed. & H.
— *Davidsoni*, „
Symphyllia Etheridgei, Duncan.
Isastræa Richardsoni, Ed. & H.
— *tenuistriatae*, M'Coy, *sp.*
— *Lonsdalei*, Ed. & H.
Thamnastræa Defranciana, Michelin, *sp.*
— *Terquemi*, Ed. & H.
— *Mettensis*, „
— *fungiformis*, „
— *Maccoyi*, „
— *Walcotti*, Duncan.
— *Manseli*, „
Gonioseris angulata, „
— *Leckenbyi*, „
Anabacia hemisphærica, Ed. & H.
Dimorphoseris Oolitica, Duncan.
Cyclolites Lyceti, „
— *Beani*, „
Podoseris constricta, „

V.—General relation of the Oolitic Coral-faunas.

The Oolitic Corals, as a whole, indicate the geographical conditions incident to reefs and atolls, and do not represent those bathymetrical states which the Upper and Middle Liassic coralliferous strata appear to have illustrated. A deep oceanic coral-fauna is not found amongst the relics of the Oolites, and the forms characteristic of the reefs are positively aggregated in an upper and lower mass at Crickley in the Inferior Oolitic beds.

Dr. Wright noticed some years since[2] an Oolitic coral-reef near Frith Quarry, on the northern spur of Brown's Hill, about two miles from Stroud. There is a corresponding

[1] The numerous forms I consider to belong to *Thecosmilia gregaria* are not mentioned or considered as species, although they have a very fair claim. There are three varieties very Symphyllian, and two very Heterogyran in their aspect, Pl. VII, figs. 12—15. There is a well-marked variety of *Montlivaltia trochoides* at Painswick in the Inferior Oolite.

[2] Dr. Wright has kindly sent me these details. See 'On Coral Reefs,' by T. Wright, M.D., F.G.S., Cotteswold Club. Transact.

2

reef on the opposite side of the valley, the whole of the intervening space having been excavated by denudation. The coral-bed consists of large masses of coralline limestone imbedded in a fine-grained cream-coloured mudstone. The corals are in a highly crystalline state, so that the genera and species are determined with difficulty. The bed is from fifteen to twenty feet in thickness, and forms one of the finest examples of fossil coral-reefs that Dr. Wright is acquainted with in thé district. The bed may be traced along the escarpment, in a north-westerly direction, for several miles, to Witcomb and Crickley on the west, and to near Cubberley and Cowley on the east, where it was worked several years ago. Judging from the thickness of the bed, and the abundance of corals it contains, it must have formed a barrier-reef of considerable magnitude in the Jurassic sea. The following is a section showing the relative position of the Lower Coral-reef.

Section of the Lower Coral-reef, in the Inferior Oolite, at the Quarry, North Frith Wood, near Brown's Hill, Gloucestershire.

Lithological Characters and Thickness.	Beds.	Organic Remains. Leading Fossils.
	UPPER FREESTONES.	
Cream-coloured Marl, with several inconstant layers of Mudstone, upper part passing into a loose, friable Freestone, with large *Terebratula fimbria*. From 20 to 25 feet.	OOLITE-MARL. MIDDLE CORAL-BED.	*Thamnastræa, Isastræa, Axosmilia, Terebratula fimbria, T. carinata, T. maxillata, Rhynchonella Lycetti, Lucina Wrighti, Lima pontonis.*
Fine-grained oolitic Limestone, very white, and emitting a metallic ring when struck with a hammer. 40 to 50 feet.	FREESTONES.	Shelly fragments, not determinable.
Coarse brown ferruginous Oolite.	LOWER RAGSTONES.	*Terebratula plicata.*
Masses of Coralline Limestone, imbedded in a light-coloured Mudstone; the Corals highly crystalline, forming the chief part of the bed. 15 to 25 feet.	LOWER CORAL-REEF.	*Latimæandra, Thamnastræa, Isastræa, Axosmilia, Thecosmilia, Pecten Dewalquei, Trichites, Lucina Wrighti.*
Brown ferruginous pisolitic rock. Pea-grit structure not much exposed.	PEA-GRIT.	*Lima sulcata, Hinnites abjectus, Ceromya Bajociana, Avicula complicata, Nerita costata, Trochotoma carinata, Pygaster, Hyboclypus, Diadema.*

The Middle Coral-bed is included in the Oolite-marl, and in some localities, as at Frith, Leckhampton, Sheepscombe, and others, it contains masses of corals.

The Upper Coral-reef occupies the horizon of the Upper Trigonia Grit, and is very well exposed in many sections. That of Cleeve Hill has yielded the best corals. The following section is open near Frith. Ascending the bank above this quarry for a short distance some fields or arable land are passed over, on which are several heaps of the Upper Ragstones, with *Trigonia costata, Gryphæa subloba,* and other shells of the higher zone. Walking in the direction of the Grove, after passing over the summit of the hill and descending a short distance, a good section of the upper reef may be seen in the Slad Valley.

Section of the Quarry at Worgin's Corner, Upper Zone of Inferior Oolite.[1]

Lithology.	Beds.	Organic Remains.
Masses of Coralline Limestone, 4 feet thick.	UPPER CORAL-REEF.	*Thamnastræa, Isastræa, Thecosmilia, Magnotia Forbesi, Stomechinus intermedius, Pecten, Trigonia costata.*
Hard shelly Limestone, full of the shells of Brachiopoda, 5 feet.	TEREBRATULA-GLOBATA BED.	*Terebratula globata, Rhynchonella spinosa, Pholadomya fidicula, P. Heraulti, Ostræa, Gervillia, Trichites.*
Hard shelly sandy Oolite, full of *Gryphæa*, 6 feet.	GRYPHÆA BED.	*Gryphæa subloba, Lima proboscidea.*

The remarkable varieties of *Thecosmilia gregaria,* which resemble the genus *Symphyllia* and *Heterogyra,* are found principally in the lower reef, but they exist in the upper also. Some species appear to be peculiar to the different reefs, but it is unsafe to form lists at present. There is evidently a considerable affinity between the faunas of the reefs, and there is nothing to indicate anything more than a temporary absence from and a return of the species to an area.

[1] See Dr. Wright's pamphlet, from which the whole of this description is abstracted.

The corals of the Great Oolite are found in the Upper Ragstones underlying the Bradford Clay. Near Bath large masses of *Calamophyllia radiata* are associated with the roots, stems, and heads of *Apiocrinites rotundus*, Mill., which flourished like a miniature forest on the reef, and luxuriated amongst the polypes until the clear water was invaded by a current charged with mud, which destroyed the Encrinites and the Corals also.[1]

The Coral Rag in Wiltshire is divisible into (1) Upper Calcareous Grit, (2) Coral Rag, (3) Clay, (4) Lower Calcareous Grit. It is in the Coral Rag proper (2) that the Coral-beds are found. Of these Mr. Lonsdale[2] remarks: "The irregular beds of Polyparia consist of nodules or masses of crystallized carbonate of lime, which afford, invariably, evidences of the labours of the Polypus; and associated with them are others of earthy limestone, which bear only partial proofs of an organic origin. The whole are connected by a pale bluish or yellowish stiff clay. It happens frequently that a bed is composed of one genus of Polyparia."

In Yorkshire the Coralline Oolite is well developed, and several reefs are found at Hackness, Ayton, Seamer, &c. John Leckenby, Esq., F.G.S., of Scarborough, gives the following details (see Dr. Wright, *op. cit.*):—

"In various parts of the district occupied by the Coralline Oolite around Scarborough are found patches of coral-reef, sometimes occupying an area of fully an acre; and, although never attaining an altitude so high as the beds on the inclined surfaces of which they rest, they are truly the uppermost beds of the formation.

"They are sometimes from ten to fifteen feet in thickness, and consist of a series of layers of crystallized coral, from eighteen to twenty-four inches in thickness, of the species *Thamnastræa concinna*, Goldf. (which is the *Th. micraston*, Phillips), each layer being separated by rubbly clay and mud, in all probability the decomposition of each successive reef. The rock is quarried to supply material for repairing the roads of the district; but it is by no means so well adapted for the purpose as the adjacent calcareous grit, which, at the cost of a little additional labour, would furnish a material much more durable. The crystalline coral-reef is quickly ground to powder, and its use affords less satisfaction to the traveller than to the geologist, as the blocks which are stored up for use along the sides of the road yield many a handsome specimen to adorn his collection.

"The largest deposit is near the village of Ayton: there are others not quite so extensive; one near the village of Seamer, another close to the hamlet of Irton, and others in the neighbourhood of Wykeham and Bromptom—the intervening distances being about a mile in every case."

Messrs. Leckenby and Cullen visited the coral-reefs of the Coralline Oolite near Scarborough with Dr. Wright, who writes as follows:—

[1] Dr. Wright, *op. cit.*

[2] "Oolitic District of Bath," 'Trans. Geol. Soc.,' 2nd ser. vol. iii, p. 261.

"One quarry, near Ayton, which may be considered as a type of the others, consisted of masses of crystalline coralline limestone, the beds having an irregular undulating appearance. The corals appear to have grown in areas of depression of the coralline sea; the rock consists of large masses of highly crystalline limestone, forming nodulated eminences and concave curves, in beds of from twelve to eighteen inches in thickness, having a stratum of yellowish clay filling up the hollows, and forming a horizontal line again to the stratification; then follows another stratum of crystalline limestone, which assumes the same nodulated condition as the one below it, the surface of the coral masses, where exposed, showing that the whole is almost entirely composed of a small-celled *Astræa, Thamnastræa concinna,* Goldf. (*Th. micraston,* Phillips), in some altered condition; the reef is exposed to about ten feet in section, and rests on another, forming the floor of the quarry, and which descends many feet deeper. The corals are bored by *Gastrochænæ,* and numerous shells were seen imbedded in the coral mass, which had nestled in the crannies of the reef."

Dr. Wright sums up with regard to the French, German, and British strata of the Étage Corallien as follows:—

"From this general view of the geographical distribution of the Coralline Zone, it would appear that this formation was composed of a series of coral-reefs in the Jurassic sea, which, during the period of their construction, occupied a large portion of the region now constituting the soil of modern Europe; and that the bed of the Jurassic sea was a slowly subsiding area of great extent, like many parts of the Coral Sea in the Indo-Pacific Ocean of our day."[1]

The restriction of species to very definite areas, and to limited zones amongst these succeeding coral-reefs, is very remarkable, and, as was noticed to occur in the Lias, the corals are occasionally persistent, and are associated with different molluscan species. But the physico-geological changes which produced new reefs must have been preceded by considerable geographical changes, for, as a rule, the species of the grand divisions of the Jurassic system are different. *Thecosmilia Wrighti* of the lower reef of the Inferior Oolite has considerable resemblance to the *Thecosmiliæ* of the Inferior Lias; but no Liassic species pass upwards into the Oolites. Only four species are common to the Inferior and Great Oolites, and one to the Coral Rag and Great Oolite; yet there was a succession of the physico-geographical conditions favorable for the formation of reefs on the same area. The existence of reefs in so high a latitude during the Oolitic Period, and their formation by polypes whose genera were all extinct during the early Caïnozoic Period, but which are clearly represented by allied genera in the existing reefs, are very suggestive. These were the last reefs of the British area; for there are no traces of agglomeration of reef-building genera in the Lower Greensand, the Gault, Upper Greensand, Chalk, or Tertiary formations. The nearest approach to a reef must have bee . in the Lower Oligocene

[1] Dr. Wright, *op. cit.*

period, when the Tabulate Corals and *Solenastraeæ* of Brockenhurst formed a small outlier of the European coral sea of the time between the Nummulitic and the earliest Falunian age.[1]

VI. Description of New Species from the Great Oolite.

MADREPORARIA APOROSA.

Family—ASTRÆIDÆ.

Genus—Thecosmilia.

1. Thecosmilia obtusa, *D'Orbigny*, sp. Pl. I, figs. 1—4.

The corallum is short.

The calices sometimes remained united in short series.

The fossula is shallow.

Some sixty septa may be counted in the series. The margin of the septa is oblique and delicately toothed; and their sides are covered with delicate striæ, which are radiating and projecting.

The English locality is in the Great Oolite, Cirencester. MM. Milne-Edwards and Jules Haime give the following French localities:—Villers (Calvados), Neuvizi (Ardennes) in the Group Oolite Moyen.

In the Collection of T. C. Brown, Esq.

Genus—Cyathophora.

2. Cyathophora insignis, *Duncan*. Pl. I, figs. 9—11.

The corallum is massive, and in layers.

The calices are unequal, not equally distant from each other, circular, and they do not project above the inter-calicular surface generally, but in some instances they form cribriform projections.

The costæ cover the inter-calicular surfaces, are sub-equal, wavy, and long.

The septa are very short, and do not reach far into the calice; there are three cycles in six systems, and the primary septa, which do not project much more than the secondary, are the largest.

[1] P. M. Duncan, "Coral Faunas of Western Europe," &c., 'Quart. Journ. Geol. Soc.,' No. 101, p. 51.

The base of the calicular fossa is formed by a broad tabulate dissepiment.

Diameter of the calices $\frac{1}{20}$th to $\frac{1}{8}$th inch.

Locality. Great Oolite, Cirencester.

In the Collection of T. C. Brown, Esq.

3. Cyathophora tuberosa, *Duncan.* Pl. III, figs. 15—18.

The corallum is tuberose, and the base is contracted and small.

The corallites are numerous, not crowded, unequal, and are separated by much exotheca.

The calices are circular, slightly crateriform, and raised, and the primary septa encroach upon the central space, which is shallow.

The costæ are unequal and long, and the calicular wall projects between the primary and secondary septa to produce tertiary costæ, which have no corresponding septa.

The septa are unequal, and there are six systems and two cycles.

Height of corallum $1\frac{1}{2}$ inch. Breadth of calices $\frac{1}{10}$th inch.

Locality. Great Oolite, Cirencester.

In the Collection of T. C. Brown, Esq.

Genus—Isastræa.

4. Isastræa gibbosa, *Duncan.* Pl. II, figs. 10, 11.

The corallum is gibbous, and the corallites are excessively crowded.

The calices are depressed, irregular in shape, and have a broad margin, and are shallow.

The septa are sub-equal, crowded, short, and marked with lateral ornamentation of a moniliform character. There are six systems and three cycles.

The central fossa is encroached upon by the larger septa, which do not meet with their central margins.

Diameter of largest calices $\frac{1}{8}$th inch.

Locality. Great Oolite, Cirencester.

In the Collection of T. C. Brown, Esq.

Family—FUNGIDÆ.

Genus—Thamnastræa.

1. Thamnastræa Browni, *Duncan*. Pl. II, figs. 1—5.

The corallum is variable in shape, and appears in two series of orms: 1st, as a nearly globular mass with a very small base; 2nd, as a pillar-shaped corallum, terminating in a knob.

The calices are large, and have wide and rounded margins; they are shallow, and do not present any appearance of columellæ.

The septa are large, unequal, broadly dentate, arched, and not crowded. There are six systems and four incomplete cycles.

The costæ pass down the base of the corallum in long, parallel, wavy lines; they are sub-equal, broadly dentated above, and most so below, where they become more equal and more level.

The epitheca is scanty, but covers the costæ here and there.

Breadth of calices $\frac{3}{10}$ths inch.

Locality. Great Oolite, Cirencester.

In the Collection of T. C. Brown, Esq., and in the British Museum.

VII. Description of New Species from the Inferior Oolite.

Family—ASTRÆIDÆ.

Genus—Montlivaltia.

1. Montlivaltia Holli, *Duncan*. Pl. I, figs. 5—8.

The corallum is cornute, tall, and slightly compressed laterally.

The epitheca is very strong and plain, but marked with transverse folds and slight costal striæ.

The calice is elliptical, rather deep, open, and has a thin margin.

The septa are very unequal as regards the higher orders, but the primary and secondary are equal, slightly exsert, and convex on the upper margin. They are moderately prominent in the calicular fossa. The other septa are much smaller. There are six systems and four cycles in each and part of the fifth. The appearance is that of twelve systems of three cycles.

Height of corallum $1\frac{1}{2}$ inch. Length of calice $\frac{6}{10}$ths inch.

Locality. Oolite-marl, Painswick.

In the Collection of Dr. Holl, F.G.S.

Calicular gemmation is frequent.

2. Montlivaltia Painswicki, *Duncan*. Pl. I, fig. 12.

The corallum is rather flabelliform, compressed, especially inferiorly, has a narrow but elongated base, with the remains of former adhesion, and an elliptical and deep calice.

The epitheca is very strong, transversely ribbed, and folded, moreover, inferiorly; there is a projection on either side of the base.

The calicular margin is broad and rounded.

The septa are numerous, unequal, not exsert, crowded, and some are attached to others near the central space.

There are six systems of septa and five cycles, with some orders of the sixth in each.

Height of corallum $\frac{7}{10}$ths inch. Length of calice $\frac{9}{10}$ths inch.

Locality. Oolite-marl, Painswick.

In the Collection of Dr. Holl, F.G.S.

3. Montlivaltia Morrisi, *Duncan*. Pl. II, fig. 13.

The corallum is turbinate, the base is slender and conical, and the calicular margin is deformed, and more or less oval. The corallum expands above.

The calice is deep; its margin is rounded, rather sharp, and there is no columella.

The septa are stout, numerous, unequal, long, and curved. The larger septa unite deep in the fossa in a kind of whorl.

There are six systems and five cycles, with part of the sixth.

The corallum is often deformed by arising close to others.

Height of corallum $\frac{9}{10}$ths inch. Breadth of calice $\frac{12}{10}$ths inch.

Locality. Inferior Oolite.

In the Collection of the Royal School of Mines.

Genus—Thecosmilia.

1. Thecosmilia Wrighti, *Duncan*. Pl. V, figs. 1—5.

The corallum is large, massive, and irregular in shape.

The corallites are cylindrical and increase very slightly in their calices during their growth. They do not remain long united after fissiparity and budding, and they form an aggregate of rather short tubes which are not united by a common epitheca.

The epitheca of each corallite is dense and marked with lateral lines, but it is usually worn off here and there so as to show the costæ which are delicate, straight, numerous, and subequal.

The calices are usually slightly elliptical and the epitheca reaches to them. They are not of greater diameter than the corallites.

The septa are few in number and probably do not attain the full complement of the fourth cycle. The primary septa are the largest, but in some calices the secondary equal them in size.

The columella is rudimentary.

Length of calices $\frac{1}{2}$ inch (largest). Height of corallite 2 inches.

Locality. Crickley. Inferior Oolite.

In the Collection of Dr. Wright, F.G.S., Cheltenham.

2. Thecosmilia gregaria, *M'Coy,* sp. Pl. VI, figs. 1—4.

This common species appears to vary greatly in some districts, and Dr. Wright, of Cheltenham, has a series which appears to gradate towards and into the genus *Symphyllia.* The figures explain this tendency, but the calices of fig. 1 are rather too much levelled internally. Fig. 2 represents the calices on the outside of the corallum.

Locality. Crickley.

In the Collection of Dr. Wright, F.G.S., Cheltenham.

Genus—Latimæandra, *Ed. & H.*

1. Latimæandra Flemingi. *Ed. & H.* Pl. V, figs. 6, 7.

A fine specimen of this Lower Oolite form is delineated in plate V. The magnified view (fig. 7) shows a calice in which gemmation has taken place very remotely from the centre. Many portions of the corallum do not present serial calices, and if such fragments were found separate they would necessarily be associated with the genus *Isastræa.* The *Latimæandræ* may be regarded as modified *Isastræa;* but most probably they descended from *Thecosmiliæ.*

Genus—SYMPHYLLIA, *Ed. & H.*

1. SYMPHYLLIA ETHERIDGEI, *Duncan.* Pl. VI, figs. 5—8.

The corallum is nodular in shape; the base is uneven, and the sides and upper surface are irregular, convex, and gibbous. The remnants of a basal epitheca exist and the costæ of the calices end in a wall with which the costæ of the base are continuous (fig. 8).

The calices are irregular in shape and size, and often form short series. The intercalicular spaces are broad, and are marked by the costæ which are continuous with the septa.

The septa are numerous, very unequal, and crowded. The larger reach to the columellary space, and the small are almost rudimentary.

In small calices the fifth cycle is incomplete.

The columella is small and not always visible.

The dissepiments are close and join the septa so as to resemble synapticulæ.

Height of corallum $1\frac{3}{4}$ inch.

Breadth of corallum $2\frac{3}{4}$ inch.

Breadth of calices $\frac{4}{10}$ths to $\frac{6}{10}$ths inch.

Locality. Crickley. Inferior Oolite.

In the Collection of Dr. Wright, F.G.S., Cheltenham.

This is the earliest representative of the genus *Symphyllia,* and its derivation from a *Thecosmilia* does not admit of much doubt.

FAMILY—FUNGIDÆ.

Genus—THAMNASTRÆA.

1. THAMNASTRÆA WALCOTTI, *Duncan.* Pl. IV, figs. 5—10.

The corallum is moderate in size and of a flat conical shape.

The apex of the cone is truncated and forms the inferior part or peduncle of the corallum which was adherent.

The base of the cone is inferior and is flat, and there is a tendency to inequality and curving of the margins.

The epitheca is well developed, rigid, and marked with transverse lines; where abraded it permits the subequal moniliform costæ to be seen and their connecting synapticulæ.

The calices are large, flat, shallow, and tolerably well defined, and are separated by much cœnenchyma covered with costæ.

The columella is distinct and formed by one pimple-shaped mass.

The septa are in very unequal systems, and there are three incomplete cycles. They are short, rather moniliform, perforated on their free margins, marked with lateral synapticulæ, and end in larger or shorter costæ, which are continuous with the septa of neighbouring calices.

The endotheca is fully developed, and is partly in the form of synapticulæ and partly of dome-shaped dissepiments.

Height of corallum $\frac{9}{10}$ths inch.

Breadth of calicular surface $2\frac{1}{2}$ inches.

Breadth of calices $\frac{3}{10}$ths inch.

Locality East Coker. Inferior Oolite.

In the Collection of W. Mansel, Esq., F.G.S.

2. Thamnastræa Manseli, *Duncan*. Pl. IV, figs. 11—14.

The corallum is small and conical, with a rounded apex, which is inferior, and a circular flat but slightly gibbous upper or calicular surface.

The epitheca is distinct and is marked with transverse lines, and where abraded permits the costæ to be seen.

The costæ are numerous, alternately large and small, slightly apart, and are connected by numerous synapticulæ.

The calices are numerous, small, nearly circular, shallow, and are separated by distinct nodular elevations of cœnenchyma.

The septa are distinct, rather moniliform, unequal, and more or less continuous with those of the neighbouring calices. They are broader externally than within the calice, and the larger unite more or less to form a false columella.

The costæ on the calicular surface are wavy and moniliform.

There are six systems of septa and usually some orders of the fourth cycle in addition to the complete but very irregularly disposed third cycle.

The endotheca is abundant and assumes the synapticular form.

Height of corallum $\frac{7}{10}$ths inch.

Breadth of calicular surfaces $1\frac{3}{10}$ths inch.

Breadth of calices about $\frac{1}{10}$th inch.

Locality. East Coker. Inferior Oolite.

In the Collection of W. Mansel, Esq., F.G.S.

Genus nov.—Gonioseris.

The corallum is simple and free.

The base is polygonal in outline and the projecting angles are formed by groups of costæ terminating in septa. Between the angles the margin is concave externally. The centre of the base is concave.

The costæ are numerous and they cover the base. Many converge at each angle along a line leading from the large septum to the centre.

The upper surface of the corallum is convex, and is divided by masses of septa which are continuous with the angles of the base, and which, after projecting there, become exsert and pass to the axial space where they meet.

There is a large, prominent, primary septum in each mass.

The calicular wall is invisible. The synapticulæ are broad and numerous.

This extraordinary genus is represented by two forms in the Inferior Oolite. Probably the normal number of projecting angles is six, but in one specimen there are five, and a careful examination of it tends to prove that there was no abortion of a septum, but that the quinary arrangement was initiated from the first.

The type is *Gonioseris angulata* nobis. Probably the small specimen delineated in the same plate is a young form of it. Plate VII, figs. 10—11. The third specimen I have called *Gonioseris Leckenbyi* after the discoverer of these fossils.

1. Gonioseris angulata, *Duncan*. Pl. VII, figs. 1—5.

The base is hexagonal, and the projecting angles are connected by marginal concavities. The space between the central concavity of the base and the margin is broad and slightly convex.

The costæ are of two kinds—those which pass from the concave margins to the concavity of the base, and those which pass from the margin near the angles to a line directed from the angle to the base. All the costæ are thin, slightly crenulate, alternately large and small; and they are all continuous with the septa. Each septal mass, which forms one of the six angles, consists of a large primary septum and several small septa associated on either side. The mass projects upwards and outwards from the base, and then curves inwards and slightly upwards to the axial space. The spaces between the six masses are convex from within outwards and concave from side to side.

There are six large primary septa; and the others are subequal, long, thin, crenulate, and uniting.

There is no columella, but the large septa and many of the small appear to unite over the axial space.

The synapticulæ are not numerous, and are delicate.

Height of corallum $\frac{6}{10}$ths inch.

Extreme length $1\frac{1}{2}$ inch.

Locality. Millepore bed, Cloughton Wyke, near Scarborough.

In the Collection of John Leckenby, Esq., F.G.S.

2. Gonioseris Leckenbyi, *Duncan.* Pl. VII, figs. 6—9.

The corallum is pentagonal.

The costæ are thick.

The concavity of the base is angular in outline.

The septal masses at the angles are formed by small septa, which converge towards the large costæ.

Height of corallum $\frac{1}{10}$ths inch.

Length $1\frac{7}{10}$ths inch.

Locality. Millepore bed, Cloughton Wyke, near Scarborough.

In the Collection of John Leckenby, Esq., F.G.S.

Genus nov.—Dimorphoseris.

The corallum is compound, turbinate, and adherent.

The epitheca is dense and faintly striated, but in no way incised or plicated.

The calicular surface is slightly concave and circular in outline.

There is a large central primary calice, and one or more concentric rows of calices at some distance from the primary.

The septa are continuous and moniliform.

There is no columella.

The secondary calices increase by fissiparous division.

1. Dimorphoseris oolitica, *Duncan.* Pl. IV, figs. 1—4.

The corallum is turbinate, and has a small peduncle and a large and slightly concave calicular surface.

The central calice is large, and about twenty-four septa enter into its composition, but there are many others just outside.

The fossa is shallow.

The septa are ornamented with elongated, bead-shaped projections, and their costal prolongations are very long, and are also ornamented in the same manner.

Some of the external costæ on the calicular surface bifurcate, and even divide into three portions. Usually the costæ are subequal and the synapticulæ are very numerous and distinct.

Height of corallum 1 inch.

Breadth of calicular surface $1\frac{3}{4}$ inch.

Locality. East Coker. Inferior Oolite.

In the Collection of W. Mansel, Esq., F.G.S.

Genus—Cyclolites.

1. Cyclolites Lyceti, *Duncan.* Pl. III, figs. 7—9.

The corallum is small, pedunculate, depressed, and nearly flat, and the calicular margin is everted and elliptical.

The epitheca is strongly marked, and is in folds.

The calicular fossa is in the centre of the calicular surface.

The septa are very numerous, alternately large and small, and are delicately ornamented with moniliform projections.

The calice is slightly convex.

Height of the corallum $\frac{4}{10}$ths inch.

Length of the calice $1\frac{2}{10}$ths inch.

Locality. Inferior Oolite.

In the Collection of Dr. Holl, F.G.S.

2. Cyclolites Beanii, *Duncan.* Pl. III, figs. 10, 11.

The corallum is turbinate and greatly expanded, and slightly concave above. It is slightly flat at the base where it adhered.

The epitheca is stout, and in transverse folds.

The calicular margin is nearly circular.

The septa are very numerous, and number about 220 They are unequal, long, and moniliform, here and there.

The synapticulæ are very numerous.

Height of corallum $\frac{1}{2}$ inch.

Breadth of calice $1\frac{1}{2}$ inch.

Locality. Lower Ragstone, Dorset.

In the Collection of Dr. Holl, F.G.S.

Genus—Podoseris.

1. Podoseris constricta, *Duncan*. Pl. III, figs. 5, 6.

The corallum is fungiform and constricted beneath the rounded calicular surface.

The base is small and presents the concave surface of a former adhesion to a foreign body.

The epitheca is delicate.

The calice is convex.

The septa are delicate, narrow, long, slightly unequal, and there are five cycles of them and part of the sixth.

The costæ are distinct and equal inferiorly where they are linear.

The synapticulæ are rare.

Height of corallum $\frac{6}{10}$ths inch.

Breadth of calice $\frac{4}{10}$ths inch.

Locality. Lower Ragstone, Dorset.

In the Collection of Dr. Holl, F.G.S.

A MONOGRAPH

OF THE

BRITISH FOSSIL CORALS.

SECOND SERIES.

BY

P. MARTIN DUNCAN, M.B.Lond., F.G.S.,

SECRETARY TO THE GEOLOGICAL SOCIETY.

Being a Supplement to the
'Monograph of the British Fossil Corals,' by MM. Milne-Edwards *and* Jules Haime.

PART IV.

Corals from the Zones of Ammonites planorbis, angulatus, Bucklandi, obtusus, and raricostatus of the Lower Lias; from the Zones of Jamesoni and Henleyi of the Middle Lias; and from the Avicula-contorta Zone and the White Lias.

Pages i, ii; 1—73. Plates I—XVII.

LONDON:

PRINTED FOR THE PALÆONTOGRAPHICAL SOCIETY.

1867—1868.

PRINTED BY
ADLARD AND SON, BARTHOLOMEW CLOSE.

CONTENTS OF SUPPLEMENT TO THE LIASSIC CORALS.

		PAGE
	PREFACE	i
I.	Introduction to the Study of Liassic Corals	1
II.	Description of the Species contained in the Zone of *Ammonites planorbis*	5
III.	Description of the Species contained in the Zone of *Ammonites angulatus* in the Sutton stone, and in deposits at Brocastle, Ewenny, and Cowbridge in Glamorganshire. List of Species	6
IV.	Description of the Species from the Zone of *Ammonites angulatus* at Marton, near Gainsborough	35
V.	Description of the Species from the Zone of *Ammonites angulatus* in the North of Ireland	38
VI.	Description of the Species from Lussay, in the Isle of Skye	41
VII.	List of the Species described and noticed from the Zones of *Ammonites planorbis* and *angulatus*	42
VIII.	Remarks upon other Species from the Zone of *Ammonites angulatus*	45
IX.	Description of the Species	46
X.	On the Corals of the British and European Lower Liassic Deposits of the Zones of *Ammonites angulatus*, *Ammonites planorbis*, and *Avicula contorta*	47
XI.	List of Species from the Continental Zone of *Ammonites angulatus*	48
XII.	List of Species of Corals from the Continental and British Strata of the Zone of *Ammonites angulatus*	49
XIII.	Description of Species from the Zone of *Ammonites Bucklandi* (*bisulcatus*)	51
XIV.	List of Species from the Zone of *Ammonites Bucklandi*	55
XV.	Description of the Species from the Zone of *Ammonites obtusus*, Sow.	56
XVI.	Sections of the Beds in Gloucestershire and Warwickshire containing Corals from the Zone of *Ammonites raricostatus*, Ziet., and Description of the Species	57
XVII.	List of Species from the Zone of *Ammonites raricostatus*	61
XVIII.	List of Species from the Zones of the Lower Lias above the Zone of *Ammonites angulatus*	61
XIX.	Corals of the Middle Lias from the Zone of *Ammonites Jamesoni*, Sow.	62
XX.	Corals of the Middle Lias from the Zone of *Ammonites Henleyi*, Sow.	63
XXI.	Enumeration of the British Liassic Species	64
XXII.	Description and Notice of Species from the Zone of *Ammonites planorbis*	65
XXIII.	List of Species from the Zone of *Ammonites planorbis*	66
XXIV.	Notice on the indeterminable Corals of the Avicula Contorta Zone and White Lias of the British Isles (Rhætic of Moore)	66
XXV.	Note on the Age of the Sutton Stone and the Brocastle Deposits	69
	INDEX OF SPECIES	72

A MONOGRAPH

OF THE

BRITISH FOSSIL CORALS.

SECOND SERIES.

PART IV. No. 1.

PREFACE.

It was noticed in the Preface to the First Part of this series that some irregularity in the succession of the Monographs would occur. According to the plan adopted by MM. Milne-Edwards and Jules Haime, the Corals of the Cretaceous rocks should have been described in this Part; but it was found advisable to take advantage of Mr. Charles Moore's splendid collection of Liassic Madreporaria, and to describe the species contained in it at once.

MM. Milne-Edwards and Jules Haime only described three species from the whole of the British Lias; and as one was probably a *remanié* fossil,[1] and another could not be determined generically,[2] only one good species, *Trochocyathus Moorei,* Ed. and H., remained to characterise this great formation.

Since those authors wrote, many careful collectors have found large numbers of Corals in the Middle and Lower Lias; and Dr. Wright,[3] the Rev. P. B. Brodie,[4] and Mr. Ralph Tate,[5] have published notices and descriptions of species.

Lately Mr. Tawney and the author brought the Corals of the Sutton Stone[6] before the notice of the Geological Society; and Mr. Charles Moore, who had long before

[1] 'Monograph. Brit. Foss. Cor.,' p. 145, Palæontograph. Soc., 1851.

[2] Ibid.

[3] 'Quart. Journ. Geol. Soc.,' 1858, p. 34.

[4] 'Reports of the British Association for 1860, Reports of Sections,' p. 73. 'Quart. Journ. Geol. Soc.,' 1861, vol. xvii, p. 151. "A Sketch of the Lias, &c.," 'Transact. Woolhope Nat. Field Club,' No. 1.

[5] 'Quart. Journ. Geol. Soc.,' vol. xx, p. 111.

[6] Ibid., vol. xxii, p. 91.

collected Sutton Stone Corals, and had discovered the highly fossiliferous deposit at Brocastle, forwarded me his specimens, which are about to be described. The above-mentioned geologists have afforded me all the information at their command; and Messrs. Kershaw, Winwood, Boyd Dawkins, Burton, Chamberlin, and Mrs. Strickland, have placed me under great obligations.

Finding that at least fifty new species would have to be added to the list of British Liassic Corals, it was thought advisable to publish the most important at once.

This Part of the Second Series will refer to the Corals of that portion of the Lower Lias which intervenes between the Rhætic strata and the beds which contain *Ammonites Bucklandi* (*bisulcatus*) and *Gryphæa incurva* (type) in abundance.[1]

The next portion of this Monograph (Part IV, No. 2) will contain the description of the Corals of the other beds of the Lower Lias, and of the forms in the Middle and Upper Liassic deposits. It is probable that several Liassic beds whose geological horizon is not yet determined may yield new species of Corals which will have to be associated with those of the zone of *Ammonites angulatus*, and they will, of necessity, be included in Part IV, No. 2, in which the lists of the species will be given, with a notice of the Corals of the Liassic strata of Continental Europe.

Owing to the paucity of specimens, it is thought advisable to defer the consideration of the species from the White Lias of the Rhætic series and from the Zone of *Avicula contorta* to a future occasion.

[1] Madreporaria of the Infra-Lias of South Wales. P. Martin Duncan, 'Quart. Journ. Geol. Soc.,' Feb., 1867.

A MONOGRAPH

OF THE

BRITISH FOSSIL CORALS.

(SECOND SERIES.)

PART IV.—No. 1.

INTRODUCTION.

THE Corals contained in the Liassic strata of Britain, France, Germany, and Italy have a very decided community of facies; at the same time it is evident that some portions of the Liassic Coral-fauna resemble Triassic types, and that another portion is allied to the Oolitic.

This was to be expected, for it is evident that the stunted *Thecosmiliæ*, and the *Astrocœniæ* of the Zone of *Ammonites angulatus*, are the descendants of the equally stunted *Thecosmiliæ* and of the *Astrocœniæ* of the Triassic age. Moreover, the descendants of the *Isastrææ*, and of the larger *Montlivaltiæ* of the Lower and Middle Lias, luxuriated in the Oolitic seas.

The bulk of the Liassic Coral-fauna is, however, characteristic of and special to the formation; and, as is the case in other great series of strata, certain assemblages of species appear to characterise certain definite geological horizons. Yet not unexceptionably, for some species range into higher zones in certain areas, whilst others, which are confined to a definite horizon in one area, are found below and above the equivalents of the horizon in a distant locality. Thus a species, which is only found in a particular bed, and is associated with a particular molluscan fauna in one locality, may be found to be associated with a molluscan fauna antecedent or posterior in its recognized succession, in another place.

It is this uncertain vertical range of species, this variation in vertical range in different geographical areas, which causes the apparent antagonism of Physical Geology (as applied to Classificatory Geology) and Palæontology.

It is this coming in of the same species at various positions in a large formation and their association with different groups of species that renders Palæontology of more or less uncertain value in the exact determination of the age of strata.

But it is this varying vertical range of species in different areas and their association with different groups of forms that points to an ever-changing life-scene, to migration of faunæ, to changes of physical conditions, to variation in the intensity of competition, to the rise of dominant and the decay of feeble forms, and to all those external agencies which affect the inherent power of variation peculiar to the animated nature of this world, where no two things are exactly alike.

The persistence of a species in a succession of strata, and its consecutive association with different groups of competitors and contemporaries, is constantly observed in the Lias, taken as a whole; and it is the strongest fact that can be adduced against the almost exploded notion of a series of cataclysmal destructions and of successive creations of beings occurring at intervals which are denoted by changes in physical geology. It is necessary to assert that those doctrines are not quite exploded, for they have a deep hold on the minds of many who have only a limited area of geological observation. The disposition to limit the possibility of the occurrence of certain specific forms to definite vertical ranges arises from a partial belief in those ideas, and they are apparently strengthened in the force of their application when physical breaks accompany palæontological changes.

Here the question concerning the physical causes which permit of and assist in the preservation of dead organisms must be considered in reference to those which have a diametrically opposite effect.

If it be admitted that when the terrestrial conditions are *in statu quo* the preservation of organic remains from destruction is hardly possible; that during elevation of areas the entombment and fossilization of organisms is equally unlikely, and that a gradual depression of the surface is in the majority of instances necessary for the preservation of deposits, it becomes evident that, whilst the physical break has a diminished value in its relation to the persistence of the life of species, the existence of a species in a considerable series of strata which could not have all been deposited during a continuous and uninterrupted sinking of their area becomes most suggestive. Taken in combination with the remarks which have preceded, it is suggestive of the evident want of relation between the formation of strata and the origin and decadence of the species of the period; and it points out that no Stratigraphical Palæontogeology can be perfect in a classificatory sense, and that zones of species may have little to do with the notion of time.

With an ever-progressing animated nature there are and have ever been associated terrestrial and inorganic changes. There is no definite connection between them, and hence our classificatory systems have an increment of error which is constantly rising to the surface when the pure physical geologist and the pure palæontologist argue upon their own bases concerning the age of strata.

The notion that the succession of strata all over the world must be upon the same plan as that of the best studied, typical or most familiar district favours this difficulty; and it is most true that the Lias has been, from the applicability of the foregoing remarks, a very debateable ground.

The relation of the Bone-bed to the Trias; the propriety of forming a Rhætic series; its relation to the Trias and Lias; the possibility of arranging the strata of the Lias in Zones of Ammonite life; the propriety of including the Liassic strata between the Keuper and the Zone of *Ammonites Bucklandi*, or between the base of the Zone of *Ammonites planorbis* and the Zone of *Ammonites Bucklandi* in a sub-group, calling it Infra-Lias; the possibility of separating the Zone of *Ammonites Bucklandi* from the Zone of *Ammonites angulatus*, and the impropriety of distinguishing Zones of Cephalopoda, Insecta, Sauria, or Madreporaria —all these have been points debated over and over again, and they will ever be so as long as artificial distinctions are placed "en rapport" with nature.

Nevertheless, carefully collected palæontological data concerning the vertical range of species are gradually deciding many of these questions, and with the effect of isolating the palæontologist more and more in his relations with the received classificatory geology.

These remarks are made because it is necessary to give the various groups of species of *Madreporaria* of the Lias places in some classification or other. It is impossible to associate them with beds determinable on purely stratigraphical or mineralogical data; and it is equally impossible to include them in Zones of special life, for *Cephalopoda* and *Saurians* are rarely in relation with Corals. The groups of *Madreporaria* have a general relation to certain zones of life and to certain strata; and if they are associated for the sake of a necessary classification with certain Ammonite-Zones, it must be understood that it is only an approximative classification, and that both the Ammonites and the Madreporaria may range out of their supposed restricted zone, or not even be represented in certain portions of its area.

If it be admitted that by a Zone of an Ammonite or of any other Mollusc the general and usual vertical limit of the species is meant, all the difficulties thrown in the way of the philosophical, but still artificial, separation of the Liassic series into Ammonite-Zones vanish.

Dr. Wright has elaborated this system in his 'Monograph of the Oolitic Asteriadæ,'[1] and had his Zone of *Ammonites angulatus* been known to have been as well developed in Glamorganshire and in Lincolnshire as it is in some of its most typical districts in France, his arrangement would have met with slight opposition. But the endeavour to give definite horizons to and to correllate *Saurian, Insect, Ostræa, Ammonite,* and *Lima* beds has resulted in the production of confusion instead of the reverse.

Whether the principle of the arrangement in Zones of Ammonites is admitted or not, it is absolutely necessary that the foreign equivalents of our Liassic subdivisions should

[1] Pal. Soc.

be studied. If this be done the association of the characteristic species of certain British beds with the characteristic species of a lower geological horizon on the Continent becomes evident, and the unphilosophical nomenclature of geologists who restrict themselves to the study of small areas is exposed.

In classifying the groups of species about to be described, in the geological scale attention will be directed to the Ammonite-Zone in which they are found and to the Mollusca associated with them.

There are a few Triassic species in the Liassic Coral-fauna, and the Branching Corals of the Sutton Stone have, generally speaking, a very Triassic facies. The majority of the Corals of the lowest members of the Lias are peculiarized by the imperfection of their septal arrangement: the distinct development of definite cycles in six systems is rarely observed, and it would appear that this high organization was not attained in the forms which had varied from Palæozoic into Mesozoic species. The *Montlivaltiæ*, *Thecosmiliæ*, and *Astrocœnia* of the Lower Lias of Glamorganshire illustrate this remark; and the first definite septal arrangement is met with in the *Montlivaltia Haimei*, Ch. et Dew, in the Zone of *Ammonites angulatus* at Marton.

The septal number is also very uncertain in the species of the above-mentioned genera in the Lower Lias, and multiseptate *Montlivaltiæ* are found in the same deposit as those possessing an unusually small number of septal laminæ. It may, in fact, be asserted that the so-called *rugose* peculiarities had hardly left their hold upon Madreporarian life at the time when the lowest members of the Lias were deposited. The genus *Elysastræa*, Laube, retains some "rugose" peculiarities, and the transition from the tabulæ and vesicular endotheca of a *Cyathophyllum* to the dissepiments and vesicular endotheca of some forms of *Montlivaltia polymorpha*, Terq. et Piette, is certainly within the bounds of possibility. Nevertheless, no Palæozoic genera of Corals have been found in the Lias except as "remanié" fossils.

The genera which are represented in those subdivisions of the Lias called the Zones of *Ammonites planorbis* and *Ammonites angulatus* are—

I. *Montlivaltia.*
II. *Rhabdophyllia.*
III. *Thecosmilia.*
IV. *Oppelosmilia*, gen. nov.
V. *Isastræa.*
VI. *Astrocœnia.*
VII. *Cyathocœnia*, gen. nov.
VIII. *Elysastræa.*
IX. *Septastræa.*
X. *Latimæandra.*

No *Tabulate* nor *Perforate* genera have been discovered; yet as they existed both in palæozoic times, and in formations more recent than the Lias, they doubtlessly will be found.

The multitude of branching *Thecosmiliæ,* stunted *Montlivaltiæ,* and small-caliced *Astrocœniæ,* give the peculiar facies to the Coral-fauna of these members of the Lower Lias.

II. Corals from the Zone of Ammonites planorbis.

The yellow shale in the section at Street which contains *Ammonites planorbis* and *Ichthyosaurus intermedius* has yielded a large and well-preserved specimen of the genus *Septastræa.*[1] At Binton there are said to be Corals in the "Guinea[2] bed," but no specimens could be obtained.

Section—*APOROSA.*

Family—ASTRÆIDÆ.

Division—Faviaceæ.

Genus—Septastræa.[3]

1. Septastræa Haimei, *Wright,* sp. Pl. I, figs. 1—5.

The corallum is massive, tall, club-shaped, and rather gibbous. The shape is generally sub-cylindrical, the base is small and conical, and the top is large and convex.

The calices cover the corallum, are very numerous, and are separated by rather thick and united walls. The calices are irregular in size, shallow, and more or less polygonal; and they have a tendency to elongate at one end, as well as to divide fissiparously.

The septa are irregular in size, shape, and number; they are small, unequal, rather distant, and the only ornamentation is an ill-defined swelling here and there. They are not exsert; the smallest reach but a slight distance from the wall, but the larger occasionally reach the centre of the calice and unite.

Fissiparity is produced by two large septa stretching across the calice and developing others from their sides. The septa vary in number, from thirty to forty, but no cyclical arrangement is distinguishable. The endotheca is rather plentiful.

[1] [2] Wright, 'Monogr. Oolitic Asteriadæ, Pal. Soc.,' pp. 56 and 60. 1863.

[3] 'Hist. Nat. des Corall.,' vol. ii.

Height of corallum, 7 inches. Breadth, 4$\frac{1}{2}$ inches. Diameter of calices, $\frac{3}{10}$ths to $\frac{4}{10}$ths inch.

Locality. Street, Somersetshire. In the Collection of Dr. Wright, F.G.S.

The genus *Septastræa* resembles *Isastræa;* but there is fissiparous growth in the calices of the first, and never in the calices of the last-named genus. The peculiar calicinal gemmation of *Isastræa* never produces septa which, crossing the calice, divide it off into separate individuals. The walls of *Septastræa* are not so perfectly united as in *Isastræa*. The genus is found in the Lias and in the Tertiary Coral-fauna.

The shape of the corallum and the septal structures and arrangement distinguish the species from *Septastræa excavata,*[1] E. de From., and *Septastræa Fromenteli,* Terquem et Piette.

III. Corals from the Zone of Ammonites angulatus.

The Sutton Stone[2] and the deposits at Brocastle, Ewenny, and Cowbridge,[3] are highly coralliferous beds in Glamorganshire. They rest on the Mountain-limestone, and are covered by members of the Lias higher in the series than the Zone of *Ammonites angulatus.* They have the homotaxis[4] of the Continental strata, which are classified within the Zones of *Ammonites angulatus* and *Ammonites moreanus,* such as the Calcaire de Valogne, the Foie de Veau, in the Côte d'Or, and the Grès Calcareux, in the Duchy of Luxembourg. Their British equivalent strata are well shown at Marton, near Gainsborough, and in Ireland[5] near Belfast, besides in the localities mentioned by Dr. Wright.[6]

Dr. Wright named this Coral *Isastræa Haimei,* and noticed its specific distinction from *Isastræa Murchisoni,* Wright. Its genus is evidently *Septastræa,* and although Dr. Wright has not published a specific diagnosis of the form, still it is just that it should retain his name. He is answerable for its discovery in the locality given above.

[2] Sir Henry de la Beche, 'Mem. Geol. Survey,' vol. i, p. 270; Mr. Tawney, and P. Martin Duncan, 'Quart. Journ. Geol. Soc.,' vol. xxii, p. 69.

[3] Mr. Charles Moore discovered the Brocastle and Ewenny deposits some years before Mr. Tawney drew attention to the Sutton Stone. He collected a vast number of fossils from them, and forwarded them to me for examination. His able essay on "Abnormal Conditions of Secondary Deposits," &c., was read before the Geological Society, March 20th, 1867. See my notice of Mr. Chas. Moore's labours, 'Quart. Journ. Geol. Soc.,' Feb. 1867, p. 13. See also "On the Lower Lias or Lias-Conglomerate of a Part of Glamorganshire," by H. W. Bristow, F.R.S.; "On the Zone of Ammonites angulatus in Britain," by R. Tate, F.G.S.

[4] "On the Madreporaria of the Infra-Lias of South Wales," by P. Martin Duncan, 'Quart. Journ. Geol. Soc.,' 1866, Feb., p. 12. See also Terquem et Piette, 'Mém. de la Soc. Géol. de la France,' 2de série, tome 8, 1868.

[5] R. Tate, 'Quart. Journ. Geol. Soc.,' vol. xx, No. 78, p. 103.

[6] Wright, 'Monogr. Ool. Aster., Pal. Soc.,' p. 63; see also Oppel's 'Juraformation.'

SECTION—*APOROSA.*

FAMILY—ASTRÆIDÆ.

Division—LITHOPHYLLACEÆ SIMPLICES.

Genus—MONTLIVALTIA.

There are seven species of the genus *Montlivaltia* in the Sutton Stone and the deposits at Brocastle, and six of them are new forms, the seventh having already been described by MM. Terquem et Piette.[1] Three of the species belong to the section of the genus which is characterised by forms having their bases and calices of equal width, and three others have a more or less turbinate shape, whilst one species is pedunculated. The discoid *Montlivaltiæ* appear to be absent, although they are largely represented in the equivalent beds in the east of England and in the north of Ireland.

1. MONTLIVALTIA WALLIÆ, *Duncan.* Pl. VIII, figs. 5, 6, 7.

The corallum is cylindro-conical in shape, the base is small, and the calice large, open, and shallow.

The calice is surrounded with a well-marked margin, which is double in some places, and the smallest or rudimentary septa, which are barely visible in the true calice, are distinct on the outer rim.

The septa are very unequal, but narrow and lamellar, and rather plain, but dentate internally. They are not exsert as regards the calicular margin, but curve upwards and then inwards, terminating by a process marked with at least two teeth.

The fourth cycle of septa is incomplete, and the fourth and fifth orders are rudimentary when they exist; so that the septal number is irregular. The rudimentary septa alternate with the larger. There are about thirty well-developed septa of unequal lengths, and between these are the rudimentary septa.

Height of the corallum, $\frac{6}{10}$ths inch. Breadth of the calice, $\frac{7}{10}$ths inch.

Locality. Brocastle. In the Collection of Chas. Moore, Esq., F.G.S., Bath.

The wide and shallow calice, the low septal number, and the capacious interseptal spaces characterise this species.

[1] Terquem et Piette, 'Mém. de la Soc. Géol. de la France,' 2de série, tom. 8, 1865.

2. Montlivaltia Murchisoniæ, *Duncan*. Pl. VIII, figs. 10, 11, 12.

The corallum is short and turbinate: the peduncle is small, and the calice is large, deep, and open.

The epitheca is distinct and swollen out in some places, being slightly constricted in others.

The calice is circular in outline, very deep, and has a sharp margin.

The septa are numerous, very distinct, and very remarkable, both in their arrangement and relation to the costæ.

The largest septa are bluntly dentate and exsert; the rest are faintly dentate, and pass deeply into the fossa, and there are a few rudimentary septa. The rounded costæ are continuous internally with the interseptal spaces, and the septa are continuous with intercostal spaces (fig. 12).

The cyclical arrangement of the septa is confused. There are forty-eight septa, but these do not appear to form four cycles in six systems, but to be arranged in four systems, there being four septa larger than the others.

The height of the corallum is $\frac{7}{10}$ths inch.

The breadth of the calice is $\frac{5}{10}$ths inch.

Locality. Brocastle.

In the Collection of Charles Moore, Esq., F.G.S., Bath.

The arrangement of the septa and the depth of the calice distinguish this species very readily. It has its mimetic *Thecosmilia* in *Thecosmilia Brodiei*, Duncan.

3. Montlivaltia polymorpha, *Terquem et Piette*.[1] Pl. VII, figs. 14, 15; Pl. VIII, figs. 1—4 and 13—15.

The corallum is simple, very variable in form, and has a thick and folded epitheca reaching to the calice, and is marked with fine and regular costæ. The corallum is rather narrowly pediculate, or adheres by a portion of its base. In shape the corallum may be conical, oblong, or flattened.

The calice is more or less deep, is either round or oval, and its margin is thin. The septa are numerous, have strong teeth on the upper margin, and are smooth laterally. There are five complete cycles, and the sixth is incomplete.

MM. Terquem et Piette do not give the measurements of the Coral, but in their plate the height varies from $\frac{3}{4}$ inch to $2\frac{1}{4}$ inches, and the calicinal diameter from $\frac{3}{4}$ inch to $1\frac{1}{3}$ inch.

[1] 'Le Lias Inférieur de l'Est de la France,' p. 127, pl. xvi, figs. 17—21.

MM. Terquem et Piette notice that the species is found in great abundance at St. Menge in a bed lower than that containing *Gryphæa incurva* and between the strata containing *Ammonites bisulcatus* (*Bucklandi*) and *A. angulatus.*

The specimens from Brocastle show much of the anatomy of the Coral; and the high septal number and dense wall of the corallites when broken off short are well seen in them.

The taller specimens are often denuded of their epitheca, and the highly developed and inclined endotheca is then well seen. One specimen had a broad base, but the others taper and become rather pedunculate.

Locality. Brocastle. In the Collection of Charles Moore, Esq., F.G.S., Bath.

4. Montlivaltia parasitica, *Duncan.* Pl. IV, figs. 13, 14.

The corallum is small, very short, has a base as broad as the calice, and is elliptical in outline.

The calice is very shallow. The septa are few in number, are very irregular; and the costæ run a short distance down the sides of the corallum.

The septa are stout and unequal in length, but not very much so in thickness. The shorter septa bend towards and usually unite themselves to a larger septum. There appear to be twelve large septa, and five of these had either one or two smaller septa joined on to them. There would appear to be two complete cycles of equal septa, and that the tertiary cycle is incomplete.

Length of the calice $\frac{1}{4}$ inch. Height of the corallum $\frac{1}{20}$th inch.

Locality. The Sutton Stone. In the Collection of Charles Moore, Esq., F.G.S., Bath.

The species is founded upon a specimen fixed upon an Astrocœnian, and the extreme shortness, the attachment to a very wide base, and the union of the tertiary to the secondary septa, are very distinctive.

5. Montlivaltia simplex, *Duncan.* Pl. III, figs. 16, 17.

The corallum is short, has a broad base, and an elliptical calice, which is very slightly broader than the base.

The epitheca is strong, does not show any costæ, and it reaches to the calicular margin.

The calice is shallow, and has rather a wide margin.

The septa are very few, very distant, slender, and curved: their arrangement is very irregular; and although there are six septa which reach nearer the calicular centre than the others, still no cyclical development can be asserted to have existed. There are sixteen septa; three are rudimentary, and there are thirteen of a larger size.

Height of corallum, $\frac{2}{10}$ths inch.

Long diameter of calice, $\frac{4}{10}$ths inch.

Locality. Brocastle. In the Collection of Charles Moore, Esq., F.G.S., Bath.

The paucity of septa and the shape distinguish this remarkable species.

6. Montlivaltia brevis, *Duncan.* Pl. VIII, figs. 8, 9.

The corallum is short and cylindrical, and has a base as broad as the calice.

The calicular margin is sharp, and the calice is rather irregular in shape: the calicular fossa is shallow, and the septa are few in number.

The septa are unequal, distant, stout, and have a large tooth at the internal end. This dentation is more distinct in the secondary and tertiary cycles than in the primary. There are three cycles of septa, but the third is incomplete. The primary septa are the longest, and reach to the central space, whilst the smallest septa end in a blunt knob, not so near the central space as the termination of the intermediate septa.

Height of the corallum, $\frac{1}{10}$th inch.

Breadth of the calice, $\frac{1}{4}$ inch.

Locality. Brocastle. In the Collection of Charles Moore, F.G.S., Bath.

The septa are very characteristic of this short and widely based Coral.

7. Montlivaltia pedunculata, *Duncan.* Pl. II, figs. 12, 13; Pl. VIII, fig. 16.

The corallum is large above, cylindro-conical midway, and finely pedunculate at the base.

The epitheca is thin, rather but finely ridged transversely, and permits the costæ which are small to be seen where it is very scanty. The calice is not symmetrical, and the septa are numerous, and apparently constitute five cycles, and part of a sixth. The peduncle is much smaller than the body of the corallum.

Height of corallum $\frac{5}{10}$ths inch. Width of the calice $\frac{4}{10}$ths inch.

Locality. In the Sutton Stone, and at Brocastle. In the Collection of Charles Moore, Esq., F.G.S., and in the Museum of Practical Geology, London.

The shape and high septal number distinguish this species.

Division.—Lithophyllaceæ Cæspitosæ.

Genus.—Thecosmilia.

The *Thecosmiliæ* of the Sutton Stone are principally capitate forms, that is to say, they spring from a peduncle and divide suddenly. The short and fissiparous species, *Thecosmilia rugosa*, is very common amongst the non-capitate forms, and so is the *Thecosmilia Michelini*, Terq. et Piette.

At Brocastle and Cowbridge the larger *Thecosmiliæ* are common, but *Thecosmilia Michelini* forms large masses at Cowbridge, and studs blocks at Laleston. Although the specimens are very numerous, still the individuals rarely attain that bush-like structure which is noticed in the Continental beds.

At Cowbridge the specimens are mostly found as casts.

1. Thecosmilia Suttonensis, *Duncan.* Pl. IV, figs. 7—9.

The corallum has a slender and nearly straight peduncle, which gives off corallites from an enlarged summit.

The peduncle is moderately marked with transverse ridges and constrictions, and does not taper symmetrically from above downwards. The epitheca is thin, and permits very numerous and fine costæ to be seen through it.

The corallites springing from the parent (the peduncle) originate by intercalicinal gemmation; they are separate as regards their walls, and differ in size, being marked with transverse epithecal folds and constrictions. The calices are not quite circular, and their septal arrangement is irregular. The septa are unequal, and one half of them extend nearly to the centre, whilst the smaller pass inwards but for a short distance. The number of septa increases with the growth of the calices. In large calices there are more than four cycles, and in the smaller less than three cycles, or three cycles.

The endotheca is highly developed.

Height of corallum $1\frac{1}{2}$ inch. Diameter of large calice $\frac{4}{10}$ths inch. Diameter of small calice $\frac{3}{10}$ths inch.

Locality. The Sutton Stone. In the Collection of Charles Moore, Esq., F.G.S., Bath.

This species has some resemblance to *Thecosmilia serialis* in its short peduncle and capitate swelling; but the retention of the circular outline by the calices is distinctive. It has some resemblance in its calice to the simple calice of *Thecosmilia rugosa*, Laube, but there is no fissiparity observed. The origin of the corallites by intercalicinal gemmation is very distinctive, as are also the thin epitheca and the columnar shape of the peduncle. The habit of the species resembles that of the majority of the stunted *Thecosmiliæ* of the period.

2. Thecosmilia mirabilis, *Duncan*. Pl. II, figs. 10, 11.

The corallum is short, very finely pedicellate, increasing rapidly in breadth, and terminating by a large upper surface on which are several circular and distinct calices. The trunk of the corallum is smooth, and is slightly marked with rounded transverse swellings, and corresponding constrictions. No costæ can be seen. The corallites are unequal in size, separate immediately, do not increase by fissiparity, and are characterised by circular calices having a very sharp margin. The calices are shallow. The septa are numerous, crowded, and very regular; they are alternately long and short, and all are marked with small lateral swellings and faint linear depressions on the upper edge. The largest calices have four cycles of septa, and nearly a complete fifth cycle, the septa numbering from seventy-six to eighty-four.

Height of the corallum $\frac{7}{10}$ths inch. Breadth of the upper surface $\frac{6}{10}$ths inch. Diameter of the largest calice $\frac{3}{10}$ths inch.

Locality. The Sutton Stone. In the Collection of Rev. H. Winwood, F.G.S., Bath.

3. Thecosmilia serialis, *Duncan*. Pl. IV, figs. 10—12.

The corallum has a narrow, curved, and rather long peduncle, which gives off several corallites from its summit.

The peduncle is strongly marked with lateral ridges and constrictions, and so are the corallites.

The epitheca is stout, and, where worn, permits the costæ to be seen.

The young corallites arising by fissiparity from the parent, which constitutes the peduncle, separate into some which remain circular in transverse outline, and into others which form short serial calices.

The circular calices present four cycles of septa, and the serial have their septa less crowded and larger. The serial calices do not present any evidences of fissiparity.

Height of corallum $1\frac{1}{10}$ inch. Diameter of circular calice $\frac{1}{5}$th inch. Length of serial calice $\frac{3}{10}$ths inch. Breadth of serial calice $\frac{1}{10}$th inch.

Locality. The Sutton Stone. In the Collection of Charles Moore, Esq., F.G.S., Bath.

This species belongs to the stunted *Thecosmiliæ* so characteristic of the Triassic and Liassic coralliferous strata; it is readily distinguished by the number of corallites springing from the peduncle, and by its long and serial calices being mixed with rounded ones.

The mineralization of the specimen gives the appearance of a columella in the elongated calice, but there is really no such a structure.

4. Thecosmilia rugosa, *Laube.*[1] Pl. II, figs. 1—6.

The corallum springs from a small base, divides soon, and the branches are covered with an exceedingly strong epitheca marked with thick folds.

The calices, one or more in number, are either nearly round, or are irregularly distorted. They are deep, and the septa are stout, straight, and not very unequal. They number from thirty-four to thirty-six.

The diameter of a tolerably regular calice is $\frac{2}{10}$ths inch, and the length of a distorted calice $\frac{4}{10}$ths inch. The height of the corallum is about an inch.

Locality. The Sutton Stone. In the Collection of Charles Moore, Esq., F.G.S., Bath.

M. Laube's description of the species from the St. Cassian beds is simple and accurate. His small *Thecosmilia* has a strong epitheca, with constrictions and swellings, and its calices are now and then fissiparous. His plate gives the idea of there being more septa; and this is the only distinction which can be made between the St. Cassian and the British species.

5. Thecosmilia Brodiei, *Duncan.* Pl. X, figs. 1, 2, 3, 4.

The corallum is rather short; the corallites are cylindrical and large in relation to their height, and they appear to divide near together, so that regular calices are rare.

The epitheca is stout and complete, being marked with slight constrictions.

The calicular margin is sharp, and the calices are deep.

The septa are numerous, and the large primary and secondary septa are equal and very dentate. The tertiary septa are very much smaller than the secondary, are not dentate, but are long; and the septa of the fourth and fifth orders are very small.

Diameter of the calices $\frac{4}{10}$ths inch.

Locality. Brocastle. In the Collection of Charles Moore, Esq., F.G.S., Bath.

The extraordinary development of the dentate first and second cycles of septa characterise the species.

[1] Laube, 'Die Fauna der Schichten von St. Cassian,' 1 Abtheil.

6. Thecosmilia Martini, *E. de From.*[1] Pl. X, figs. 6—9.

The corallum is bush-shaped, and is formed by dichotomous cylindrical corallites, which are covered with a strong folded and complete epitheca.

The corallites separate rapidly, and remain free for some distance before fissiparous growth occurs again.

The dissepiments are very developed, and are inclined.

The calices are circular, or slightly oval.

The septa are very thin and distant. There are thirty-two large septa, one half of which reach the centre, and there are forty-eight small, or rudimentary septa.

The calices are about $\frac{5}{8}$ths inch in diameter.

Localities. Brocastle, Ewenny, Cowbridge. In the Collection of Charles Moore, Esq., F.G.S., Bath.

This species is distinguished by its size, high septal number, and highly developed endotheca.

In the British specimens the septa are stouter, and the calices are often larger than in the French; moreover the larger septa are often raised to the number of forty-eight. The rudimentary septa are not shown in M. Martin's plate.

The localities whence the species has been derived have been the middle and upper beds of the Zone of *Ammonites Moreanus*, at Semur, and Vic de Chassenay Côte d'Or.

It is found in the limestone of Charleville, with *Ammonites bisulcatus*, in the sandstone containing *Ammonites bisulcatus*, at Saul, and in the Hettangian sandstone containing *Ammonites angulatus*. The species had thus a considerable range both in space and time; and it followed the usual habit of widely wandering species, in varying from the true specific type.

7. Thecosmilia Michelini, *Terq. et Piette.*[2] Pl. VII, figs. 10—13; and Pl. X, figs. 10—14.

The corallum is bush-shaped, and is formed by numerous, close, dichotomous, subcylindrical, long and slender corallites, which are surrounded with a thick, folded, smooth, complete, and persistent epitheca.

The calices are nearly on the same level, are rounded or oval, and the fossa is not very deep. The septa are forty in number, and are alternately large and small.

The endothecal dissepiments are very close.

[1] Martin, 'Pal. Strat. de l'Infra-Lias,' 1860, pl. viii, figs. 8, 9.

[2] 'Le Lias Inférieur de l'est de la France,' p. 127, pl. xvii, figs. 7, 8.

The height of the corallum may reach 6 inches.

Diameter of a corallite $\frac{1}{3}$rd inch.

Localities. Brocastle, Cowbridge, Laleston, the Sutton Stone, and Ewenny. In the Collection of Charles Moore, Esq., F.G.S., Bath.

MM. Terquem et Piette ('Le Lias Inférieur de l'est de la France,' p. 127, Pl. xvii, figs. 7, 8) have described this well-marked species in their usually concise manner. The smaller size of the corallites, the septal number, and the small amount of endotheca, distinguish the species, which is very common in the Glamorganshire beds. The bush shape of the corallum may be imagined from the grouping of the casts of the species in the limestone at Cowbridge; and the dichotomous and slender form of the corallum is common at Laleston and in the Sutton Stone.

The rounded swellings and intermediate constrictions of the plain epitheca are very characteristic.

The French specimens are derived from the beds at Aiglemont, in the zone of *Ammonites angulatus*.

8. Thecosmilia irregularis, *Duncan*. Pl. III, figs. 1—6; and Pl. X, fig. 5.

The corallum is small, short, and has a broad base. It consists of a short and rather cylindrical peduncle with a broad base, a very strongly marked and ridged epitheca, and of an upper part whence the calices spring by fissiparity.

The calicular surface is considerably broader than the peduncle, and overhangs.

The calices are small, shallow, irregular in shape, and have a distinct margin.

The septa are few in number, large, unequal, and very irregular in their arrangement. They have large rounded teeth upon their upper margins, and the larger septa occasionally unite by their inner margins, which are toothed.

There are about twenty septa, and several others which are rudimentary.

There are no costæ.

Height of corallum $\frac{4}{10}$ths inch. Diameter of calices $\frac{2}{10}$ths inch.

Locality. Brocastle. In the Collection of Charles Moore, Esq., F.G.S., Bath.

Thecosmilia irregularis, *Duncan*. (A variety.) Pl. III, figs. 14, 15.

The calices are deeper, the septa longer and more slender, and the dentations sharper and more distinct than in the type.

Locality. Brocastle. In the Collection of Charles Moore, Esq., F.G.S., Bath.

9. THECOSMILIA TERQUEMI, *Duncan*. Pl. III, figs. 7—12.

The corallum has a fine pedicle, which increases in breadth very rapidly, and produces a large upper surface, upon which are the calices, or one corallite may spring from the edge of the upper surface, and give rise to others in the same manner. (Plate X, fig. 4.)

The epitheca is strong, folded, and constricted; where worn, the costæ and exothecal dissepiments appear.

The calices are irregular in shape, size, and distance.

The septa are unequal in size and are bluntly dentate, their arrangement is irregular, and a quaternary disposition of the laminæ is very evident, and they may number sixteen, twenty, or thirty-two. The larger septa do not indicate an hexameral arrangement. All are thick, distant, and pointed internally.

Height of the corallum $\frac{5}{10}$ths inch. Diameter of the calices $\frac{3}{20}$ths—$\frac{5}{20}$ths inch.

Locality. Brocastle. In the Collection of Charles Moore, Esq., F.G.S., Bath.

10. THECOSMILIA AFFINIS, *Duncan*. Pl. III, figs. 18—20.

The corallum is short, and the corallites separate soon after leaving a short, conical peduncle.

The calices are deep and open.

The septa are irregular, unequal, distant, often curved, dentate at their inner margin, and about sixteen in number.

The epitheca is moderately strong.

The height of the corallum is $\frac{8}{10}$ths inch. The diameter of the calices is $\frac{4}{10}$ths inch.

Locality. Brocastle. In the Collection of Charles Moore, Esq., F.G.S., Bath.

11. THECOSMILIA DENTATA, *Duncan*. Pl. III, figs. 21—23.

The corallum has a broad base, and the corallites separate soon, and diverge; they are subcylindrical, and their epitheca is smooth.

The calices have a very distinct margin; they are slightly deformed, not very deep, and contain numerous septa.

The septa are unequal; alternately large and small, irregular, and present distinct and numerous blunt dentations. The smallest septa are simple dentations, and the different sizes of the septa and dentations are very remarkable.

There is no exact arrangement of the septa in cycles, and their number varies from thirty to thirty-two and thirty-six.

Height of corallum $\frac{7}{10}$ths inch; breadth of calice $\frac{4}{10}$ths inch.

Locality. Brocastle. In the Collection of Charles Moore, Esq., F.G.S., Bath.

12. THECOSMILIA PLANA, *Duncan.* Plate III, figs. 24, 25.

The corallum is short, the calices separate rapidly, and soon attain a considerable size.

The epitheca is strong, and constricted here and there.

The calices are large, shallow, oval, and are deeper at the centre than elsewhere; their margin is indistinct, and the septa are rounded, faintly dentate, distant, and very irregular. There are about thirty septa.

Height of corallum $\frac{4}{10}$ths inch; breadth of calice $\frac{8}{10}$ths inch.

Locality. Brocastle. In the Collection of Charles Moore, Esq., F.G.S., Bath.

These species of the genus *Thecosmilia* may be arranged for the purposes of diagnosis as—

Long and more or less bush-shaped .	*Th. Martini.* *Th. Michelini.*
Pedunculate and capitate . . .	*Th. Suttonensis.* *Th. mirabilis.* *Th. Terquemi.* *Th. serialis.*
Short and stunted	*Th. irregularis.* *Th. rugosa.* *Th. plana.* *Th. Brodiei.* *Th. dentata.* *Th. affinis.*

Taken as a series, the species are very characteristic of the Coral-fauna of the period.

Genus—RHABDOPHYLLIA.

1. RHABDOPHYLLIA RECONDITA, *Laube.*[1] Plate II, figs. 7—9.

The corallum is pedunculated, has very fine costal markings, which are flat, and a delicate epitheca.

The corallites separate rapidly at the extremity of the peduncle.

The calice (section of) is almost circular, and is crowded with rather stout septa.

The septa are unequal, longer at the calicular margin than elsewhere, and either reach the columella or enlarge at their free extremity at different distances from it.

[1] Laube, 'Die Fauna der Schichten von St. Cassian,' 1 Abtheil.

There are four cycles of septa in six systems. The primary reach the columella. The tertiary, which are longer than those of the higher orders, join the secondary septa.

The columella is well defined, and is circular in its transverse outline.

The diameter of the corallite is about $\frac{1}{4}$ inch.

The specimens are usually covered with parasitic corals or sponges.

Locality. The Sutton Stone. The St. Cassian beds. In the Museum of Practical Geology, London.

Laube's[1] description of this species is very faithful, and it is readily recognised by the curious septal arrangement. The specimens are rare in the Sutton Stone, and the sections showing the septa require very careful examination before they can be understood.

FAMILY—ASTRÆACEÆ.

Genus—ASTROCŒNIA.

1. ASTROCŒNIA GIBBOSA, *Duncan*. Pl. V, figs. 2, 3, 4, 12; Pl. IV, fig. 3; Pl. VI, figs. 1, 2, 3, 4.

The corallum is large, and covered with rounded eminences of various sizes.

The calices are large, polygonal, close, irregularly placed, irregular in size, and shallow.

The septa are usually twenty in number, are joined to small narrow club-shaped straight costæ, are very unequal in size, and usually one half of them reach to the columella. The smaller and shorter septa unite in many instances to the larger septum between them, but not very close to the columella. The septa are finely dentate laterally, and there is a trace in some of the longest of a swelling close to the columella. Their development is very irregular.

The columella is moderately prominent and large.

The cœnenchyma is not strongly developed, but in sections the presence of ornamentation in the form of round processes is observable. The endotheca is occasionally noticed in the calicular fossa, and extends from septum to septum.

Three large calices, with their cœnenchyma, occupy the length of nearly $\frac{3}{10}$ths inch.

Locality. The Sutton Stone, and Brocastle.

In the Museum of Practical Geology, London; and in the Collections of Charles Moore, Esq., F.G.S., and Rev. W. Winwood, F.G.S., Bath.

[1] Laube, op. cit.

2. Astrocœnia plana, *Duncan*. Pl. V, fig. 1.

The corallum is large, flat, and short.

The calices are small, very regular in their linear arrangement, polygonal, and nearly equal: they are rather deep and rather distant.

The septa appear to be from eight to ten in number, and reach the columella.

The costæ are very indistinct.

The columella is large.

The cœnenchyma is well developed, and becomes divided into rounded eminences between the calices; and where four of these are together, the intervening cœnenchyma is decidedly peaked.

Three of the largest calices, with the intervening cœnenchyma, cover a length of $\frac{2}{10}$ths inch.

Locality. The Sutton Stone. In the Museum of Practical Geology, London.

3. Astrocœnia insignis, *Duncan*. Plate IX, figs. 1 and 2.

The corallum is large; it is flat on the upper surface, and is short.

The calices are somewhat regular in their linear arrangement; they are unequal, and are irregular as regards their outline and distance. They are shallow, and are large in comparison with those of most of the species of the genus.

The septa are large, and nearly equal in size at the calicular margin, but all do not reach the columella. Generally five primary septa extend to the columella, and there are three which only reach a little way into the calicular fossæ between the longer primary. The central of these smaller septa is often longer than those on each side of it. The septal number is irregular, but twenty is the usual number. In some calices the shorter septa are decidedly smaller than the others.

The costæ are large, broad, straight, nearly equal, club-shaped, close, and are oblique in some, but flat in other calices. They extend over the cœnenchyma when it exists, do not coalesce with those of other calices, and are often separated by a ridge. Neither septa nor costæ appear to be spined or dentate, but a very slight unevenness of the margin may be noticed in well-preserved specimens.

The columella is small, sharp, and prominent.

The size of the calices varies, and in large specimens, where there is some cœnenchyma, three calices and their cœnenchyma occupy rather more than $\frac{3}{10}$ths inch. The smallest calices, with a small quantity of cœnenchyma, do not occupy one half of that space.

Locality.—Brocastle. In the Collection of Charles Moore, Esq., F.G.S., Bath.

4. Astrocœnia reptans, *Duncan*. Plate IV, figs. 4, 5, 6, 15.

The corallum is short, convex, and very irregular; it is moderately large for an Astrocœnian, and is covered with numerous and closely packed calices.

The calices are polygonal and shallow, and are separated by very distinct, plain, cœnenchyma, which is obtusely ridged, and prominent here and there.

The septa are twenty in number; ten reaching the columella, and ten joining five of the longer, in pairs.

The septal arrangement is very marked.

The columella is small and the costæ are rudimentary.

The length of three calices, with the cœnenchyma, is about $\frac{3}{20}$ths inch.

Locality. The Sutton Stone, Brocastle, and at Ewenny. In the Collection of Charles Moore, Esq., F.G.S., Bath.

5. Astrocœnia parasitica, *Duncan*. Plate V, figs. 5, 6.

The corallum encrusts other *Madreporaria*, such as dendroid Astræaceæ or "remanié" Lithostrotions; it is small and short, and possesses much cœnenchyma.

The calices are very small, distant, and shallow.

The septa appear to be ten in number.

The columella is well marked.

The cœnenchyma is plain.

The diameter of the calices is about $\frac{1}{20}$th inch.

Locality. The Sutton Stone. In the Museum of Practical Geology, London.

6. Astrocœnia pedunculata, *Duncan*. Plate V, figs. 7, 8, 9.

The corallum is small, pedunculate, and fungiform; it has an epitheca and much cœnenchyma.

The peduncle is short, small, and rounded, and joins the expanded discoid epithecate base of the true corallum near its centre.

The discoid base has an epitheca, and its edges are slightly rounded.

The convex upper part of the corallum is covered with unequal, shallow, and distant calices.

The calices are irregular in size, and are small.

The septa are small, alternately long and short, and are granulated laterally. There are twenty of them, and the smallest are rudimentary.

The cœnenchyma is abundant, and is elevated between some calices and flat between others.

Height of the corallum $\frac{4}{10}$ths inch.

Locality.—Brocastle. In the Collection of Charles Moore, Esq., F.G.S., Bath.

7. ASTROCŒNIA COSTATA, *Duncan.* Pl. IX, figs. 15, 16, 17.

The corallum is small, irregular in shape, and rounded above.

The calices are numerous, and rather deep; they are either very close together, or they are separated by more or less cœnenchyma, whose upper surface is marked by wavy costæ.

The septa are usually twenty in number, and their costal ends are nearly equal.

The costæ are either very small, small and curved, or large and more or less curved as they approach the costæ of neighbouring corallites.

The columella is small.

The space occupied by three large calices, separated by much cœnenchyma, is $\frac{2}{10}$ths inch.

Locality. Brocastle. In the Collection of Charles Moore, Esq., F.G.S., Bath.

8. ASTROCŒNIA FAVOIDEA, *Duncan.* Pl. IX, figs. 12, 13, 14.

ASTRÆA FAVOIDES? *Quenstedt,* Der Jura, 1858.

The corallum is more or less globose, and the calices are very small, very deep, and are separated by sharp ridges. The cœnenchyma is rudimentary.

The septa are twenty in number, the smaller being very rudimentary.

The costæ are rudimentary.

The columella is small, and is situated at the base of the very deep calice.

Localities. Brocastle. In the Collection of Charles Moore, Esq., F.G.S., Bath. Also in the Arieten-Kalk of Germany.

9. ASTROCŒNIA SUPERBA, *Duncan.* Pl. IX, figs. 3, 4, 5.

The corallum is small, and irregular in shape.

The calices are shallow, wide apart, and usually circular in outline.

The septa are usually twenty in number, are small near the columella, and thicker at the costal end. About one half of them reach the columella. They are dentate.

The costæ are highly developed, and cover the cœnenchyma, which is also spiny between the costal ends. Nearly all the costæ are equal; they are straight in some places and wavy in others, but all are strongly dentate and well marked.

The columella is small.

Three calices occupy about $\frac{1}{4}$ inch in length.

Locality. Brocastle. In the Collection of Charles Moore, Esq., F.G.S., Bath.

10. Astrocœnia dendroidea, *Duncan*. Pl. IX, figs. 9, 10.

The corallum is in small branches, with blunt extremities.

The cœnenchyma is highly developed and plain.

The calices are wide apart in some places, but close in others; they are shallow, small, and more or less circular.

The septa are very irregular in their number, and their costal ends are club-shaped and rounded.

The columella is small.

The branches rarely exceed $\frac{1}{2}$ inch in length.

Locality. Brocastle, and at Ewenny. In the Collection of Charles Moore, Esq., F.G.S., Bath.

11. Astrocœnia minuta, *Duncan*. Pl. IX, figs. 18, 19, 20.

The corallum is large, flat, and thin. It is more or less encrusting in its habit.

The calices are very small, rather deep and close: they are more or less circular in outline, and are separated by a small quantity of cœnenchyma.

The septa are usually twenty in number, and many of them have a paliform tooth close to the columella. The costæ are small.

The columella is small.

Locality. Brocastle. In the Collection of Charles Moore, Esq., F.G.S., Bath.

12. Astrocœnia Sinemuriensis, *D'Orb.*, sp.[1]

The corallum is in the shape of a rounded mass, which is formed of superimposed layers. The calices are small, and tolerably regularly polygonal. The columella is stout, and projects. The septa are rather thick, unequal, and slightly close. There are twenty

[1] Martin, op. cit., pl. vii, figs. 26, 27.

in each calice, ten large and ten small. The internal ends of the large septa are rounded and swollen out.

Diameter of the calices $1\frac{1}{2}$ inch.

M. D'Orbigny considered the swollen ends of the principal septa to be pali, and placed the species in the genus *Stephanocœnia*, but M. de Fromentel determined the correct position of the form to be amongst the "decemeral" Astrocœniæ.[1]

The species does not appear to have been formed from very perfect specimens, and in M. J. Martin's admirable plate the septa are all equal in length and thickness, and the calices are close together. It is impossible to determine from either the description or the plate whether the calices are deep, whether there is any ornamentation, or whether the cœnenchyma is marked in any way. There are many species of Astrocœnia which are massive, and their formation from superimposed layers is the result more of a process of mineralization than of growth. The form is readily recognisable in its strata, because it is rare and as yet the only species discovered; but placed in comparison with others from distant localities it is hardly to be distinguished, on account of its defective specific distinctive peculiarities. The Astrocœniæ of the Sutton Stone, and from Brocastle, show the smaller septa joining the larger more or less, but this does not appear to be the case in *A. Sinemuriensis*. The enlarged state of the septal ends is common to several Astrocœniæ. It is very probable that with more complete specimens, the occasional union of the septa will be observed in *A. Sinemuriensis*, for in some specimens of most of the species this non-union is seen in certain calices.

Specimens of some Astrocœniæ in the Sutton Stone and Brocastle beds put on all the appearances of this species when worn. It is therefore introduced here; but not figured.[2]

The genus *Astrocœnia* was formerly included in the *Eusmilinæ aggregatæ*, but Reuss[3] pointed out the fact that the upper margins of the septa of the species falling under his observation were dentate and not smooth.

M. de Fromentel[4] discovered in the Neocomian formation some species which had dentate septa; and after acknowledging Reuss's discovery, he placed the genus amongst the XXVth family of his classification, the "*Astréens.*" This family corresponds in part to the *Astræaceæ* of Milne-Edwards and J. Haime, and the genus may be considered to form a part of the *Astræaceæ*.

In the Introduction to the British Fossil Corals,[5] *Astrocœnia*, being placed amongst the *Eusmilinæ*, follows the genus *Stylocœnia*, and was evidently considered to be closely allied to it. The following is the generic diagnosis by MM. Milne-Edwards and J. Haime:

[1] Pal. Strat. de Infra-lias, p. 94. J. Martin, 1860.

[2] See "Remarks on *Astrocœnia Sinemuriensis* and *Astrocœnia Oppeli*," Laube, in my essay on the "Madreporaria of the Infra-Lias of South Wales," 'Quart. Journ. Geol. Soc.,' Feb. 1867, p. 25 (note).

[3] Reuss. 'Beiträge, zur Charakteristik der Kriedeschichten.'

[4] E. de Fromentel, 'Introd. à l'Étude des Polyp. Foss.'

[5] 'Introd. to Brit. Foss. Corals: Palæontogr. Soc.'

"Corallum very dense, and not bearing columnar processes, as in the preceding genus. Calices polygonal, columella styliform, not projecting much. No pali. Septa thick, apparently eight or ten systems, two or four of the secondary septa being as much developed as the six primary ones. Walls thick and united as in *Stylocœnia*."

M. de Fromentel separated the genera *Astrocœnia* and *Stylocœnia,* and retained the latter amongst the *Eusmilinæ aggregatæ.* There was no reference made, therefore, in his generic diagnosis of *Astrocœnia* to the genus *Stylocœnia.* M. de Fromentel's description of the generic peculiarities of *Astrocœnia* are as follows: "Corallum massive, composed of corallites united by their walls, which are prismatic in shape; the calices are polygonal; the columella is styliform, and more or less projecting; the septa are tolerably thick, are few in number, and are dentate, especially near the columella; there are no pali."

Whilst investigating the Madreporaria of the Maltese rocks in 1865, I found that the septa of the common species *Stylocœnia lobato-rotundata,* Mich. sp., were dentate.[1] The species occurs also in the Chert of Antigua, and presents there the usual plain septa considered to mark the family of the genus. If fossilization can remove the dentations of the septa of one *Stylocœnian,* it can do so in others, and it may be safely asserted that all the *Stylocœnians* had dentate septa.

This dentate condition of the septa brings the genera *Astrocœnia* and *Stylocœnia* together again, although it removes them from the *Eusmilinæ* into the *Astræaceæ.*

MM. Milne-Edwards and J. Haime's generic description can thus stand, and its concluding sentence respecting the thick walls of the genus which was omitted by M. de Fromentel is very important.

In some species, as in *A. pulchella,* Ed. & H., the calices are so wide apart in some specimens, and in certain spots in all the specimens, that there is evidently here and there a cœnenchyma between the walls of the corallites. The surface of the cœnenchyma, which appears to arise from an hypertrophied condition of the adjacent corallite walls, is usually ornamented either with prolongations of the costæ, or with small papillose granules. This is observed in other species, and it is noticed that the amount of cœnenchyma varies according to the shape of the corallum, and the rapidity of the multiplication of the corallites. The presence of scattered granules, or of small papillæ on the cœnenchymal surface, and between the external terminations of the costæ, is observed in some specimens of a species, and not in others; but the costæ, although they may extend far over the inter-calicular spaces (or, in other words, over the surface of the cœnenchyma), never unite, and run into those of adjoining corallites. There are modifications in the length and straightness of the costæ, and where there is no cœnenchyma, and the walls of the corallites are thin, they may be so reduced in size as to appear to be simple terminations of septa.

In many species the cœnenchyma, when non-costulated, and not ornamented with granules, becomes slightly ridged, and foreshadows the condition which peculiarises the genus *Stylocœnia.*

[1] 'Ann. Mag. Nat. Hist.,' April, 1865.

The reproduction by gemmation cannot occur from the walls of the corallites, except at the edge of the corallum. The close contact of the walls, and the existence of dense cœnenchyma, prevent any budding from the wall; but where the outside corallites are partly free, there gemmation may occur outside and below the calicular margin.

Fissiparity does not occur, but the young buds arise either from the top of the calicular edge or margin, or just within the calice. When there is some distance between the calices on account of the thickened walls or cœnenchyma, buds may arise on the cœnenchymal or inter-calicular surface.

Many of the species have an epitheca, some are pedunculated, and others are massive, encrusting, or dendroid.

The septa vary greatly in their numbers and cyclical arrangement, and very often they have a large paliform tooth close to the columella. There are no pali.

A styloid columella projecting more or less, is an essential generic requisite. The endotheca is scanty, but it always exists.

The calices are small, and vary in depth; but, as a rule, they are arranged with great symmetry, and are polygonal in outline. Transverse sections show the complete consolidation of the walls, and the space between the costal ends, in these sections, is often marked with granules.

The species without any cœnenchyma, and whose walls are thin, are distant in their alliance to *Stylocœnia*, and had they no columella, they would be considered to belong to the genus *Isastræa*. The genus *Cyathocœnia* (Duncan) comprehends *Astrocœniæ* without columellæ.

The fossil condition of the specimens must be considered during the specific determination of *Astrocœniæ*. Usually, the columella is represented by a flat, central, and more or less circular mass with the ends of thick septa adherent to it. In these instances a calcareous deposition has occurred around the columella and between the septal ends, the columella having been broken off. It happens, however, that the columella may be broken off without the deposition having taken place, and either the structure retains its normal size at the point of fracture, or is absent altogether.

On examining doubtful specimens which have lost their columellæ, much attention should be paid to longitudinal sections produced by weathering, fracture, or by artificial means. A small projection at the base of the calice is more readily determined to exist in longitudinal views than in those which simply show the open calice.

There are eleven species of the genus *Astrocœnia* special to the Welsh Lias, and one species found with these has been described by D'Orbigny as *Stephanocœnia Sinemuriensis*. M. D'Orbigny obtained his specimens from the Lower Liassic deposits of France. M. de Fromentel and MM. Terquem and Piette have found the species in several localities, and the first-named palæontologist has determined it to belong to the genus *Astrocœnia*.

The Liassic *Astrocœniæ* occur as large and massive, small and dendroid, or as irregular and sometimes as encrusting forms. All are very irregular in their septal arrangement, and none of them present definite and clear cyclical sequences.

Some of the species have the cœnenchyma between the calices irregularly ridged, so as to present the first traces of that cœnenchymal development which characterises the genus *Stylocœnia*. The columella is very distinct in all the species, and the junction of the largest septa to it is marked in some forms by a paliform swelling, but there are no pali. In many species the smaller septa unite more or less to the larger, and in others the dentate condition of the septal edge is very marked. The costæ are either rudimentary or well developed in different species; they may be straight, spined, and wavy.

The size of the corallum, its shape and its habit, with the size of the calices, and the character of the costæ and of the cœnenchyma, appear to separate certain forms from others and enable eleven new species to be classified with the *Astrocœniæ*.

The following scheme of the structural peculiarities of the new *Astrocœniæ* will show how readily their specific distinctions may be recognized :

ASTROCŒNIA.

Corallum . .	large . .	gibbous and tall . . .	*Astrocœnia gibbosa.*
		flat and short	— *plana.*
			— *insignis.*
		short, and irregular in outline .	— *reptans.*
	small . . .	encrusting	— *parasitica.*
		pedunculate, with an epitheca .	— *pedunculata.*
		dendroid	— *dendroidea.*
		flat and narrow . . .	— *superba.*
		globose	— *favoidea.*
		irregular	— *costata.*
		flat and semi-encrusting . .	— *minuta.*

Corallum having the cœnenchyma . .	scanty	*Astrocœnia favoidea.*
		— *minuta.*
	abundant	— *parasitica.*
		— *dendroidea.*
		— *superba.*
		— *pedunculata.*
		— *insignis.*
	moderately developed .	— *reptans.*
		— *costata.*
		— *gibbosa.*
		— *plana.*

The surface of the cœnenchyma	ornamented .	costæ well developed	and straight .	*Astrocœnia insignis.*
			spined .	— *superba.*
			wavy .	— *costata.*
			. . . .	— *gibbosa.*
	ridged			— *gibbosa.*
				— *plana.*
				— *minuta.*
				— *reptans.*
				— *dendroidea.*
	plain			— *parasitica.*
				— *pedunculata.*
	rudimentary			— *favoidea.*

Genus nov.—Cyathocœnia.

This genus has been determined for species which, had they columellæ, would belong to the genus *Astrocœnia*.

The walls of the corallites of the species are joined, and there is more or less cœnenchyma. The costæ are not confluent, and the septa are finely dentate. There are no pali, nor is there a columella. There is no fissiparity, and the gemmation is either from the intercalicular surface, or from the calicular margins.

There is always some cœnenchyma present, and this distinguishes the new genus from *Isastræa*, the only genus with which it can be confounded.

The following is the generic formula:

Cyathocœnia.—The corallum is compound. The corallites are united by their walls and by more or less cœnenchyma; they are more or less polygonal, but are often cylindrical. The calices are small, the costæ are not confluent, and the septa are finely dentate. There is no columella. There are no structures on the cœnenchyma between the calices except granules and costæ. The gemmation is superior and marginal.

1. Cyathocœnia dendroidea, *Duncan*. Plate IX, figs. 6, 7, 8, 9.

The corallum is large and tall, forming fasciculate masses. The corallites are more or less crowded on the surface of stems, which branch rarely, and which are close and more or less parallel. The transverse outline of the stems is irregular, from variability in their thickness, and also from the presence of superficial calices. The stems consist of calices separated by cœnenchyma whose amount varies.

The calices are distant when there is much cœnenchyma, but occasionally they are close, and their margin then becomes round; they are small, are irregularly placed, and are rather deep.

The septa are dentate, distinct, distant, unequal, stout, not exsert, and pass obliquely downwards and inwards, so that they do not encroach much upon the calicular fossa. There are eighteen in some and twenty-four in the largest calices. Three cycles appear to be the normal number. There is no columella.

The costæ either reach on to the surface of the cœnenchyma or end abruptly at the calicular margins, and they never become continuous with those of other calices.

The stems are several inches in height, and are from $\frac{3}{20}$ths to $\frac{5}{20}$ths of an inch in diameter.

The calices rarely exceed $\frac{1}{12}$th of an inch in diameter.

Locality. Brocastle.

In the Collection of Charles Moore, Esq., F.G.S.

The peculiar mineralization of the specimen prevents the structure of the central parts of the stems being distinguished.

There is a dendroid Astrocœnian in the Brocastle beds which has some resemblance to this species, but the well-developed columella of the first distinguishes it at once.

MM. Terquem and Piette have described a species *Microsolena Fromenteli*,[1] whose bush-like form and parallel constricted and irregular stems resemble *Cyathocœnia dendroidea*, but the calices have a columella, and the costæ are continuous; nevertheless, the "habit" of both species is very similar.

2. Cyathocœnia incrustans, *Duncan*. Pl. IV, figs. 1, 2.

The corallum is very thin and encrusts portions of the shells of Bivalve Mollusca.

The calices are unequal, circular or subpolygonal, rather close and very shallow. The septa are few in number, are very small, and are marked with distinct and almost moniliform processes. They are thickest at the margin of the calices. The larger septa usually alternate with smaller, but, as a rule, the largest are the most numerous.

The septa cannot be recognized as following a cyclical arrangement, and they vary in number from fifteen to twenty.

There is no columella. The cœnenchyma is scanty and is marked with large granules, which are the representatives of costæ. The gemmation occurs between the calices.

Diameter of calices $\frac{1}{20}$th to $\frac{1}{10}$th of an inch.

Locality. The Sutton Stone, encrusting an *Ostrea*.

In the Collection of Charles Moore, Esq., F.G.S.

The papillate septa and encrusting habit distinguish the species from *C. dendroidea*, and *C. costata*.

[1] Op. cit., pl. xvii, figs. 11, 12.

3. Cyathocœnia costata, *Duncan*. Pl. V, figs. 10, 11.

The corallum is flat, and presents slightly rounded eminences; it is short, and has an irregular base where it is attached to foreign bodies.

The calices are numerous, nearly equal, and distant. The margins of the calices are flat, and are continuous externally with the cœnenchyma, whose upper surface is covered by the costæ.

The calicular fossæ are deep.

The septa are small, unequal as regards length, but rather equal in their thickness; they vary in number from twenty to twenty-four.

There is no columella.

The costæ are large, slightly rounded, not continuous, and occasionally slightly wavy.

Three calices occupy a length of $\frac{2}{10}$ths inch.

Locality. Brocastle.

In the Collection of Charles Moore, Esq., F.G.S., Bath.

Genus—Elysastræa.[1]

1. Elysastræa Fischeri, *Laube*.[2] Pl. VI, figs. 5—9.

The corallum is massive; the corallites are close and united above and near the calices, but separate and more or less covered with epitheca below.

The corallites are unequal in size, tall, and more or less cylindrical below and polygonal above.

The calices are very variable in shape and size, and the margin is broad and distinct.

The septa are numerous, often wavy, unequal in length, and near to the centre of the calice a new set appears to come in, in some calices. There are no pali.

The number of septa depends upon the size of the calice, and it may vary from forty to sixty.

The septal laminæ are thin, and faint traces of costæ may be seen where the walls are not fused together.

The gemmation is extra-calicular, but the bud probably springs from the centre of a corallite, and works its way outwards.

The columella is rudimentary.

Diameter of calices, $\frac{2}{10}$ths to $\frac{4}{10}$ths inch.

Locality. The Sutton Stone. The St. Cassian beds.

In the Museum of Practical Geology, London.

1 Laube, op. cit., and 'Intro. Brit. Foss. Corals,' 2nd series, part i.

2 Laube, op. cit.

2. Elysastræa Moorei, *Duncan*. Pl. VI, figs. 10—15.

The corallum is massive, and the upper surface is very irregular.

The corallites are joined by their walls in many places, but are free in others, both superiorly and lower down in the corallum.

The corallites vary greatly in size, and the smallest are usually joined by their walls, and are more or less angular in outline. The largest corallites are circular in outline.

The calices are irregular in their depth, and are either circular or polygonal. They are close, even when not adherent.

The septa are alternately large and small, are faintly dentate, and are very variable in number. There are forty-eight septa in the largest calices.

The costæ are continuous with the septa in the separate corallites, but do not exist when the walls are united.

The columella is deficient.

The endotheca is very abundant.

The diameter of the calices is from $\frac{2}{10}$ths—$\frac{4}{10}$ths inch.

Locality. The Sutton Stone, and at Brocastle.

In the Collection of Charles Moore, Esq., F.G.S., Bath.

The genus *Elysastræa* is very remarkable; it has affinities with *Isastræa* and with the very close bush-shaped *Thecosmiliæ*. The bush-shaped *Thecosmiliæ* are noticed to become united by their walls in some specimens, and the walls of *Septastræœ* and *Prioastrææ* are occasionally not united inferiorly.

The species *Elysastræa Moorei* has its corallites more distinctly separate than the St. Cassian form, which is, however, clearly represented in the Sutton Stone.

The appearance of septa near the centre of the calice is very characteristic of the genus.

Genus—Isastræa.

1. Isastræa Sinemuriensis, *E. de Fromentel.*[1] Pl. VII, figs. 1—9.

The calices are polygonal, and tolerably deep.

The septa are very numerous, spined, close, and unite occasionally by their inner

[1] Martin, 'Pal. Strat. de l'Infra-Lias du dép. de la Côte d'Or,' 1860, pl. vii, figs. 16, 17.

border. There are seventy-eight septa in the largest calices, and they are unequal. The calices are from $\frac{3}{12}$ths—$\frac{4}{12}$ths inch in diameter.

To this specific determination of M. de Fromentel the following may be added, as better specimens have been derived from the Brocastle bed than elsewhere.

The corallum is massive, and irregular in shape, but often assumes a subglobular form. When this is the case there is an epitheca, which is strongly folded, but which is lost as the calices are developed.

The size of the calices is very irregular, and marginal gemmation is very common.

The septa are crowded and distinct, and in the largest calices there are many of the fifth cycle, but there is great irregularity in the septal number. The septa are often not quite straight, and present swellings at several points.

Locality. Brocastle. Menetreux, near Samur.

In the Collection of Charles Moore, Esq., F.G.S., Bath.

2. Isastræa globosa, *Duncan.* Pl. VIII, figs. 17, 18.

The corallum is nearly spherical in shape; it has a cylindrical but short peduncle, covered with epitheca, and a rounded upper surface marked with very numerous and closely placed small calices.

The calices are shallow, faintly polygonal, and crowded with septa. The septa are unequal, not very thin, and have now and then an enlargement at the inner end. The smaller septa frequently unite to the larger. All are very distinct. A cyclical arrangement of the septa cannot be distinguished, and the septal number varies from twenty, twenty-four, to thirty-six.

There is no columella.

The diameter of the calices is about $\frac{1}{10}$th inch.

Locality. Brocastle.

In the Collection of Charles Moore, Esq., F.G.S., Bath.

The largest specimens of this fossil are usually much worn, and some care must be taken in examining the perfect calices, for their mineralization often suggests a columella.

Genus—LATIMÆANDRA.

LATIMÆANDRA DENTICULATA, *Duncan*.

One or two calices of a *Latimæandra* occur in several of the hand-specimens in the Collection of Charles Moore, Esq., F.G.S. The calices are long and are straight; they are separated by sharp walls, and the larger septa have a high paliform tooth close to their inner end. This structure of the septa distinguishes the species; but as no very satisfactory views can be obtained of a series of calices in the specimens, it has not been thought worth while to have the incomplete structures drawn.

Locality. Brocastle.

In the Collection of Charles Moore, Esq., F.G.S., Bath.

Division—FAVIACEÆ.

Genus—SEPTASTRÆA.

SEPTASTRÆA EXCAVATA, *E. de Fromentel*.[1] Pl. I, figs. 6, 7.

The corallum is rather tall and rounded.

The corallites are intimately united by their walls, which, although very thin, have a slight line of separation between them.

The calices are polygonal, irregular, and deep.

The septa are thin, distant, and strongly dentate, especially near the centre.

Fissiparity occurs, and the longest calices may have three calicinal centres.

In simple calices there are from thirty-six to forty-two septa, which are unequal. The hexameral type is very distinct.

The diameter of simple calices is from $\frac{2}{10}$ths to $\frac{3}{10}$ths inch.

In the specimens from Brocastle the abrupt rise of the septa near the calicular margin is very well seen. The calices are very irregular, and the longitudinal sections show constrictions and irregular swellings, which are very characteristic. Most of the calices have forty-eight septa or more, especially those about to divide.

Locality. Brocastle; and Pont d'Aisy, Côte d'Or.

In the Collection of Charles Moore, Esq., F.G.S., Bath.

[1] Martin, op. cit., pl. viii, figs. 1—5.

The remarkable *Thecosmilia rugosa*, *Rhabdophyllia recondita*, and *Elysastræa Fischeri*, from the lower part of the Sutton Stone, have been described and drawn by Laube from the St. Cassian beds of the Trias. The fauna with which they are associated in the Trias has not been described; but the presence of the species in the Angulatus-Zone of the Lower Lias or the Infra-Lias is very interesting.

The species described by Terquem and Pictte, E. de Fromentel, and D'Orbigny, viz.—

Isastræa Sinemuriensis, E. de From.,
Septastræa excavata, E. de From.,
Montlivaltia polymorpha, Terquem et Piette,
Astrocœnia Sinemuriensis, D'Orb., sp.,
Thecosmilia Martini, E. de From.,
„ *Michelini*, Terquem et Piette,

are associated in the Continental Liassic strata with many of the species of Mollusca which are noticed in the Sutton Stone and in the deposits at Brocastle in Glamorganshire.

In the *Lumachello* of the upper series of the Infra-Lias of Normandy[1] (the Calcaire de Valogne) *Septastræa excavata* is found to be associated with the following species, found also with it in the Glamorganshire Lias, which rests on Carboniferous Limestone:

Cerithium acuticostatum, Terquem.
Turritella Dunkeri, Terquem.
„ *Zenkeni*, Dunker, sp.
Phasianella Morencyana, Piette.
Ostrea anomala, Terquem.
Cardinia regularis, Terquem.

At Vic de Chassenay[2] *Thecosmilia Martini*, E. de From., and *Astrocœnia Sinemuriensis*, D'Orb., sp., are associated with—

Ammonites Moreanus, D'Orb.
Littorina clathrata, Desh.
Cerithium Semele, D'Orb.
„ *gratum*, Terquem.
„ *acuticostatum*, Terquem.

The middle bed of the Grès calcareux described by Terquem ('Pal. de Hettange,'

[1] Deslongchamps, 'Mém. Soc. Linnéenne de Normandie,' vol. xiv, 1865. p. 1.
[2] Martin, 'Pal. Strat. de l'Infra-Lias,' &c., 1860.

5

1855), contains *Isastræa Sinemuriensis* and the following species, in addition to the Mollusca just mentioned from the Côte d'Or and the Calcaire de Valogne:

Neritopsis exigua, Terquem.
Gervillia acuminata, Terquem.
Lima tuberculata, Terquem.
Plicatula intusstriata, Emm.
Ostrea irregularis, Münst.

These species, common in the French beds which are included in the Zone of *Ammonites angulatus*, and which form part of the Lower Lias of some and of the Infra-Lias of other geologists, are those that are associated with the great Coral-fauna of the Sutton Stone and of the equivalent deposits at Brocastle, in Glamorganshire.

The following Table shows the community of some well-known species in the coralliferous Liassic beds of Glamorganshire, and those of France and the Duchy of Luxembourg.

	Sutton Stone.	Southerndown.	Brocastle.	*Ammonites planorbis* Zone.[1]	Calcaire de Valogne.	*Ammonites Burgundiæ* Zone.[2]	*Ammonites angulatus* Zone.[2]	Luxembourg, Grès calcareux.
Septastræa excavata, E. de From.	...	...	*	...	*	...	*	...
Montlivaltia polymorpha, Terq.	*	...	*	...	...	...	*	...
— *pedunculata*, Dunc.	*	...	*	...	...	...	...	...
Isastræa Sinemuriensis, E. de From.	...	...	*	...	...	...	*	...
Thecosmilia Martini, E. de From.	...	...	*	...	*	...	*	..
— *Michelini*, Terq.	*	...	*	...	...	...	*	...
Ammonites angulatus, Schl.	...	*	...	...	...	...	*	...
Cerithium acuticostatum, Terq.	...	...	*	*	*	...	*	*
— *gratum*, Terq.	...	...	*	...	...	*	*	*
— *Semele*, D'Orb.	...	...	*	...	...	*	*	...
Turritella Dunkeri, Terq.	...	...	*	...	*	...	*	*
— *Zenkeni*, Dunk., sp.	...	...	*	...	*	...	*	*
Littorina clathrata, Desh.	...	*	*	...	...	...	*	*
Phasianella Morencyana, Piette	...	...	*	...	...	...	*	...
Neritopsis exigua, Terq.	*	...	*	...	...	...	...	*
Gervillia acuminata, Terq.	...	...	*	...	...	...	*	*
Ostrea irregularis, Münster (*O. liassica*, Strickland)	*	*	*	*	...	*	...	*
— *multicostata*, Münst.	*	*	*	...	...	...	...	...
— *anomala*, Terq.	*	...	*	...	*	...	...	...
Lima tuberculata, Terq.	*	*	*	...	...	*	*	*
Cucullæa Hettangiensis, Terq.	...	...	*	...	...	...	*	...
Cardita Heberti, Terq.	...	...	*	...	...	...	*	...
Lima exaltata, Terq.	*	...	...	...	...	...	*	*
— *dentata*, Terq.	*	*	...	...	...	...	...	*
Cardinia regularis, Terq.	*	...	...	...	*	*	*	*
Plicatula intusstriata, Emm.	*	*	...	*	...	*	*	*

[1] Terquem et Piette, op. cit. [2] Côte d'Or.

The range in space and in time of some of these species is very remarkable. Several of them range from the Italian to the Welsh Lias, and from the Zone of *Avicula contorta* to that of *Ammonites Bucklandi;* but the general grouping of the Gasteropoda, Lamellibranchiata, and Madreporaria indicates a Zoological Province which flourished anterior to the characteristic fauna of the time of *Gryphæa incurva* and *Ammonites Bucklandi.*[1]

The richness of the Glamorganshire beds beneath the arenaceous deposits containing *Gryphæa incurva* in species and specimens is very evident. The Madreporaria are rare in the equivalent strata on the Continent.

IV. Description of the Species from the Zone of Ammonites Angulatus at Marton, near Gainsborough.

At Marton,[2] on the line of railway from Gainsborough, in Lincolnshire, to Lincoln, there are dull blue earthy and shelly limestones, which are very fossiliferous. These beds have been carefully searched for fossils, and a very rich and interesting fauna has been collected.[3]

They occupy a position above the White Lias and below the blue compact limestones of the *Ammonites Bucklandi* series.

The fauna is very characteristic, but the Madreporaria are allied rather to those of the equivalent beds of the Lower Lias in the North of Ireland and of the East of France than to the species at Brocastle and in the Sutton Stone.

1. Montlivaltia Haimei, *Chapuis et Dewalque.* Pl. X, figs. 24—32.

"The corallum is simple, discoidal, and depressed; the base is very slightly pedicillate; the epitheca is very thin, ridged, and extends to the calicular margin.

"The calice is circular in outline, slightly or not at all convex, and the central fossa is small and circular.

"The septa are numerous, and form six cycles in six systems. The primary and secondary septa nearly reach the centre of the calice, and barely differ from those of the third cycle. The septa of the sixth cycle are very small. All the septa are thin, and their margin is strongly crenulate; those of the first and second cycles become thicker near the centre of the calice, and thinner at the periphery, where all the septa are about the same thickness." ('Descript. des Foss. des Terr. Second. du Luxembourg,' Chapuis et Dewalque, p. 268.)

The resemblance of the species to a Cyclolite is noticed by MM. Chapuis and

[1] P. Martin Duncan, 'Quart. Jour. Geol. Soc.,' Feb., 1867.

[2] F. M. Burton, Esq., F.G.S., and the Rev. B. Chamberlin, F.G.S., have given me information on this section.

[3] Ralph Tate, Esq., F.G.S., "On the Fossiliferous Development of the Zone of *A. angulatus*, &c.," an unpublished paper from which I have obtained much information, and all my knowledge of the Molluscan fauna of Marton.

Dewalque; and they remark that the base is ordinarily slightly convex, but sometimes perfectly horizontal; moreover, they observe that the calice is more convex when the base is horizontal.

MM. de Fromentel and de Ferry have divided the species *Montlivaltia Haimei* into three:

1. *Montlivaltia Haimei,* Ch. et Dew.
2. „ *tenuisepta,* From. et Ferry.
3. „ *granigera,* From. et Ferry.

He doubts the propriety of admitting so great a variation in septal number and in septal ornamentation as must be tolerated if the species were left entire.

A very considerable series of specimens of the species has been examined, and the distinctness of such forms as those considered worthy of the specific names *tenuisepta* and *granigera* has not been satisfactorily determined. Like the recent simple corals, *Montlivaltia Haimei* may have had a great variability. It was a very common species, and therefore all the more likely to vary in its shape, septal number, and ornamentation.

It is evident that there are forms of the species which are either concave or horizontal at the base; and others which are barely convex at the base, and which may become conical, sensibly taller than usual, and even cylindro-conical in shape. The convexity of the calice, or rather the exsertness of the septa, is often, but not always, correlative to this development of the base, and concave calices are not uncommon in the tallest corallites. The septal number varies in the development of part of the seventh cycle, and the dentate or crenulate condition of the septal edge is very variable.

The diameter of the calices and the height of the corallum depend upon the age of the individual.

It would appear that no British specimen exactly resembles the type from Jamoigne, but a variety from the Irish Lias at the Island Magee is nearer to it than any of the British forms.

Localities. Marton, near Gainsborough; Newark, Notts; east shore of Island Mageė, North of Ireland. In the Collections of Rev. P. B. Brodie, Mr. Burton, Rev. Mr. Chamberlin, the Geological Society, and the British Museum.

2. Montlivaltia papillata (sp. nov.). Pl. X, figs. 15—18.

The corallum is Cyclolitoid in shape, the base is slightly concave, and the calice is convex, there being a circular depression at the centre.

The epitheca of the base reaches to the calicular margin; it is very thin, is marked with concentrical shallow depressions and elevations, and the costæ are seen through it faintly.

The calice is nearly circular.

The septa are exsert, and the larger have very large dentations or papillæ on them. The papillæ are small at the margin and at the columellary space, but midway there are six or more of them which are very prominent. There are twenty-four septa, which reach the margin of the columellary space, and they are strongly papillated. Between two of the longest septa there are three others, one, the central, is longer than the others, which are almost rudimentary; all are papillate. There are thus five complete cycles of septa, in six systems.

The columellary space presents several small papillæ, but they are septal. There is no columella.

Diameter of calice, $\frac{10}{12}$ths inch. Height of corallum, $\frac{8}{10}$ths inch.

Locality. Marton, near Gainsborough; east shore of Island Magee, in the North of Ireland. In the Collections of F. M. Burton, Esq., F.G.S., Gainsborough, and R. Tate, Esq., F.G.S.

3. MONTLIVALTIA PAPILLATA (sp. nov.). A variety. Pl. X, figs. 19—21.

The corallum is smaller than the type, and the papillæ are smaller and sharper.

Locality. Marton, near Gainsborough. In the Collection of Rev. B. Chamberlin, F.G.S.

4. SEPTASTRÆA FROMENTELI, *Terquem et Piette.* Pl. XI, fig. 5.

The corallum is massive, and resembles a flattened cone in shape.

The corallite walls are very thin, and are fused together.

The calices are polygonal, irregular in shape, and deep.

The septa are thin, finely dentate, and rather wavy; they number from twenty-four to twenty-six in small calices, and from fifty-two to sixty-two in the larger.

The fissiparous division of the calices is very constant, and occurs both in the midst of the calices and at their angles. It is very rare to observe calices which do not present evidences of fissiparity, so that the calices are almost always double.

Diameter of the calices, about $\frac{1}{3}$rd to $\frac{2}{3}$rds inch.

Locality. Marton, near Gainsborough; Harbury, Warwick; east shore of Island Magee, North of Ireland. In the Collections of F. M. Burton, Esq., F.G.S., and Ralph Tate, Esq., F.G.S.

The shape of the corallum is subject to variation, and the Marton specimens are massive and flat, whilst that from Harbury, belonging to Rev. P. B. Brodie, is very gibbous and irregular in shape. The specimens from the North of Ireland are also irregular in shape. The species has a considerable range, and it has been found by MM. Terquem et Piette[1] in the "Calcaire à *A. planorbis* de Volfsmuhl, près de Mondorf et de Beaufort." But in

[1] Terquem et Piette, op. cit., p. 129.

England and Ireland it occupies a higher zone, and is accompanied by *Montlivaltia Haimei* and its varieties.

The following Cephalopoda, Gasteropoda, and Lamellibranchiata accompany the Madreporaria just described in the section at Marton:[1]

Ammonites Johnstoni, Sow.
,, *angulatus*, Schl.
Nautilus striatus, Sow.
Cerithium Semele, D'Orb.
Phasianella Morencyana, Piette.
Turbo subelegans, Münst.
Turritella Dunkeri, Terquem.
Cucullæa Hettangiensis, Terquem.
Anomia pellucida, Terquem.
Cardinia Listeri, Sow.
,, *ovalis*, Stutch.
Cardita Heberti, Terquem.
Lima tuberculata, Terquem.
,, *punctata*, Sow.
Pecten punctatissimus, Quenst.

The following is a list of the Madreporaria from the zone of *Ammonites angulatus* at Marton:

Montlivaltia Haimei, Ch. et Dew.
,, ,, 2 varieties.
,, *papillata*, Duncan.
,, ,, a variety.
Septastræa Fromenteli, Terquem et Piette.

V. Description of the Species from the Zone of Ammonites angulatus in the North of Ireland.

In the subdivision of the Lias at Waterloo, Larne, where the Cephalopoda and Mollusca about to be mentioned are found, there is a very remarkable coral which cannot be classified with any of the genera of the Astræidæ. I have founded the new genus Oppelismilia to receive this species and another which belongs to the Lias at Harbury, and which will be described in the next part of this Monograph.

In the *Ammonites angulatus* Zone on the east shore of Island Magee there are several species of Madreporaria.

[1] List furnished by Ralph Tate, Esq., F.G.S., as was also that at p. 40.

Montlivaltia Haimei, Ch. et Dew., is found there, and the form has a greater resemblance to the Belgian type than to the specimens from Marton. The multiseptate and granular varieties of the species are also found.

Montlivaltia papillata, Duncan, is noticed amongst the Irish coral-fauna, and *Septastræa Fromenteli*, Terquem et Piette, also.

The Coral-fauna of the Zone of *Ammonites angulatus* of Lincolnshire is clearly strongly represented in the North of Ireland, and the Mollusca which accompany the Corals of the first locality are noticed to be associated with those of the last.

There is a Montlivaltia of the *Montlivaltia papillata* type which is special to the Irish Lias. *Oppelismilia gemmans* is not found in any other locality than Waterloo, Larne.

Genus—Oppelismilia.

The corallum is simple, attached, and conical. The epitheca is well marked, and reaches to the calicular margin.

The calice is shallow, and the septa are numerous and close. There are no costæ, and there is no columella. Gemmation occurs within the calice, and the bud, which has an epitheca, grows with the parent.

The genus thus includes Montlivaltiæ with calicular gemmation.

Oppelismilia gemmans (sp. nov.). Pl. X, figs. 33, 34.

The corallum is short; it has a broad and flat calice, an oval space at the base where it was once adherent; a strong epitheca, with circular markings, and there are no costæ.

The calice is flat and shallow, and its margin is sharp.

The septa are very numerous and unequal.

The bud on the calice has an epitheca, and its septa are faintly dentate.

Height of the corallum, $\frac{5}{10}$ths inch. Width of the calice, $\frac{9}{10}$ths inch.

Locality. Waterloo, Larne, North of Ireland. In the Collection of Ralph Tate, Esq., F.G.S.

The following new Montlivaltia is also from the Lias of Ireland:

Montlivaltia Hibernica (sp. nov.). Pl. X, figs. 22, 23.

The corallum is discoidal, the base is flat, and the calice is convex.

The epitheca of the base is strongly marked concentrically.

The septa are numerous, close, unequal, and are marked by small papillæ, which are very close together, and by flat eminences, which are also very close together. There appear to be nearly five cycles of septa, and the largest septa are papillose.

The diameter of the calice is $\frac{4}{10}$ths inch, and the height of the corallum $\frac{1}{10}$th inch.

Locality. The eastern shore of Island Magee, in the North of Ireland. In the Collection of Ralph Tate, Esq., F.G.S.

List of Species of Madreporaria from the Zone of Ammonites angulatus in the North of Ireland.

1. *Oppelismilia gemmans*, Duncan.
Montlivaltia Haimei, Ch. et Dew.
,, ,, varieties.
,, *papillata*, Duncan.
,, *Hibernica*, Duncan.
Septastræa Fromenteli, Terquem et Piette.

The following Cephalopoda, Gasteropoda, and Lamellibranchiata were found associated with the Madreporaria in the zone of *Ammonites angulatus* of the North of Ireland:

Ammonites Johnstoni, Sow.
,, *angulatus*, Sch.
Nautilus striatus, Sow.
Actæonina fragilis, Dunk.
Cerithium Semele, D'Orb.
,, *gratum*, Terquem.
Phasianella Morencyana, Terquem.
Pleurotomaria cœpa, Terquem.
Turbo subelegans, Münst.
Turritella tenuicosta, Portl.
Pecten calvus, Goldf.
Plicatula Hettangiensis, Terquem.
,, *intusstriata*, Emm.
Terebratula perforata, Piette.
Avicula Sinemuriensis, D'Orb.
Cardinia Listeri, Sow.
,, *ovalis*, Stutch.
Cardium Philippianum, Dunk.
Lima acuticosta, Münst.
,, *tuberculata*, Terquem.
Ostrea irregularis, Münst.

VI. Description of the Species from Lussay in the Isle of Skye.

Dr. T. Wright has described the Coral-bed of the Lower Lias of Skye, and the species of *Isastræa* which, grouped in masses, appears to be the only Coral found there. It is most probable, from the position of this coral-bed,[1] and the association of *Ostrea arietis* and *Cardinia concinna* with it (in the bed beneath), that *Isastræa Murchisoni* belongs to the same geological horizon as the Liassic deposit at Brocastle and the Sutton Stone.

Isastræa Murchisoni, *Wright.*[2] Pl. XI, figs. 1—4.

Dr. Wright's description of this species gives the following characters:

Corallum large, massive, convex. Calices unequal, deep, polygonal; sides unequal. Margin thin. Septa, 30 to 36, and even 40 or more in the large calices; unequal in length, thin, waved, granulated superiorly. Columella absent; point of convergence of septa excentral. Diameter of calices, $\frac{3}{10}$ths to $\frac{4}{10}$ths inch. Depth of fossa, $\frac{3}{20}$ths inch.

Locality. Lussay, Skye.

The surface of the type specimen is very uneven; the calices are very irregular in size, shape, and depth, and the margins are not even. Thus one calice may be on a higher level than those to which it is attached, and often so much so that there is a faint trace of a subsequent growth of wall. The septa are very irregular in their number, and the longest have one or more teeth at their inner end. There is often a ridge between the margin of the calice and the centre, indicating calicinal gemmation, but the gemmation of the corallum usually takes place at the margin, and there is no fissiparity. No cyclical arrangement of the septa can be distinguished.

The large and shallow calices, thin septa, the peculiar relation of contiguous calices, and the sharp elevated margins, distinguish this species, which is allied rather to a new genus from the Middle Lias of Pabba, *Lepidophyllia* (Duncan), than to any of the Liassic Isastrææ.

[1] See Mr. Geikie's memoir "On the Geology of Strath, Skye;" with "Descriptions of Fossils," by Dr. T. Wright, 1857, 'Quart. Journ. Geol. Soc.,' vol. xiv, pp. 1 *et seq.* There is a most interesting description of the Coral-bed in the Isle of Skye by Hugh Miller, in his "Essay on the Corals of the Oolitic System of Scotland," read before the Royal Physical Society of Edinburgh, and published in 'The Old Red Sandstone,' 7th edition, 1859.

[2] 'Quart. Journ. Geol. Soc.,' vol. xiv, p. 34. 1858.

VII. List of the Species described in this Part from the Zones of Ammonites planorbis and Ammonites angulatus.

1. *Oppelismilia gemmans*, Duncan.
2. *Montlivaltia Walliæ*, „
3. — *Murchisoniæ*, „
4. — *polymorpha*, Terquem et Piette.
5. — *parasitica*, Duncan.
6. — *simplex*, „
7. — *brevis*, „
8. — *pedunculata*, „
9. — *Haimei*, Chapuis et Dewalque.
10. — *papillata*, Duncan.
11. — *Hibernica*, „
12. *Thecosmilia Suttonensis*, „
13. — *mirabilis*, „
14. — *serialis*, „
15. — *rugosa*, Laube.
16. — *Brodiei*, Duncan.
17. — *Martini*, E. de Fromentel.
18. — *Michelini*, Terquem et Piette.
19. — *irregularis*, Duncan.
20. — *Terquemi*, „
21. — *affinis*, „
22. — *dentata*, „
23. — *plana*, „
24. *Rhabdophyllia recondita*, Laube.
25. *Astrocœnia gibbosa*, Duncan.
26. — *plana*, „
27. — *insignis*, „
28. — *reptans*, „
29. — *parasitica*, „
30. — *pedunculata*, „
31. — *costata*, „
32. — *favoidea*, „
33. — *superba*, „
34. — *dendroidea*, „

35. *Astrocœnia minuta*, Duncan.
36. — *Sinemuriensis*, D'Orbigny, sp.
37. *Cyathocœnia dendroidea*, Duncan.
38. — *incrustans*, „
39. — *costata*, „
40. *Elysastræa Fischeri*, Laube.
41. — *Moorei*, Duncan.
42. *Septastræa excavata*, E. de Fromentel.
43. — *Haimei*, Wright, sp.
44. — *Fromenteli*, Terquem.
45. *Latimæandra denticulata*, Duncan.
46. *Isastræa Sinemuriensis*, E. de Fromentel.
47. — *globosa*, Duncan.
48. — *Murchisoni*, Wright.

Varieties of *Thecosmilia irregularis*, Duncan.
— *Montlivaltia Haimei*, Chapuis et Dewalque.
— — *papillata*, Duncan.

A MONOGRAPH

OF THE

BRITISH FOSSIL CORALS.

(SECOND SERIES.)

PART IV.—NO. 2.

VIII. CORALS FROM THE ZONE OF AMMONITES ANGULATUS.
(*Continued.*)

THERE are some Coralliferous deposits belonging to the Lower Lias at Inkbarrow, at Chadbury, in Worcestershire, and Fladbury, near Evesham, whose exact geological horizon has not been determined. They are low down in the Lower Lias, but their commonest Corals do not identify them with the Coralliferous beds of Brocastle. The genus *Isastræa* is dominant in these localities, and its species are unlike any which have been described. The Corals will not do more than associate these beds on one horizon. There is a great probability, from the presence of small Gasteropoda, whose shells are left in the calices of the Corals, that careful search will yield a sufficient number of fossils to determine whether these deposits are below the Zone of *Ammonites Bucklandi.* Our present knowledge does not justify the association of these *Isastrææ* with the Coral-fauna of the Zone of *Ammonites angulatus.*

The Coralliferous deposits at Abbott's Wood, Harbury, Aston Magna, and Down Hatherly may belong to more than one zone; but, from the association of *Thecosmilia Michelini, Thecosmilia Martini,* and *Septastræa Fromenteli,* the presence of the Zone of *Ammonites angulatus* may be satisfactorily asserted.

There is an *Isastræa* found in the Lower Lias of Lyme Regis, which is said to belong to the Zone of *Ammonites angulatus,* but the mineralization of the specimen and its affinities are sufficiently distinct to associate it with the beds containing *Ammonites Bucklandi.*

IX. Description of the Species.

Section—*APOROSA.*

Family—ASTRÆIDÆ.

Division—Lithophyllaceæ simplices.

Genus—Montlivaltia.

1. Montlivaltia Ruperti, *Duncan.* Pl. XII, figs. 3, 4, 5; Pl. XV, fig. 15.

The corallum is turbinate; it is truncated at the base, and is widest at the calice.

The epitheca is strong, and is marked transversely with ridges, prominent lines, and constrictions; the longitudinal markings are faint, but there is a tendency to split in their direction.

The calice is moderately deep, and is circular in outline.

The septa are crowded, unequal, long, and irregular; the longest are thick internally, and reach so far inwards as to give the appearance of a false columella; all are slightly spined.

There are five cycles of septa, in six systems, and those of the highest orders are small, whilst the primary and secondary are equal and very long.

The wall is thick, and the epitheca does not project upwards as a ridge around the circular margin. The endotheca is abundant.

The costæ are small, and are rarely visible beneath the epitheca.

Height of the corallum $\frac{5}{10}$ths inch.

Breadth of the calice $\frac{7}{10}$ths inch.

Locality. Down Hatherly.

In the Collection of R. Tomes, Esq.

Division—Astræaceæ.

Genus—Isastræa.

1. Isastræa Tomesii, *Duncan.* Pl. XV, fig. 20.

The corallum is massive, large, and irregular in shape. The upper surface is subgibbous.

The calices are irregular in size, are separated by very thin walls, and are rather deep and polygonal, quadrangular, or more or less circular.

The septa are very thin, and are faintly dentate; they often curve and unite. They reach well into the axial space, and are united by dissepiments. They are subequal, but many rudimentary septa exist. There are not four complete cycles of septa.

Diameter of calices $\frac{2}{10}$ths—$\frac{4}{10}$ths inch.

Locality. Long Coppice, near Binton, Warwickshire.

In the Collection of R. Tomes, Esq.

The delicacy and subequal character of the septa, their deficiency in decided dentations, and the dissepiments between the septa, characterise this species.

There is an immense *Isastræa* at Inkbarrow, with small calices and thick walls; unfortunately it is not determinable specifically, but the honeycomb appearance and subgibbous upper surface, and the low septal number, may distinguish it. A specimen is in the collection of the Rev. P. B. Brodie, F.G.S.

Isastræa Murchisoni, Wright, is found attached to the Inkbarrow specimen, and thus this Scottish Coral has also an English habitat.

X. On the Corals of the British and European Lower Liassic Deposits of the Zones of Ammonites angulatus, Ammonites planorbis, and Avicula contorta.

The strata of the Lower Lias evidently contain more than one Coral-fauna, and there is a strong distinction between the assemblage of species of the Zone of *Ammonites Bucklandi* and those of the zones below. The Corals of the White Lias are few in number, and probably belong to the genus *Montlivaltia*, but they cannot be distinguished specifically. The *Avicula contorta* series of France and England are uncoralliferous, but the Italian beds at Azzarola, which probably are on that horizon, contain a very remarkable Coral-fauna.

The extent of the area of Coralliferous beds described by Stoppani as the Azzarola series is very considerable. The "Madrepore-bed," as it is termed by Stoppani, is seen above the Azzarola beds, with *Cardium Rhæticum, Myophoria inflata, Mytilus psilonoti, Avicula contorta, Terebratula gregaria,* &c., wherever the succession of the rocks can be made out, either on the south-eastern slopes of the Alps, as on the Lake of Como, or on the north-western slopes to the south of the Lake of Geneva.[1] The Madrepore-bed is described, moreover, as occurring below and in the midst of the Azzarola beds, and as forming a dense layer of eight to ten yards in thickness. The prevailing Coral is *Rhabdophyllia Langobardica,* Stop., and the genus is represented by three other species. The *Rhabdophylliæ* resemble in their habit of growth many *Thecosmiliæ*, and form in the Azzarola beds great masses of tangle, like *Thecosmilia Martini* in the Coralliferous beds of the Côte d'Or and of Cowbridge in South Wales. Stoppani describes a *Stylina* from some casts which

[1] Stoppani, 'Monog. des Foss. de l'Azzarola.'

resemble those of *Astrocœnia gibbosa*, nobis, from the Sutton Stone. The species I select as distinguishable in the Azzarola deposits are—

Rhabdophyllia Sellæ, Stopp.
" *Langobardica*, Stopp.
" *Meneghini*, Stopp.
" *De-Filippi*, Stopp.
Montlivaltia Gastaldi, Stopp.
Stylina Savii, Stopp.
Thamnastræa Batarræ, Stopp.
" *Escheri*, Stopp.
" *Meriani*, Stopp.
" *rectilamellosa*, Winkl.

These *Rhabdophylliæ*, *Stylinæ*, and *Thamnastrææ*, are represented in the lowest zones the British Lower Lias by *Thecosmiliæ*, *Rhabdophylliæ*, and *Astrocœniæ*.

The strata between the Trias and the Zone of *Ammonites Bucklandi* in Germany are very uncoralliferous, and the determinations of the species given by Quenstedt are not sufficiently exact. MM. D'Orbigny, Terquem et Piette, and De Fromentel, have noted and described the following species from the Lower Lias, below the Zone of *Ammonites Bucklandi*, *Gryphæa incurva*, &c., in France and Luxembourg, and by omitting *Isastræa basaltiformis*, De From., which belongs to the Zone of *Ammonites planorbis*, the following table will give all the species from the Continental Zone of *Ammonites angulatus*:—

XI. List of Species from the Continental Zone of Ammonites angulatus.

1. *Montlivaltia Sinemuriensis*, D'Orb.
2. " *dentata*, De From. et Ferry.
3. " *Martini*, De From.
4. " *Rhodana*, De From. et Ferry.
5. " *discoidea*, Terquem et Piette.
6. " *Haimei*, Chapuis et Dewalque.
7. " *Guettardi*, Chapuis et Dewalque.
8. " *polymorpha*, Terquem et Piette.
9. " *denticulata*, De From. et Ferry.
10. *Thecosmilia Martini*, De From.
11. " *Michelini*, Terquem et Piette.
12. " *coronata*, Terquem et Piette.
13. *Septastræa Fromenteli*, Terquem et Piette.

14. *Septastræa excavata,* De From.
15. *Isastræa Condeana,* Chapuis et Dewalque.
16. „ *Sinemuriensis,* De From.
17. *Stylastræa Sinemuriensis,* De From.
18. „ *Martini,* De From.
19. *Astrocœnia Sinemuriensis,* D'Orb.
20. „ *clavellata,* Terquem et Piette.

Probably some of the species of *Montlivaltia* will have to be absorbed by others, but this list, when added to the Table of British Corals from the Zone of *Ammonites angulatus,* proves that, instead of the Lias being an uncoralliferous series, it was quite the contrary. The great development of Coral-life in the Azzarola series, the scanty remains of it in the western and north-western European *Avicula contorta* Zones and in the White Lias and in the Zone of *Ammonites planorbis,* and the luxuriance of the species in the Zone of *Ammonites angulatus,* are very significant facts; and the significance is not diminished when the paucity of the species of the *Ammonites Bucklandi* Zone and their distinctness from those of the Zone of *Ammonites angulatus* is considered.

XII. List of Species of Corals from the Continental and British Strata of the Zone of Ammonites angulatus.

1. *Oppelismilia gemmans,* Duncan. Ireland.
2. *Montlivaltia Walliæ,* „ South Wales.
3. — *Murchisoniæ,* „ „
4. — *Ruperti,* „ England.
5. — *parasitica,* „ South Wales.
6. — *simplex,* „ „
7. — *brevis,* „ „
8. — *pedunculata,* „ „
9. — *polymorpha,* Terquem et Piette. South Wales; East of France.
10. — *Haimei,* Chapuis et Dewalque. England; Ireland; Luxembourg; France.
11. — *Hibernica,* Duncan. Ireland.
12. — *papillata,* „ England; Ireland.
13. — *Sinemuriensis,* D'Orb. France.
14. — *dentata,* De From. et Ferry. „
15. — *denticulata,* „ „
16. — *Rhodana,* „ „
17. — *Martini,* „ „

18. *Montlivaltia discoidea*, Terquem et Piette. France.
19. — *Guettardi*, Blainville. Luxembourg and England.
20. *Thecosmilia Suttonensis*, Duncan. South Wales.
21. — *mirabilis*, „ „
22. — *serialis*, „ „
23. — *irregularis*, „ „
24. — *Terquemi*, „ „
25. — *affinis*, „ „
26. — *dentata*, „ „
27. — *plana*, „ „
28. — *Brodiei*, „ „
29. — *Martini*, E. de From. South Wales; England; France; Luxembourg.
30. — *Michelini*, Terquem et Piette. „ „ „ „
31. — *coronata*, „ France
32. — *rugosa*, Laube. South Wales; St. Cassian.
33. *Rhabdophyllia recondita*, „ „ „
34. *Astrocœnia Sinemuriensis*, D'Orb., sp. South Wales; France.
35. — *clavellata*, Terquem et Piette. France.
36. — *gibbosa*, Duncan. South Wales; Azzarola?
37. — *plana*, „ „
38. — *insignis*, „ „
39. — *reptans*, „ „
40. — *parasitica*, „ „
41. — *pedunculata*, „ „
42. — *costata*, „ „
43. — *favoidea*, „ „
44. — *superba*, „ „
45. — *dendroidea*, „ „
46. — *minuta*, „ „
47. *Cyathocœnia dendroidea*, „ „
48. — *incrustans*, „ „
49. — *costata*, „ „
50. *Elysastræa Fischeri*, Laube. „ St. Cassian.
51. — *Moorei*, Duncan. „
52. *Septastræa excavata*, E. de From. „ France.
53. — *Fromenteli*, Terquem. „ England; Ireland; France.
54. *Stylastræa Sinemuriensis*, E. de From. „
55. — *Martini*, „ „
56. *Latimæandra denticulata*, Duncan. „

57. *Isastræa Sinemuriensis*, E. de From. South Wales ; France.
58. — *Condeana*, Chapuis et Dewalque. Luxembourg ; France.
59. — *globosa*, Duncan. South Wales.
60. — *Murchisoni*, Wright. Isle of Skye ; Inkbarrow, England.
61. — *Tomesii*, Duncan. Worcestershire.

Of these 61 species 50 are found in the British Isles.

XIII. Corals from the Zone of Ammonites Bucklandi (bisulcatus).

Corals are not numerous in this zone, and the commonest species of the Zone of *Ammonites angulatus* are not found in any of its strata. It is probable that *Thecosmilia Martini*, E. de From., which in France ranges from the beds containing *Ammonites Moreanus* into those in which *Ammonites bisulcatus* is found, has a corresponding vertical distribution in England. *Thecosmilia Michelini*, Terq. et Piette, appears to be present in the Zone of *Ammonites Bucklandi*, but only in the form of casts, which resemble those found at Abbott's Wood, in the Zone of *Ammonites angulatus*. These casts and some of *Thecosmilia Martini* have been assigned to the genus *Cladophyllia*, but without sufficient reason. *Thecosmilia* is a large genus, and the species contain individuals of all sizes, so that to give to very small cylindroid *Thecosmiliæ* the term *Cladophylliæ* is too artificial a distinction. The septa of *Thecosmiliæ* are generally, but by no means universally, regularly toothed, granular, and slightly exsert ; and the septa of *Cladophylliæ* are said to be small, not numerous, and slightly dentate ; moreover, the endotheca is scanty in *Cladophyllia*, but abundant in *Thecosmilia*. These are not generic distinctions, and it is very probable that one genus will absorb the other.

Section—*APOROSA.*

Family—ASTRÆIDÆ.

Division—Lithophyllaceæ simplices.

Genus—Montlivaltia.

1. Montlivaltia Guettardi, *Blainville*, 1830. Pl. XII, figs. 10—14. + 6-7

The following is the specific diagnosis given by MM. Chapuis and Dewalque.[1]

Corallum simple and rather variable in shape; often conical, more or less depressed, rarely cylindro-conical; the base is slightly pedicillate.

[1] Chapuis et Dewalque, 'Descript. des Foss. des Terr. Second. du Luxembourg, 1854.'

Epitheca thick, ridged, and extending to the calicular border.

Calice circular, ordinarily concave, shallow.

Septa usually thin, granular; strongly toothed on their arched margins.

Five cycles, the first three nearly equal.

This Coral varies greatly in its height and basal flatness. It may be sub-turbinate, or even discoidal; and the specimen from Bottesford, in Lincolnshire, is flat below and very convex above, but it presents an axial depression. The Continental specimens appear to be found in a lower horizon of the Lias than that in which the specimen figured in Pl. XII was found.

Locality. Bottesford, Lincolnshire.

In the Collection of Rev. T. C. B. Chamberlin, F.G.S.

There are specimens, which I believe are young forms, that were found at Fenny Compton and Aston Magna. Pl. XII, figs. 6 and 7.

There is a microscopic Coral at Willsbridge, in the Lima-series (Pl. XV, fig. 9), but the species is not distinguishable. It is figured, as perhaps a larger form may be discovered. Small and young *Montlivaltiæ* are very common on Gryphææ and on large Corals.

Family—ASTRÆIDÆ.

Division—Faviaceæ.

Genus—Septastræa.

1. Septastræa Eveshami. *Duncan*. Pl. XIII, figs. 5—7.

The corallum is large, tall, and flabelliform, and the surface is subgibbous. The base is small, and the corallites radiate and elongate rapidly.

The calices are very irregular in shape and size, and many are twisted and irregular; all are shallow, and those which are fissiparous are narrow. Some calices are polygonal, but fissiparity can be distinguished in most.

The septa are small, dentate, and very irregular in size and arrangement. There are between thirty and forty septa in regular calices, but in the elongated there are many more. The calicular wall is very thin, but where it has been worn a groove is noticed. The endotheca is rather scanty.

Diameter of a polygonal calice $\frac{2}{10}$ths inch, and of elongated calices from $\frac{1}{10}$th to $\frac{2}{10}$ths inch.

Locality. Evesham.

In the Collection of the Rev. P. B. Brodie, F.G.S.

Division—Astræaceæ.

Genus nov.—Lepidophyllia.

The corallum is compound, and the corallites are joined by their walls. The gemmation is calicular and gives an overlapping appearance both to the sides and the upper part of the corallum.

The epitheca is distinct. There is no columella.

The septa are dentate. The calicular gemmation and Astræacean characters distinguish the genus.

There are two species; one is found at Chadbury, and the other in the Island of Pabba, in the Middle Lias.

1. Lepidophyllia Stricklandi, *Duncan*. Pl. XII, fig. 15.

The corallum is tall, and is composed of two sets of corallites.

The calicular gemmation is very frequent and successive. The calicular margins are sharp and wavy; and they are free, except where the corallites join.

The fossa is deep.

The costæ are distinct.

The epitheca is scanty.

Height of corallum 1 inch. Breadth of calice $\frac{9}{10}$ths inch.

Locality. Chadbury, Worcestershire.

In the Collection of Mrs. Strickland. The specimens were collected by the late Hugh Strickland, F.G.S.

Genus—Isastræa.

1. Isastræa endothecata, *Duncan*. Pl. XII, figs. 17—21.

The corallum is large, and either massive and flat, or tall and arising from a small base.

The calices are very irregular in shape, size, and depth. The largest calices are very deep.

The septa are small, and often wavy. They are not exsert, but are very irregular. They are faintly dentate, wide apart, and project slightly from the calicular wall.

The cyclical arrangement cannot be determined by the number of the septa; there are between four and five cycles. The largest septa reach the floor of the calice, where they join.

The endotheca is greatly developed, and it forms small dissepiments, and others which stretch across the corallites almost like tabulæ.

The marginal gemmation is frequent.

Length of the largest calice $\frac{1}{2}$ inch. Depth $\frac{1}{5}$—$\frac{1}{4}$ inch.

Locality. Lyme Regis.

In the Collection of R. Tomes, Esq.

2. Isastræa insignis, *Duncan*. Pl. XIII, figs. 10, 11.

The corallum is massive and forms a flat mass. The corallites are very equal in size and regular in shape.

The calices are placed very regularly in linear series; they are shallow, open, and are separated by a stout wall. The calices are generally hexagonal, but many are square.

The septa are small, project but slightly from the wall, are dentate and unequal. There are four cycles of septa in six systems in the largest calices. The primary and secondary septa are nearly equal; the tertiary are decidedly smaller, and the rest are the smallest.

The endotheca is close.

There is no columella.

Diameter of largest calices $\frac{3}{10}$ths inch, and of the usual size $\frac{2}{10}$ths inch.

Locality. Lyme Regis.

In the Collection of R. Tomes, Esq.

This is a very well-marked species, and belongs to a section which comprises *Isastræa Henocquei*, Ed. and H., from the Lower Lias of Hettange, *Isastræa polygonalis*, Michelin, sp., of the Muschelkalk, and *Isastræa Lonsdalei*, Ed. and H., of the British Inferior Oolite.

3. Isastræa Stricklandi, *Duncan*. Pl. XIII, figs. 1—4.

The corallum is very tall, has a small base, and is expanded superiorly.

The corallites are unequal in size and length, and vary much in shape.

The calices are very irregular in form and depth; their walls are thick, and the septa stout and very dentate. The dentations are blunted and are very large, and more so internally than near to the calicular margin.

The septal number varies, and 32—40 appear to be the usual number. The laminæ are stout, and the primary and secondary septa reach downwards to the base of the fossa and are dentate. The others, which are short, are also stout.

The endotheca is greatly developed, and consists of small vesicular dissepiments, and of larger masses which stretch across the corallites like tabulæ and close in the calicular fossa.

Height of corallum 6 inches. Breadth of largest calice $\frac{3}{10}$ths inch.

Locality. Chadbury, Evesham.

In Mrs. Strickland's Collection.

Genus—Cyathocœnia.

1. Cyathocœnia globosa, *Duncan,* Pl. XIII, figs. 8, 9.

The corallum is nearly spherical.

The calices are numerous, small, and shallow. They are rarely circular, and are generally rather polygonal in outline, and they are separated by a small amount of cœnenchyma. There are no costæ.

The septa are stout at the wall and taper off inwardly; they are subequal, distant, and form three more or less perfect cycles in six systems.

Diameter of the calices $\frac{1}{10}$th inch.

Locality. Fladbury, in Drift with *Gryphæa incurva.*

In the Collection of R. Tomes, Esq.

The shape of the corallum, the absence of costæ, and the shallow calices, distinguish this species from *Cyathocœnia dendroidea,* nobis, *C. costata,* nobis, and *C. incrustans,* nobis, from Brocastle and the Sutton Stone.

The following analysis of the genus will enable the diagnosis of the species to be determined readily.

Cyathocœniæ with the corallum	branching, having costæ	*C. dendroidea.*
	encrusting, without costæ, cœnenchyma granular .	*C. incrustans.*
	flat, having large costæ and a deep calice . .	*C. costata.*
	globular, without costæ, cœnenchyma plain .	*C. globosa.*

XIV. List of Species from the Zone of Ammonites Bucklandi.

1. *Montlivaltia Guettardi,* Blainville.
2. *Septastræa Eveshami,* Duncan.
3. *Lepidophyllia Stricklandi,* „
4. *Isastræa endothecata,* „
5. „ *insignis,* „
6. „ *Stricklandi,* „
7. *Cyathocœnia globosa,* „

XV. Corals from the Zone of Ammonites obtusus, Sow.

Some worn and light-coloured simple Corals of the genus *Montlivaltia* are found at Pebworth, five miles south-west of Stratford-on-Avon, in a bed with *Ammonites Sauzeanus*, D'Orb., and *Ammonites semicostatus*. One of the species (*Montlivaltia mucronata*, Duncan) will be described amongst those of the next zone, in which it is common. The specimens are worn, the calices especially, and all the spines are broken off. The columellary space is occasionally occupied by the prominent ends of the principal septa, the laminæ having been worn away in their middle course. The longitudinal series of costæ are rarely visible, and there are many examples of deformed corallites.

Family—ASTRÆIDÆ.

Division—Lithophyllaceæ simplices.

Genus—Montlivaltia.

1. Montlivaltia patula, *Duncan*. Pl. XV, figs. 6, 7, 8.

The corallum is turbinate, depressed, and slightly longer than broad.

The calice is large, elliptical, very shallow, and open; its margin is sharp, and the wall shelves very gradually inwards, giving to the calice a very open appearance.

The septa are unequal and numerous, and the largest are very long and dentate; the tooth nearest the axial space points inwards and is rounded, and those of the longest septa form an irregular circle around the space. The smallest septa are very rudimentary, but the next in size have, in common with all the others, an internal oval tooth. All the septa are delicate, and they are not crowded. There are five cycles of septa in six systems. The primary, secondary, and tertiary are nearly equal in length. The septa are not exsert, but all are lower than the calicular margin.

Length of the calice $\frac{8}{15}$ths inch. Breadth $\frac{6}{10}$ths inch.

Locality. Walford Hill, Stratford-on-Avon, with *Ammonites semicostatus* and *Ammonites Sauzeanus*. In the Collection of R. Tomes, Esq.

XVI. Corals from the Zone of Ammonites raricostatus, Ziet.

The brick-fields in the vicinity of Cheltenham present dark-coloured clay beds, which have the following succession (see Wright, Fossil Oolitic Asteriadæ,' p. 25).

Marle Hill Section.

NO.		FT.	IN.
1.	*Gryphæa-bed;* a hard, ferruginous clay, which broke into fragments, and contained *Gryphæa obliquata*, Sow. . .	3 to 4	0
2.	*Coral-band;* a thin seam of lightish-coloured unctuous clay, containing a great many small sessile Corals, *Montlivaltia rugosa*, Wright and Duncan, most of which appeared to have been attached to the curved valves of the Gryphææ .	1 in. to $1\frac{1}{2}$	
3.	*Hippopodium-bed*	10	0
4.	*Ammonite-bed*		?

In Warwickshire the railway-cutting at Honeybourne presented the same beds, and the Coral-band contained a considerable number of the *Montlivaltiæ.*

A section on the line of railway at Fenny Compton, in Oxfordshire, near the station, presents the following beds in descending order; the bed No. 2 is highly coralliferous.[1]

Fenny Compton Section.

NO.		FT.	IN.
1.	White clay, containing *Gryphæa obliquata* (*Maccullochii ?*), *G. incurva, Belemnites acutus, Hippopodium ponderosum, Pleurotomaria similis,* &c.	4	0
2.	Blue clay, with included hard blue calcareous bands, containing Corals and the Mollusca mentioned in Bed No. 1 .	2	0
3.	Blue shale	10	0

Middle Lias clays and shales, with *Ammonites Henleyi,* are superimposed on Bed No. 1; and the blue shale (3) rests on a clay with calcareous masses, the "Cardinia-zone."

The Coral-bands at Marle Hill and Honeybourne are upon the same geological horizon as bed No. 2 of the Fenny Compton section. These beds contain some of the finest specimens of *Montlivaltia* ever discovered.

[1] The Rev. P. B. Brodie, M.A., F.G.S., has given me great assistance, and has furnished me with this section.

Family—ASTRÆIDÆ.

Division—Lithophyllaceæ simplices.

Genus—Montlivaltia.

The *Montlivaltiæ* from Fenny Compton, Honeybourne, and Cheltenham, belong to several species, and two of these are singularly polymorphic. Shape has not very much to do with the specific diagnosis of some recent simple Corals, and it is necessary to assert this in collecting under one fossil species Corals of very diverse external forms. Singularly enough, the Liassic *Montlivaltiæ* from the Zone of *Ammonites raricostatus* are common and are extraordinarily well preserved, although a few years ago a Liassic Coral was excessively rare. Even the ornamentation upon the dentations of the septa is preserved, and the longitudinal striations of the epitheca also. The Fenny Compton Coral-bed contains specimens of the species of all sizes, and this is the case with the deposits containing the so-called *Thecocyathus rugosus*, Wright, at Honeybourne and Cheltenham. At Pebworth the Fenny Compton species are not found in a dark blue matrix, but in a white deposit; moreover, the specimens are usually worn, and they appear to have grown under less favorable conditions than the others.

Thecocyathus rugosus is referable to a group of forms specimens of which are very common; it does not belong, however, to the same family that contains the *Thecocyathi*. Dr. Wright gave the species a name in his MS., but the description and diagnosis have never been published. The Corals have been associated so long with the name of Dr. Wright, that it is just to append his name to the species.

1. Montlivaltia rugosa, *Duncan and Wright*. Pl. XIV, figs. 1, 2. 3; Pl. XV, figs. 14, 16, 17; Pl. XVI, figs. 5—15.

Thecocyathus rugosus, *Wright*, MS.

The corallum is very variable in its shape; it may be tall, conico-cylindrical, and curved, sub-turbinate and curved, short and cylindrical, short and turbinate and curved, or straight. It is pedunculate, and the scar by which it adhered to foreign substances, such as shells, is large and oval, or small and very irregular in shape.

The epitheca is stout and identified with the wall; it is strongly ridged transversely and folded as well as grooved. It is rarely marked by longitudinal lines, and is usually deficient in ornamentation. When the epitheca is worn, the external ends of the septa are seen like costæ, and the oblique external terminations of the endotheca are very apparent, sometimes having a herring-bone pattern. Young corallites are often adherent to the epitheca, they are therefore not buds, but accidentally attached Corals. When more than one Coral is attached to the same shell the bases appear to join.

The calice is shallow, and its margin is formed by the epitheca, which often intrudes upon its periphery ; it is circular or slightly deformed, and it may be either contracted or very open.

The septa are numerous and unequal; they are irregular in size and in their arrangement; they are dentate, and the teeth are regular, rounded above, and ornamented with waving lines and are largest near the axial space. The worn septa show their bases in the form of oval swellings, and when these are of full size the appearance of a columella and pali is simulated. There are four perfect cycles of septa, and the fifth is very irregularly developed, the higher orders being often rudimentary. In some large calices the fifth cycle is complete.

The endotheca is strong and well developed, and its dissepiments are numerous, oblique, and arched.

Height of Coral $1\frac{8}{10}$ths inch, 2 inches, $1\frac{5}{10}$ths inch, $\frac{5}{10}$ths inch.

Breadth of calice $\frac{6}{10}$ths inch, $\frac{8}{10}$ths inch, $\frac{8}{10}$ths inch, $\frac{7}{10}$ths inch.

Locality. Hippopodium-bed and Coral-bed of Marle Hill, Honeybourne, and Fenny Compton.

In the Collections of the Geological Society, British Museum, Dr. Wright, F.G.S., Charles Moore, Esq., F.G.S., R. Tomes, Esq., and Rev. P. B. Brodie, F.G.S.

The ornamentation of the teeth of the septa is very well seen in some specimens, but usually it is worn off and the teeth also. The Coral, although very polymorphic, is very easily distinguished from all others by its septa, epitheca, and base.

2. MONTLIVALTIA MUCRONATA, *Duncan*. Pl. XIV, figs. 4—11 and 14—18; Pl. XV, figs. 10—13.

The corallum is very variable in shape; it has a small peduncle, and a small and more or less circular flat scar. The corallum is turbinate and symmetrical, and widely open at the calice; or more or less compressed and subturbinate; or cylindrical and compressed. When turbinate and with a circular calice, the calice is singularly shallow; but when cylindrical and compressed, or in the young state, the calice is deeper.

The epitheca is strong and rises up with the wall to produce a sharp margin to the calice. The transverse markings are very distinct, and there are constricting ridges and folds. The longitudinal markings are very distinct, ornamental, and symmetrical; they are in groups which are smallest at the base, where they are most distinct and rounded, but they are less distinct at the calice where they are flat. The groups dichotomize, so that there are usually 12 at the base and 24 at the calice; they are separated by well-marked grooves and consist of bundles of longitudinal epitheca swellings and costæ.

The calice is either very shallow and circular, or deep and circular, or deep and elliptical. The septa are strongly spined, and the spines are very large, and are ornamented with granules. The largest septa are exsert and largely spined, and the spines near the axial space are so distinct from the septa as to simulate a columella. The first, second, and third cycles of septa are nearly equal in size, and the rest are much smaller. The septa of the fourth cycle are spined, and are larger than those of the fifth cycle. There are a few rudimentary septa of the sixth cycle in large specimens.

The endotheca is abundant.

Height of the corallum 1 inch, $\frac{8}{10}$ths inch, $\frac{6}{10}$ths inch, $\frac{9}{10}$ths inch.

Breadth of calice $1\frac{2}{10}$ths inch, $1\frac{1}{10}$th inch, $\frac{9}{10}$ths inch, $\frac{8}{10}$ths inch.

Locality. Fenny Compton and Pebworth.

In the Collection of the British Museum, Charles Moore, Esq., F.G.S., R. Tomes, Esq., and Rev. P. B. Brodie, F.G.S.

This beautiful Coral is readily distinguished by its peduncle, longitudinal markings, many septa, and ornamented spines; it is very variable in shape, and some very distinct varieties occur, as well as deformed and monstrous shapes.

Variety I.—With curved peduncles, elliptical calice. Pl. XIV, figs. 5 and 15, 16, 17.

Variety II.—Coral cylindro-conical, tall and compressed. Pl. XIV, figs. 14, 18.

Height of Coral $1\frac{3}{10}$ths inch.

Breadth of calice $\frac{7}{10}$ths inch.

Length of calice $\frac{9}{10}$ths inch.

Variety III.—Coral conical and wide at the calice. Pl. XV, figs. 12, 13, 14.

Variety IV.—Coral cornute, slightly curved, and the longitudinal markings very indistinct. No transverse corrugations of the epitheca. Pl. XV, figs. 10, 11.

A deformed corallum is figured in Pl. XIV, fig. 10.

The young corallites are often slightly curved, and their septa are very numerous.

Very probably a Coral with strong transverse markings, but much worn, and which is figured Pl. XIV, fig. 11, is a variety.

3. Montlivaltia nummiformis, *Duncan*. Pl. XIV, figs. 12, 13.

The corallum is nummiform. The base is perfectly flat, and is covered with epitheca strongly marked with concentric lines.

The epitheca does not extend to the septal edges, and these project out from the periphery of the base.

The calice is flat, and has a central depression.

The septa are numerous, crowded, convex externally, and less so superiorly. The larger septa are spined, and the septa of the highest orders are small and rudimentary. There are five cycles of septa and a part of a sixth.

The breadth of the calice is $\frac{7}{10}$ths inch, and the height $\frac{8}{10}$ths inch.

Locality. Fenny Compton.

In the Collection of R. Tomes, Esq.

The shape and the peculiar base distinguish this species.

4. Montlivaltia radiata, *Duncan*. Pl. XV, figs. 1—5.

The corallum is small, circular, discoid, and very flat. The wall is almost horizontal, and has a central depression. The calicular margin presents radiating septo-costal ends, which project.

The calice is very shallow, circular, and presents four principal septa.

The septa are unequal. The longest reach into the axial space, and are more exsert than the others. There are thirty-six septa, and a few rudimentary laminæ. The quaternary arrangement is obvious. The septa are stout, straight, and granular.

The costæ are very distinct, and are covered with a pellicular endotheca, which presents transverse ridges between the costæ. No dissepiments can be discovered.

The diameter of the corallum is $\frac{5}{10}$ths inch, the height $\frac{1}{10}$th inch.

Locality. Fenny Compton.

In the Collection of R. Tomes, Esq.

This is a very abnormal species, and retains the quadrate septal arrangement which is faintly preserved in many Liassic Corals, but which is so characteristic of many Palæozoic forms. The pellicular epitheca is very remarkable.

XVII.—List of Species from the Zone of Ammonites raricostatus.

Montlivaltia rugosa, Wright, and many varieties.
— *mucronata,* Duncan, and many varieties.
— *nummiformis,* ,,
— *radiata,* ,,

XVIII. List of Species from the Zones of the Lower Lias above the Zone of Ammonites angulatus.

1. *Montlivaltia Guettardi,* Blainville.
2. — *patula,* Duncan.
3. — *rugosa,* Wright, sp.
4. — *mucronata,* Duncan.
5. — *nummiformis,* ,,

6. *Montlivaltia radiata*, Duncan.
7. *Septastræa Eveshami*, „
8. *Lepidophyllia Stricklandi*, „
9. *Isastræa endothecata*, „
10. — *insignis*, „
11. — *Stricklandi*, „
12. *Cyathocœnia globosa*, „

CORALS FROM THE MIDDLE LIAS.

XIX. Corals from the Zone of Ammonites Jamesoni, Sow.

Dr. Wright notices that this zone is well developed in the Island of Pabba, near Skye, in the Hebrides, and the remarkable Coral about to be described appears to form a bed there of some extent.[1]

Family—ASTRÆACEÆ.

Genus—Lepidophyllia.

1. Lepidophyllia Hebridensis, *Duncan*. Pl. XVI, figs. 1—4.

The corallum is flat, and the corallites are short.

The calices vary in size and number; they are open and shallow, and are crowded with delicate, unequal, and not prominent septa.

The septal arrangement is very irregular. The laminæ are dentate and narrow, and the largest approach the axial space. In calices of ordinary size there are four cycles of septa, and part of a fifth in some systems, whilst in the largest calices the fifth cycle is complete.

The epitheca on the free wall of the corallites, where they overlap those below them in the general imbrication, is smooth. The calicular gemmation occurs centrally, and also near the margin.

Height of the corallum $\frac{8}{10}$ths inch.

Breadth of the calices $\frac{4}{10}$ths—$\frac{6}{10}$ths inch.

Locality. Pabba shale.

In the Collection of the School of Mines, Jermyn Street.

[1] See note 1, page 41, Part IV, No. 1.

XX. Corals from the Zone of Ammonites Henleyi, Sow.

A great number of specimens of all sizes of a very polymorphic *Montlivaltia* have been found on the surface of the fields at Cherrington, near Skipton, and in a water-course or ditch section of the Middle Lias close by. *Ammonites Henleyi, Ammonites Chiltensis, Cardinia attenuata,* and *Cardinia elongata,* were found with the Corals.

Family—ASTRÆIDÆ.

Division—Lithophyllaceæ simplices.

Genus—Montlivaltia.

1. Montlivaltia Victoriæ, *Duncan.* Pl. XVII, figs. 1—10.

The corallum grows to a great size, and generally presents a scar where it was formerly attached. The shape of the corallum is very variable, and it may be short, turbinate or sub-turbinate, or long and conical, or rudely cylindrical. The corallum is rarely straight, and generally there is a very decided curve in it and a twist also; moreover, there is frequently a constriction just below the calice, and at this point also there is generally a curve.

The calice is either widely open or contracted and small; it is never very deep, but may be characterised either by exsert and rounded septa or by septa which dip at once into a concave fossa. The outline of the calice is usually circular, and slightly compressed. The margin is sharp and is formed by the epitheca.

The septa are numerous, crowded, long, and the principal often extend to and across the axial space, which is rather elongated.

The laminæ are not much thicker at the wall than elsewhere, and the dentation is more distinct close to the wall.

There are six cycles of septa, in six systems, the highest orders being very small.

The epitheca is very dense and is strongly marked with transverse elevations and depressions; where it is worn away, the septal ends are seen, like costæ with transverse dissepiments connecting them. The wall is very thin, and appears to be identified with the epitheca.

The endotheca is very abundant, thick, curved, and branching.

Height of various specimens 5 inches, $3\frac{8}{10}$ths inches, $2\frac{7}{10}$ths inches. Breadth of specimens 2 inches, $2\frac{8}{10}$ths inches, 2 inches.

Locality. Cherrington, Skipton-on-Stour.

In the Collections of the Geological Society, British Museum, Rev. P. B. Brodie, F.G.S., R. Tomes, Esq., &c.

This is the largest simple Coral of the British Fossil Coral-fauna, and is readily distinguished. Its variability of shape almost equals that of *Montlivaltia rugosa,* Wright.

There are some fragmentary Corals in the Marlstone, but their genera are doubtful; and the cast of a *Montlivaltia* was found by Mr. Charles Moore at Wells, but I cannot determine the species.

The Corals from the Middle Lias are—

1. *Lepidophyllia Hebridensis,* Duncan.
2. *Montlivaltia Victoriæ,* „

XXI. Total number of Species of Madreporaria which can be distinguished in the Lias of the British Islands.

In the zone of *Ammonites planorbis* . . .	2[1]
„ „ — *angulatus* . . .	50
„ „ — *Bucklandi* . . .	7
„ „ — *obtusus*	1[2]
„ „ — *raricostatus* . .	4
„ „ — *Jamesoni* . . .	1
„ „ — *Henleyi*	1
	66
From the Upper Lias described by MM. Milne-Edwards and Jules Haime[3] . . .	1
Total . .	67

Lower Lias . . .	64
Middle Lias . .	2
Upper Lias . . .	1
Total . .	67

[1] See page 65.

[2] Some are common to this and the next zone.

[3] 'Brit. Foss. Corals,' Pal. Soc.

XXII. Corals from the Zone of Ammonites planorbis.[1]

Division—Astræaceæ.

Genus—Isastræa.

1. Isastræa latimæandroidea, *Duncan*. Pl. XV, figs. 18, 19.

The corallum is massive, and has an angular and rather gibbous upper surface.

The corallites are long, and their united walls are thick.

The calices are very irregular in shape, and although some are small and polygonal, others are more like the serial calices of the genus *Latimæandra*. The calices are deep, and gemmation takes place quite on the margin.

The septa are numerous, very unequal, and there is a very small septum between the larger. The larger septa are very dentate, and the tooth near the axial space is very distinct, especially in the long calices. The larger septa are not very unequal, do not project much into the calice, and the axial space is left very free, but is closed by endotheca. The existence of the small rudimentary septa makes the septal number very irregular, and the long serial calices contain very variable numbers of septa.

The endotheca is strongly developed, is vesicular, and closes in the corallites.

Diameter of ordinary calices $\frac{2}{10}$ths inch to $\frac{3}{10}$ths inch.

Diameter of serial calices $\frac{2}{10}$ths inch. Length of serial calices $\frac{5}{10}$ths—$\frac{6}{10}$ths inch.

Locality. "No. 3" bed in the Street section.

In the Collection of Dr. Wright, F.G.S.

This is a most remarkable species, and the existence of serial calices with an abundant marginal gemmation is very suggestive. It renders the genus *Latimæandra* of rather doubtful value. The new species is readily distinguished by the calices and the dentate septa.

It is erroneously named *Isastræa Murchisoni* by some collectors.

The locality whence the specimen was derived is the same which yielded *Septastræa Haimei*, Wright, sp.

Genus—Thecosmilia.

Some small stunted Corals have been found in the "Guinea bed" at Binton, in Worcestershire. Only one specimen is fairly preserved, and its calice is so like that of *Thecosmilia Terquemi*, Duncan, from Brocastle, that it must be referred to that species.

[1] The specimens from the Zone of *Ammonites planorbis* were not forwarded for description until after the first part of this Monograph was finished.

Fig. 1. *Thecosmilia Terquemi*, from the " Guinea Bed" at Binton.

The drawing of *Thecosmilia Terquemi* (Pl. III, fig. 11) greatly resembles the form from Binton, fig. 1, in the large septum which passes across the calice.

There are small Corals, probably *Thecosmiliæ*, in the " Guinea bed," at Wilmcote, but, as may be decided from the accompanying drawing, fig. 2, they are not determinable specifically.

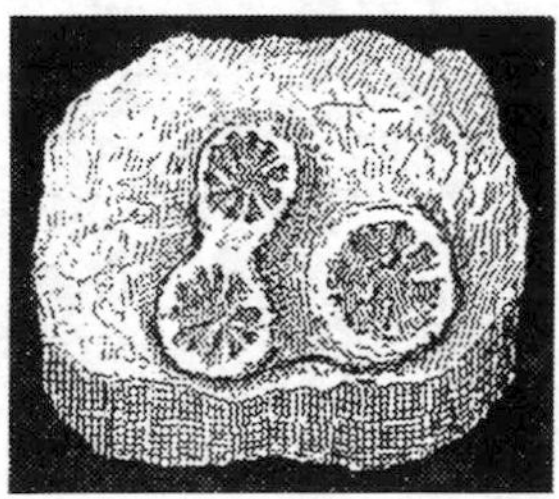

Fig. 2. *Thecosmiliæ*, from the " Guinea Bed " at Wilmcote.

XXIII. List of Corals from the Zone of Ammonites planorbis.

1. *Septastræa Haimei*, Wright, sp.
2. *Isastræa latimæandroidea*, Duncan.
3. *Thecosmilia Terquemi*, ,,

XXIV. Corals from the Zone of Avicula contorta and the White Lias.
(The Rhætic series, *Moore.*)

It has been noticed that but one fossil which could be referred to a Coral has been discovered in the Zone of *Avicula contorta* in England. The specimen is said to belong to the genus *Montlivaltia*, and to have a high septal number. The deposits containing

Avicula contorta in England, Wales, and Ireland, are not of that character in which Corals would be usually found; but the Azzarola beds of Lombardy are, as has been already noticed, highly coralliferous. The *Montlivaltia* from the British *Avicula contorta* series is, however, of some importance as a species, for it is the oldest Secondary form, there being no *Madreporaria* between it and the Carboniferous fauna except the few species of the Permian.

The White Lias, which was deposited under very different conditions to the *Avicula contorta* series, contains two genera of Corals, but the species are indeterminable, on account of the specimens being either in the form of casts or so altered by a destructive mineralization as only to present sections of their septa and part of the epithecal covering.

The White Lias of Watchet contains *Montlivaltiæ* and stunted conico-cylindrical *Thecosmiliæ*. A species of this last genus has its wall and epitheca very well shown (fig. 3).

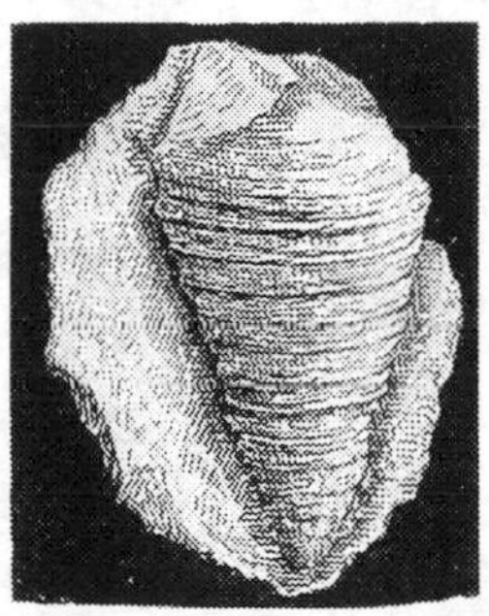

Fig. 3. *Thecosmilia*, from the White Lias of Watchet.

No Thecosmilia from the White Lias can be determined to belong to the species *Michelini* or *Martini*, but there is a cast of a Coral in the White Lias of Sparkfield which has some resemblance to casts of *Thecosmilia Terquemi*, Duncan.

Fig. 4. Cast of a *Thecosmilia*, from the White Lias of Sparkfield.

Several specimens, probably, of *Montlivaltiæ*, from the White Lias of Warwickshire, are only distinguishable by the radiating septal laminæ (fig. 5).

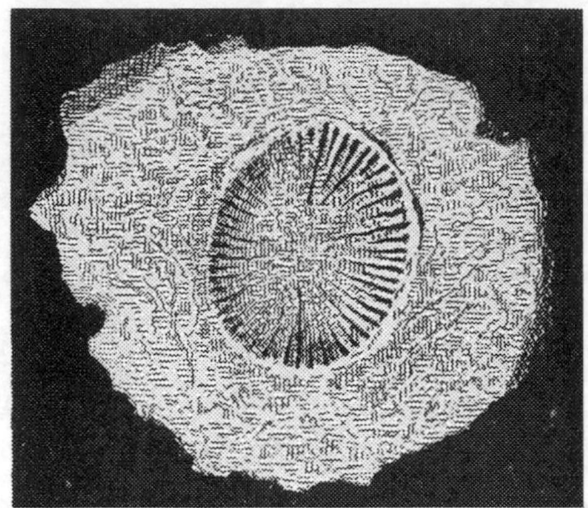

Fig. 5. *Montlivaltia,* from the White Lias of Warwickshire.

There is a great *Montlivaltia* in the Leamington beds, which is elliptical and very large at the calice. It is only found in the form of casts, one of which is here figured.

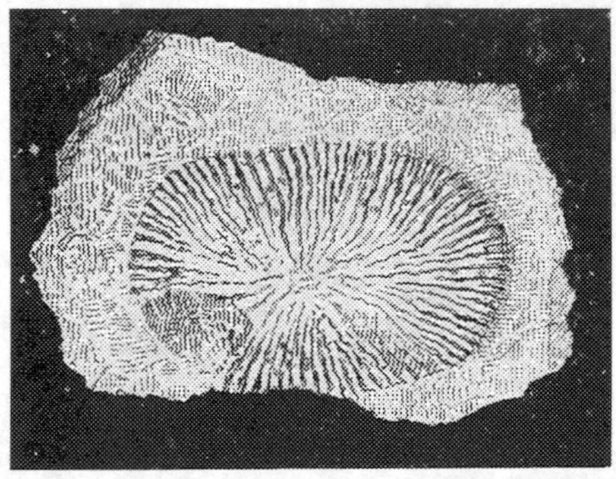

Fig. 6. Cast of a *Montlivaltia* from Leamington.

A cast of a multiseptate discoidal *Montlivaltia* is found at Punt Hill, Warwickshire, and I believe it to belong either to *Montlivaltia Haimei,* Chapuis et Dewalque, or to one of its varieties which have been noticed in the part No. 1 of this description of the Corals of the Lias. It is figured below.

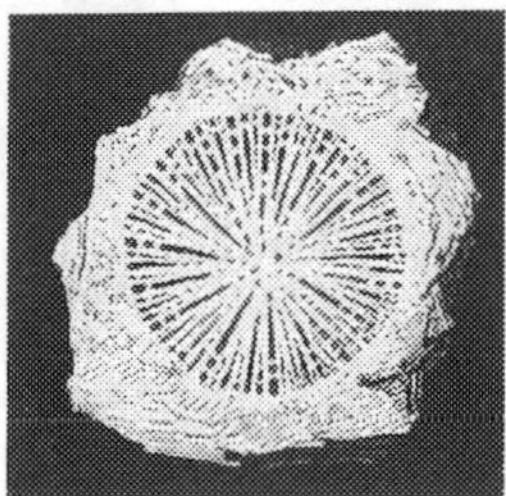

Fig. 7. *Montlivaltia,* from Punt Hill, Warwickshire.

The great vertical range of this *Montlivaltia* has already been noticed. When the local nature of the White Lias is appreciated, and it is acknowledged as "a passage-bed" between the Zone of *Avicula contorta* and the beds containing *Ammonites planorbis,* the discovery of these Corals, which in the East of France and in Luxembourg are found in the *Ammonites planorbis* series, and in that of *Ammonites angulatus,* will not be exceptional.

XXV. APPENDIX.

Note on the Age of the Sutton Stone and Brocastle, &c., Deposits.

A long and very elaborate essay, by Mr. Charles Moore, F.G.S., has been read before the Geological Society, and published in the 'Quarterly Journal' of that society, with the title, "On Abnormal Conditions of Secondary Deposits when connected with the Somersetshire and South Wales Coal-Basin, and on the Age of the Sutton and Southerndown Series," which suggests that it is more or less controversial; but although this is the case, still it has great intrinsic merits.

Mr. Bristow, F.R.S., read a paper before the Geological Society, which appeared in its 'Quarterly Journal,' "On the Lower Lias or Lias-conglomerate of a part of Glamorganshire." Like Mr. Charles Moore's communication, it is very valuable, besides being controversial. Lately also Mr. R. Tate, F.G.S., in his essay "On the Fossiliferous Development of the Zone of *Ammonites angulatus*, Schlot., in Great Britain," has produced a palæontological criticism which refers in one part to the "abnormal deposits" and "the Lias-conglomerate."

Each of these essays refers to the characters and to the age of the Sutton Stone, whose Madreporaria have been described in this Part. Mr. Bristow considers the Southerndown series of Mr. Tawney[1] to be a portion of the Sutton Stone or "Lias-conglomerate," and asserts that Mr. Tawney has made a great error in his section of the sea face of the deposit by giving it too great an elevation. Mr. Bristow also considers the Sutton Stone to be Lower Lias, and that the usual *Gryphæa incurva* occurring in abundance renders his opinion incontrovertible. Mr. Moore, on the contrary, admits the correctness of Mr. Tawney's section, but considers that insufficient altitude has been given. He considers that, as *Ostrea Liassica* (*O. irregularis*) occurs high up in the series as a "zone," and as *Ammonites planorbis* is wanting, the Sutton Stone is in the "Ostrea division" of the *Ammonites planorbis* Zone. Mr. Moore places the Brocastle deposit in the Ostrea series. He insists upon the presence of *Gryphæa incurva* in the Sutton Stone and in the deposit at Brocastle "in abundance," and localizes the deposits in the Lower Lias.[2]

Mr. Tate proves what I had already demonstrated[3]—that Mr. Tawney placed the Sutton Stone too low down in the geological scale; and, after a survey of the beds above the White Lias in Ireland and England, he considers that the *Ammonites planorbis* Zone is

[1] Tawney, 'Quart. Jour. Geol. Soc.,' vol. xxii, p. 91.

[2] A palæontological combination of the forms of the lower part of the Zone of *Ammonites planorbis* with *Gryphæa incurva* would indeed be incredible.

[3] P. Martin Duncan, 'Quart. Jour. Geol. Soc.,' Feb., 1867; and in the 1st No. of this Part.

so mixed up with that of the *Ammonites angulatus* that it had better disappear from British geology.

Mr. Tate, however, supports indirectly the geological position I have given, from the study of the Madreporaria, to the Sutton Stone and Brocastle deposits.

I agree with Mr. Bristow, or rather he agrees with me, as I was first in the field, that the Sutton Stone is what is usually called Lower Lias;[1] but I dispute the possibility of associating the Sutton Stone, Brocastle, and other equivalent deposits, including, of course, the Coral-bed of Cowbridge, with the strata composing the *Ammonites Bucklandi* Zone in the same division of a great formation.

The word "Infra-Lias," which refers to the deposits below the *Ammonites Bucklandi* series, does not assume separation from the Lias, and, although Low, Lower, and Lowest will apply to some places, it would rather confuse a geological series.

To combine in one division of the Lias, under the term Lower, such zones as those of *Ammonites raricostatus* and *Ammonites planorbis* is to associate widely different faunæ. There are many species which have a great range in this division of the Lias, but there is a clear palæontological distinction to be drawn in the British Isles, in France, Luxembourg, and in Germany, between the faunæ of the Zone of *Ammonites Bucklandi* and of those below.

Ostrea irregularis (*O. Liassica*) is a shell so widely distributed, and has so great a vertical range, that it is of no value in fixing a geological horizon. It must be considered in relaton to the fauna associated with it; and the forms found in the Sutton Stone in company with this variable Oyster are not those which elsewhere characterise the Ostrea beds of the *A. planorbis* Zone.

I have examined the *Gryphææ*, and do not consider them typical *incurvæ*. The characters of the Molluscan and Madreporarian fauna which I have already pointed out, and the affinities and grouping of the species, induce me to retain my opinion that the Sutton Stone, the Brocastle, Ewenny, and Cowbridge deposits are on one geological horizon, and still to assert that they are the equivalents of the French and Luxembourgian Zones of *Ammonites angulatus*.

The deposits have a different Coral-fauna to the corresponding beds in the east of England, where simple *Montlivaltiæ* indicate different external conditions, but not a difference in time.

Corals from the Upper Lias.

MM. Milne-Edwards and Jules Haime described *Thecocyathus Moorei*, Ed. and H., from the Upper Lias of Ilminster. Mr. Charles Moore has sent me specimens from Lansdown, near Bath. The same excellent collector has a fossil, probably a Sponge, with

[1] Sir Henry de la Beche was the first to pronounce the Sutton Stone to belong to the Lower Lias.

markings upon it like those of a cast of the calice of a Coral; it is from Ilminster. *Trochocyathus primus*, Ed. and H., is too doubtful a species to be admitted into the Liassic Coral-fauna at present.

I have to acknowledge with many thanks the great assistance I have received in completing this Monograph of the Corals of the Lias from Mr. H. Woodward, of the British Museum, Mr. R. Etheridge, of the School of Mines, and Mr. R. Tomes, besides those gentlemen whose collections have been placed at my service. (See Preface to Part IV, No. 1.)

ERRATUM.

In the Preface to Part IV, No. 1, "*Trochocyathus Moorei*, Ed. and H.," should be "*Thecocyathus Moorei*, Ed. and H."

INDEX OF THE SPECIES, ETC.

DESCRIBED AND NOTICED IN

PART IV, No. 1 AND No. 2,

CORALS FROM THE LIAS.

PAGES

Astrocœnia costata . . 21, 50.
„ dendroidea . . 22, 50.
„ favoidea . . 21, 50.
„ gibbosa . . 18, 48, 50.
„ insignis . . 19, 50.
„ minuta . . 22, 50.
„ parasitica . . 20, 50.
„ pedunculata . 20, 50.
„ plana . . 19, 50.
„ reptans . . 20, 50.
„ Sinemuriensis . 22, 33, 34, 49, 50.
„ superba . . 21, 50.
Astrocœniæ, notes on the . 26.
„ scheme of the . 23.

Cyathocœnia costata . . 29, 50.
„ dendroidea . 27, 50.
„ globosa . . 55, 62.
„ incrustans . 28, 50.

Elysastræa Fischeri . . 29, 50.
„ Moorei . . 30, 50.

Isastræa Condeana . . 49, 51.
„ endothecata . . 53, 55, 62.
„ globosa . . 31, 48, 51.
„ insignis . . 54, 55, 62.
„ latimæandroidea . 65, 66.
„ Murchisoni . . 41, 47, 51.
„ Sinemuriensis . . 30, 33, 34, 51.
„ Stricklandi . . 54, 55, 62.
„ Tomesii . . 46, 51.

PAGES

Latimæandra denticulata . 32, 43, 50.
Lepidophyllia, genus of . . 41, 53.
„ Hebridensis . 62, 64.
„ Stricklandi . 53, 55, 62.

Montlivaltia brevis . . 10, 49.
„ dentata . 48, 49.
„ denticulata . 48, 49.
„ discoidea 48, 50.
„ Gastaldi . . 48.
„ Guettardi . . 48, 50, 51, 55, 61.
„ Haimei . . 35, 40, 48, 49.
„ Hibernica . 39, 40, 49.
„ Martini . . 48, 49.
„ mucronata . . 59, 61.
„ Murchisoniæ . 8, 49.
„ nummiformis . 60, 61.
„ papillata . . 36, 38, 40, 49.
„ parasitica . . 9, 49.
„ patula . . 56, 61.
„ pedunculata . 10, 49.
„ polymorpha . 8, 33, 48, 49.
„ radiata . . 61, 62.
„ Rhodana . . 48, 49.
„ rugosa . . 57, 58, 61.
„ Ruperti . . 46, 49.
„ simplex . . 9, 49.
„ Sinemuriensis . 48, 49.
„ Walliæ . . 7, 49.
„ Victoriæ . . 63, 64.

Oppelismilia gemmans . . 39, 40, 41, 49.

	PAGES
Rhabdophyllia De-Filippi .	48.
,, Langobardica	47, 48.
,, Menighini	48.
,, recondita	17, 42, 50.
,, Sellæ	48.
Septastræa Eveshami	52, 55, 62.
,, excavata	32, 33, 49, 50.
,, Fromenteli	37, 40, 48, 50.
,, Haimei	5, 6, 66.
Stylastræa Martini	49, 50.
,, Sinemuriensis	49, 50.
Stylina Savii	48.
Thamnastræa Batarræ	48.
,, Escheri	48.
,, Meriani	48.
,, rectilamellosa	48.

	PAGES
Thecocyathus Moorei	i, 71.
,, rugosus	57, 58, 61.
Thecosmilia affinis	16, 42, 50.
,, Brodiei	13, 42, 50.
,, coronata	48, 50.
,, dentata	16, 42, 50.
,, irregularis	15, 42, 50.
,, Martini	14, 33, 45, 48, 50, 51.
,, Michelini	14, 33, 45, 48, 50, 51.
,, mirabilis	12, 42, 50.
,, plana	17, 42, 50.
,, rugosa	13, 50.
,, serialis	12, 42, 50.
,, Suttonensis	11, 42, 50.
,, Terquemi	16, 42, 50, 65, 66.
Trochocyathus primus	71.

INDEX

TO THE

BRITISH TERTIARY CORALS

DESCRIBED IN THE MONOGRAPH

BY

H. MILNE EDWARDS,

PROFESSOR AT THE MUSEUM OF NATURAL HISTORY, PARIS, ETC.,

AND

JULES HAIME,

AND IN THE SUPPLEMENTARY MONOGRAPH

BY

P. MARTIN DUNCAN, M.B. LOND., F.R.S., F.G.S., &c.,

PROFESSOR OF GEOLOGY TO KING'S COLLEGE, LONDON.

LONDON:

PRINTED FOR THE PALÆONTOGRAPHICAL SOCIETY.

1850—1866.

Dates of Publication of the various portions of the Monographs on the Tertiary Corals.

Pages i—lxxxv, 1—43, Plates I—VII, of the Monograph by H. Milne Edwards and Jules Haime (in the volume of the Palæontographical Society issued for the year 1849), were published *August*, 1850.

Pages i—iii, 1—66, Plates I—X, of Part I, by P. Martin Duncan (in the volume for the year 1865), were published *December*, 1866.

INDEX TO THE TERTIARY CORALS.*

PAGE

Classification and Structure . . . Edw. and Haime, p. i; Duncan, Pt. i, 1

Corals from the Crag . . . . . . Edw. and Haime, 1
„ Brockenhurst and Roydon . . . . Duncan, Pt. i, 40, 52, 65
„ London Clay . . . Edw. and Haime, p. 12; Duncan, Pt. i, 54, 65

*** *The synonyms are printed in italics.*

ALCYONARIA, *Dana* . . . . Edw. and Haime, pp. lxxiv, 41
Alveolites Parisiensis, Michelin; *see* Holaræa Parisiensis.
ASTRÆACEÆ, *Edw.* and *Haime* . . . . Duncan, Pt. i, 41
ASTREA, *Lamarck* . . . . . . Edw. and Haime, xxxix
Astrea cylindrica, Defrance; *see* Stylocœnia emarciata.
„ *decorata*, Michelin; *see* Stylocœnia emarciata.
„ *emarciata*, Lamarck; *see* Stylocœnia emarciata.
„ *hystrix*, Defrance; *see* Stylocœnia monticularia.
„ *stylopora*, Goldfuss; *see* Stylocœnia emarciata.
„ *Websteri*, Bowerbank; *see* Litharæa Websteri.
ASTREIDÆ, *Dana* . . Edw. and Haime, pp. xxiii, 8, 30, 47; Duncan, Pt. i, 41
ASTREINÆ, *Edw.* and *Haime* . . Edw. and Haime, pp. xxxi, 8; Duncan, Pt. i, 41
ASTROCŒNIA, *Edw.* and *Haime* . . Edw. and Haime, pp. xxx, 33
Astrocœnia pulchella, *Edw.* and *Haime* . . . . . Edw. and Haime, 34
AXOPORA, *Edw.* and *Haime* . . . . . . Duncan, Pt. i, 50, 64
Axopora Fisheri, *Duncan* . . . . . . Duncan, Pt. i, 64
„ Michelini, *Duncan* . . . . . Duncan, Pt. i, 50

BALANOPHYLLIA, *Wood* . Edw. and Haime, pp. lii, 9, 35; Duncan, Pt. i, 47
„ calyculus, *Wood* . . . Edw. and Haime, 9
„ granulata, *Duncan* . . . . Duncan, Pt. i, 47
„ desmophyllum, *Edw.* and *Haime* . . . Edw. and Haime, 35

CARYOPHYLLIACEÆ, *Edw.* and *Haime* . . . Duncan, Pt. i, 57

---

* The words "Edw. and Haime" preceding the numerals, refer to the Pages in the Monograph of British Fossil Corals by MM. Milne Edwards and Jules Haime; and the word "Duncan" to the Pages in the Supplementary Monograph by Professor Duncan.

PAGE

Cellastrea emarciata, Blainville; *see* Stylocœnia emarciata.
„ *hystrix*, Blainville; *see* Stylocœnia monticularia.
Cladocora cariosa, Lonsdale; *see* Cryptangia Woodii.
CRYPTANGIA, *Edw.* and *Haime* . . . . Edw. and Haime, pp. xliv, 8
Cryptangia Woodii, *Edw.* and *Haime* . . . . Edw. and Haime, 8
CYATHININÆ, *Edw.* and *Haime* . . . Edw. and Haime, pp. xii, 21

DASMIA, *Edw.* and *Haime* . . . . Edw. and Haime, pp. xix, 25
Dasmia Sowerbyi, *Edw.* and *Haime* . . . . Edw. and Haime, 25
DENDRACIS, *Edw.* and *Haime* , . Edw. and Haime, p. xxiii; Duncan, Pt. I, 62
Dendracis Lonsdalei, *Duncan* . . . . . Duncan, Pt. I, 62
DENDROPHYLLIA, *Blainville* . Edw. and Haime, pp. liii, 36; Duncan, Pt. I, 61
Dendrophyllia dendrophylloides, *Edw.* and *Haime* . . Edw. and Haime, 36
„ elegans, *Duncan* . . . Duncan, Pt. I, 61
DIPLOHELIA, *Edw.* and *Haime* . . Edw. and Haime, pp. xxi, 28
Diplohelia papillosa, *Edw.* and *Haime* . . . Edw. and Haime, 28

EUPSAMMIDÆ, *Edw.* and *Haime* . . Edw. and Haime, pp. li, 9, 34
EUPSAMMINÆ, *Edw.* and *Haime* . Duncan, Pt. I, pp. 47, 61
EUSMILINÆ, *Edw.* and *Haime* . . . Edw. and Haime, pp. xxiii, 30

FLABELLUM, *Lesson* . . . . Edw. and Haime, pp. xviii, 6
Flabellum Woodii, *Edw.* and *Haime* . . . . . Edw. and Haime, 6
FUNGIA, *Lamarck* . . . . . . Edw. and Haime, xlvi
Fungia semilunata, Wood; *see* Flabellum Woodii.

GORGONIDÆ, *Dana* . . . . Edw. and Haime, pp. lxxix, 42
GRAPHULARIA, *Edw.* and *Haime* . . . Edw. and Haime, pp. lxxxiii, 41
Graphularia Wetherelli, *Edw.* and *Haime* . . . . Edw. and Haime, 41

HOLARÆA, *Edw.* and *Haime* . . . . Edw. and Haime, pp. lvi, 40
Holaræa Parisiensis, *Edw.* and *Haime* . . . Edw. and Haime, 40

ISINÆ, *Dana* . . . . Edw. and Haime, pp. lxxxi, 42

LEPTOCYATHUS, *Edw.* and *Haime* . . . Edw. and Haime, pp. xiv, 21
Leptocyathus elegans, *Edw.* and *Haime* . . . Edw. and Haime, 21
LITHARÆA, *Edw.* and *Haime* . Edw. and Haime, pp. lv, 38; Duncan, Pt. I, 49
Litharæa Brockenhursti, *Duncan* . . . . . Duncan, Pt. I, 49
„ Websteri, *Edw.* and *Haime* . . . . Edw. and Haime, 38
LOBOPSAMMIA, *Edw.* and *Haime* . . Edw. and Haime, p. liii; Duncan, Pt. I, 48
Lobopsammia cariosa, *Goldfuss* , . . . Duncan, Pt. I, 48

MADREPORA, *Lin*, ... Duncan, Pt. I, 51
Madrepora Anglica, *Duncan* . . Duncan, Pt. I, 51
„ Roemeri, *Duncan* . . . . Duncan, Pt. I, 51
„ Solanderi, *Defrance* . . . . . Duncan, Pt. I, 51
MADREPORIDÆ, *Edw.* and *Haime* . . . . Duncan, Pt. I, pp. 47, 51, 61, 64

PAGE

MILLEPORIDÆ, *Edw.* and *Haime* . . . . Duncan, PT. I, pp. 50, 64
MOPSEA, *Lamouroux* . . . . . Edw. and Haime, pp. lxxxi, 42
Mopsea costata, *Edw.* and *Haime* . . . . . Ewd. and Haime, 42

OCULINA, *Lamarck* . . . Edw. and Haime, pp. xix, 27 ; Duncan, PT. I, 60
Oculina conferta, *Edw.* and *Haime* . . . . . Edw. and Haime, 27
„ *dendrophylloides*, Lonsdale ; *see* Dendrophyllia dendrophylloides.
„ incrustans, *Duncan* . . . . . Duncan, PT. I, 60
„ Wetherelli, *Duncan* . . . . . Duncan, PT. I, 60
OCULINACEÆ, *Edw.* and *Haime* . . . . . Duncan, PT. I, 60
OCULINIDÆ, *Edw.* and *Haime* . . Edw. and Haime, pp. xix, 27 ; Duncan, PT. I, 60

PARACYATHUS, *Edw.* and *Haime* . . . Edw. and Haime, p. 23 ; Duncan, PT. I, 58
Paracyathus brevis, *Edw.* and *Haime* . . . . . Edw. and Haime, 25
„ caryophyllus, *Edw.* and *Haime* . . . . Edw. and Haime, 24
„ crassus, *Edw.* and *Haime* . . . . . Edw. and Haime, 23
„ cylindricus, *Duncan* . . . . . Duncan, PT. I, 58
„ Haimei, *Duncan* . . . . . Duncan, PT. I, 59
PENNATULIDÆ, *Fleming* . . . . . Edw. and Haime, pp. lxxxii, 41
PORITES, *Edw.* and *Haime* . . . Edw. and Haime, p. lv ; Duncan, PT. I, 63
Porites panacea, *Lonsdale* . . . . . . Duncan, PT. I, 63
PORITIDÆ, *Edw.* and *Haime* . . Edw. and Haime, pp. lv, 38 ; Duncan, PT. I, p. 49, 63
PORITINÆ, *Edw.* and *Haime* . . Edw. and Haime, pp. lv, 38 ; Duncan, PT. I, 63
PSEUDOTURBINOLIDÆ, *Edw.* and *Haime* . . . Edw. and Haime, pp. xix, 25

Siderastræa Websteri, Lonsdale ; *see* Litharæa Websteri.
SOLENASTRÆA, *Edw.* and *Haime* . . . . . Duncan, PT. I, 41
Solenastræa Beyrichi, *Duncan* . . . . . Duncan, PT. I, 44
„ cellulosa, *Duncan* . . . . . Duncan, PT. I, 41
„ gemmans, *Duncan* . . . . Duncan, PT. I, 44
„ granulata, *Duncan* . . . Duncan, PT. I, 45
„ Koeneni, *Duncan* . . . . . Duncan, PT. I, 42
„ Reussi, *Duncan* . . . . . Duncan, PT. I, 43
SPHENOTROCHUS, *Edw.* and *Haime* . . . Edw. and Haime, pp. xvi, 2
Sphenotrochus intermedius, *Edw.* and *Haime* . . . Edw. and Haime, 2
STEPHANOPHYLLIA, *Michelin* . . . Edw. and Haime, pp. liii, 34
Stephanophyllia discoides, *Edw.* and *Haime* . . . . Edw. and Haime, 34
STEREOPSAMMIA, *Edw.* and *Haime* . . . Edw. and Haime, pp. liii, 37
Stereopsammia humilis, *Edw.* and *Haime* . . . . Edw. and Haime, 37
STYLOCŒNIA, *Edw.* and *Haime* . . . . Edw. and Haime, pp. xxix, 30
Stylocœnia emarciata, *Edw.* and *Haime* . . . . Edw. and Haime, 30
„ monticularia, *Edw.* and *Haime* . . . . Edw. and Haime, 32
Stylopora monticularia, Schweigger ; *see* Stylocœnia monticularia.

TROCHOCYATHACEÆ, *Edw.* and *Haime* . . . . Duncan, PT. I, 57
TROCHOCYATHUS, *Edw.* and *Haime* . Edw. and Haime, pp. xiv, 22 ; Duncan, PT. I, 57
Trochocyathus Austeni, *Duncan* . . . . . Duncan, PT. I, 57
„ insignis, *Duncan* . . . . . Duncan, PT. I, 57

PAGE

Trochocyathus sinuosus, *Edw.* and *Haime* . . . . . Edw. and Haime, 22
TURBINARINÆ, *Edw.* and *Haime* . . . . . . Duncan, Pt. I, 62
TURBINOLIA, *Lamarck* . . . Edw. and Haime, pp. xvi, 13; Duncan, Pt. I, 54
Turbinolia affinis, *Duncan* . . . . Duncan, Pt. I, 54
,, Bowerbankii, *Edw.* and *Haime* . . . . . Edw. and Haime, 17
,, *caryophyllus*, Lamarck; *see* Paracyathus caryophyllus.
,, Dixonii, *Edw.* and *Haime* . . . . . . Edw. and Haime, 15
,, *dubia*, Defrance; *see* Trochocyathus sinuosus.
,, exarata, *Duncan* . . . . Duncan, Pt. I, 55
,, firma, *Edw.* and *Haime* . . . . . . Edw. and Haime, 20
,, Forbesi, *Duncan* . . . . Duncan, Pt. I, 55
,, Fredericiana, *Edw.* and *Haime* . . . . . Edw. and Haime, 17
,, humilis, *Edw.* and *Haime* . . . . . . Edw. and Haime, 18
,, *intermedia*, Münster; *see* Sphenotrochus intermedius.
,, *Milletiana*, Wood; *see* Sphenotrochus intermedius.
,, minor, *Edw.* and *Haime* . . . . . . Edw. and Haime, 19
,, Prestwichii, *Edw.* and *Haime* . . . . Edw. and Haime, 20
,, *sinuosa*, Brogniart; *see* Trochocyathus sinuosus.
,, sulcata, *Lamarck* . . . . Edw. and Haime, 13
,, *turbinata* (pars), Lamarck; *see* Trochocyathus sinuosus.
TURBINOLIDÆ, *Edw.* and *Haime* . . Edw. and Haime, pp. xi, 2, 13; Duncan, Pt. I, 54
TURBINOLINÆ, *Edw.* and *Haime* . . Edw. and Haime, pp. xvi, 2, 13; Duncan, Pt. I, 54

WEBSTERIA, *Edw.* and *Haime* . . . Edw. and Haime, pp. lxxxiv, 43
Websteria crisioides, *Edw.* and *Haime* . . . Edw. and Haime, 43

ZOANTHARIA, *Gray* . . Edw. and Haime, pp. ix, 2, 13; Duncan, Pt. I, 54

INDEX

TO THE

BRITISH SECONDARY CORALS

DESCRIBED IN THE MONOGRAPH

BY

H. MILNE EDWARDS,

PROFESSOR AT THE MUSEUM OF NATURAL HISTORY, PARIS, ETC.,

AND

JULES HAIME,

AND IN THE SUPPLEMENTARY MONOGRAPH

BY

P. MARTIN DUNCAN, M.B. Lond., F.R.S. F.G.S., &c.,

PROFESSOR OF GEOLOGY TO KING'S COLLEGE, LONDON.

LONDON:

PRINTED FOR THE PALÆONTOGRAPHICAL SOCIETY.

1850—1872.

DATES OF PUBLICATION of the various portions of the MONOGRAPHS on the SECONDARY CORALS.

Pages i—lxxxv, 44—71, Plates VIII—XI, of the Monograph by H. Milne Edwards and Jules Haime (in the volume of the Palæontographical Society issued for the year 1849), were published *August*, 1850.

Pages 73—145, Plates XII—XXX, by H. Milne Edwards and Jules Haime (in the volume for the year 1851), were published *June*, 1851.

Pages 1—26, Plates I—IX, of the Supplementary Monograph, Part II, by P. Martin Duncan (in the volume for the year 1868), were published *February*, 1869.

Pages 27—46, Plates X—XV, of Part II, by P. Martin Duncan (in the volume for the year 1869), were published *January*, 1870.

Pages 1—24, Plates I—VII, of Part III, by P. Martin Duncan (in the volume for the year 1872), were published *October*, 1872.

Pages i—ii, 1—43, Plates I—XI, of Part IV, by P. Martin Duncan (in the volume for the year 1866), were published *June*, 1867.

Pages 45—73, Plates XII—XVII, of Part IV, by P. Martin Duncan (in the volume for the year 1867), were published *June*, 1868.

INDEX TO THE SECONDARY CORALS.*

PAGE

Classification and Structure . . . Edw. and Haime, p. i; Duncan, Pt. i, 1

Corals from Upper Chalk . . Edw. and Haime, p. 44; Duncan, Pt. ii, pp. 2, 16, 17, 44
,, Lower Chalk . . . Edw. and Haime, p. 53; Duncan, Pt. ii, pp. 2, 16, 44
,, Upper Greensand . . Edw. and Haime, p. 57; Duncan, Pt. ii, pp. 18, 23, 31
,, ,, of Haldon . . . . Duncan, Pt. ii, 27
,, Gault . . . Edw. and Haime, p. 61; Duncan, Pt. ii, pp. 31, 38, 45
,, Red Chalk . . . . . Duncan, Pt. ii, pp. 23, 45
,, Lower Greensand . . Edw. and Haime, p. 70; Duncan, Pt. ii, pp. 39, 43, 45
,, Portland Oolite . . Edw. and Haime, p. 73; Duncan, Pt. iii, 7
,, Coral Rag . . . Edw. and Haime, p. 75; Duncan, Pt. iii, 7
,, Great Oolite . . . Edw. and Haime, p. 104; Duncan, Pt. iii, pp. 7, 14
,, Inferior Oolite . . Edw. and Haime, p. 125; Duncan, Pt. iii, pp. 8, 16
,, Lias . Edw. and Haime, p. 144; Duncan, Pt. iv, pp. 5, 42, 51, 56, 65, 66
,, Zone of Ammonites angulatus . . Duncan, Pt. iv, pp. 6, 35, 38, 45, 47, 49
,, ,, Am. Bucklandi . . . . Duncan, Pt. iv, pp. 51, 55
,, ,, Am. Jamesoni and Am. Henleyi . . . Duncan, Pt. iv, 62
,, ,, Am. planorbis . . . . Duncan, Pt. iv, pp. 5, 17, 47, 65
,, ,, Am. obtusus . . . . . Duncau, Pt. iv, 56
,, ,, Am. raricostatus . . Duncan, Pt. iv, pp. 57, 58, 61
,, ,, Avicula contorta . . . . . Duncan, Pt. iv, 66
,, Skye . . . . . . Duncan, Pt. iv, 41
,, Sntton Stone . . . . . . Duncan, Pt. iv, 69
Remarks on Coral Zones . . . . . . Duncan, Pt. iv, 1
,, Oolitic Coral Faunas . . . . . Duncan, Pt. iii, pp. 2, 9

*** *The synonyms are printed in italics.*

Agaricia lobata (pars), Morris; *see* Stylina tubulifera.
,, ,, *see* Thamnastrea concinna.
Alveopora microsolena, M'Coy; *see* Microsolena regularis.
ANABACIA, D'Orbigny . . . . Edw. and Haime, pp. xlvii, 120, 142
Anabacia Bajociana, D'Orbigny; *see* Anabacia orbulites.

* The words "Edw. and Haime" preceding the numerals refer to the Pages and Plates in the Monograph by MM. H. Milne Edwards and Jules Haime; and the word "Duncan" to the Pages and Plates in the Supplementary Monograph by Prof. Duncan.

PAGE

Anabacia hemispherica, *Edw.* and *Haime* . . . . Edw. and Haime, 142
„ orbulites, *Edw.* and *Haime* . . . . Edw. and Haime, pp. 120, 142
Anthophyllum obconicum, Goldfuss; *see* Montlivaltia dispar.
Aplocyathus Magnevilliana, D'Orbigny; *see* Trochocyathus Magnevillianus.
ASTRÆACEÆ, *Edw.* and *Haime* . . Duncan, Pt. II, pp. 22, 29; Pt. IV, pp. 46, 53, 62, 65
Astrea arachnoides, Fleming; *see* Thamnastrea arachnoides.
„ *concinna*, Goldfuss; *see* Thamnastrea concinna.
„ *Defranciana*, Michelin; *see* Thamnastrea Defranciana.
„ *elegans*, Fitton; *see* Holocystis elegans.
„ *explanata*, Goldfuss; *see* Isastræa explanata.
„ *explanulata*, M'Coy; *see* Isastræa explanata.
„ *favosioides*, Phillips; *see* Isastræa explanata.
„ *helianthoides*, M'Coy; *see* Isastræa explanata.
„ inæqualis, *Phillips* . . . . . Edw. and Haime, 104
„ *limitata*, Lamoureux; *see* Isastræa limitata.
„ *micraston*, Phillips; *see* Thamnastrea concinna.
„ *tenuistriata*, M'Coy; *see* Isastræa tenuistriata.
„ *Tisburiensis*, Fitton; *see* Isastræa oblonga.
„ *tubulifera*, Phillips; *see* Stylina tubulifera.
„ *tubulosa*, Morris; *see* Stylina tuberosa.
„ *varians*, Roemer; *see* Thamnastrea concinna.
ASTRÆIDÆ, *Dana* Edw. and Haime, pp. xxiii, 47, 57, 68, 73, 76, 105, 110, 128; Duncan, Pt. II, pp. 5, 11, 21, 27, 41; Pt. III, pp. 14, 16; Pt. IV, pp. 5, 7, 46, 51, 52, 56, 58, 63
ASTREINÆ, *Edw.* and *Haime* . . Edw. and Haime, pp. xxxi, 59; Duncan, Pt. II, 42
ASTROCŒNIA, *Edw.* and *Haime* . . . Duncan, Pt. II, p. 29; Pt. IV, pp. 18, 23
Astrocœnia costata, *Duncan* . . . . . Duncan, Pt. IV, 21
„ decaphylla, *Edw.* and *Haime* . . . . Duncan, Pt. II, 29
„ dendroidea, *Duncan* . . . . . Duncan, Pt. IV, 22
„ favoidea, *Duncan* . . . . . Duncan, Pt. IV, 21
„ gibbosa, *Duncan* . . . . . Duncan, Pt. IV, 18
„ insignis, *Duncan* . . . . . Duncan, Pt. IV, 19
„ minuta, *Duncan* . . . . . Duncan, Pt. IV, 22
„ parasitica, *Duncan* . . . . . Duncan, Pt. IV, 20
„ pedunculata, *Duncan* . . . . . Duncan, Pt. IV, 20
„ plana, *Duncan* . . . . . Duncan, Pt. IV, 19
„ reptans, *Duncan* . . . . . Duncan, Pt. IV, 20
„ Sinemuriensis, *D'Orbigny* . . . . Duncan, Pt. IV, 22
„ superba, *Duncan* . . . . . Duncan, Pt. IV, 21
AXOSMILIA, *Edw.* and *Haime* . . . . Edw. and Haime, pp. xxvi, 128
Axosmilia Wrighti, *Edw.* and *Haime* . . . . Edw. and Haime, 128

BATHYCYATHUS, *Edw.* and *Haime* . Edw. and Haime, pp. xiii, 67; Duncan, Pt. II, 35
Bathycyathus Sowerbyi, *Edw.* and *Haime* . . . . Edw. and Haime, 67
BRACHYCYATHUS, *Edw.* and *Haime* . . . . Duncan, Pt. II, 40
Brachycyathus Orbignyanus, *Edw.* and *Haime* . . . . Duncan, Pt. II, 40

CALAMOPHYLLIA, *Guettard* . . . Edw. and Haime, pp. xxxiii, 89, 111

PAGE

Calamophyllia prima, D'Orbigny ; *see* Cladophyllia Babeana.
Calamophyllia Stokesi, *Edw.* and *Haime* . . . . . Edw. and Haime, 89
Calamophyllia radiata, *Lamouroux* . . . . . . Edw. and Haime, 111
CARYOPHYLLIA, *Lamarck* . . . . . Duncan, PT. II, pp. 2, 31
Caryophyllia annularis, Fleming ; *see* Thecosmilia annularis.
,, *centralis*, Flem. ; *see* Parasmilia centralis.
,, *cespitosa*, Conybeare and Phillips ; *see* Cladophyllia Conybearii.
,, *conulus*, Phillips ; *see* Trochocyathus conulus.
,, convexa, *Phillips* . . . . . . Edw. and Haime, 143
,, cylindracea, *Reuss* . . . . . . Duncan, PT. II, 3
,, *cylindrica*, Phillips ; *see* Thecosmilia annularis.
,, *fasciculata*, De Blainville ; *see* Lithostrotion irregularis.
,, Lonsdalei, *Duncan* . . . . . . Duncan, PT. II, 3
,, Tennanti, *Duncan* . . . . . . Duncan, PT. II, 4
CARYOPHYLLIACEÆ, *Edw.* and *Haime* . . . . Duncan, PT. II, pp. 2, 31, 40
CARYOPHYLLINÆ, *Edw.* and *Haime* . . . . Duncan, PT. II, pp. 31, 40
CLADOPHYLLIA, *Edw.* and *Haime* . . . . Edw. and Haime, pp. 91, 113
Cladophyllia Babeana, *D'Orbigny* . . . . . . Edw. and Haime, 113
,, Conybearii, *Edw.* and *Haime* . . . . . Edw. and Haime, 91
CLAUSASTREA, *D'Orbigny* . . . . . . Edw. and Haime, 117
Clausastrea Pratti, *Edw.* and *Haime* . . . . . . Edw. and Haime, 117
CŒLOSMILIA, Edw. and Haime . . Edw. and Haime, pp. xxv, 52 ; Duncan, PT. II, pp. 5, 8
Cœlosmilia laxa, *Edw.* and *Haime* . . . . . . Edw. and Haime, 52
COMOSERIS, *D'Orbigny* . . . . Edw. and Haime, pp. 101, 102, 122, 143
Comoseris irradians, *Edw.* and *Haime* . . . . Edw. and Haime, 101
,, vermicularis, *M'Coy* . . . . Edw. and Haime, pp. 122, 143
CONVEXASTREA, *D'Orbigny* . . . . . . Edw. and Haime, 109
Convexastrea Waltoni, *Edw.* and *Haime* . . . . . Edw. and Haime, 109
Coralloidea columnaria, Parkinson ; *see* Isastræa oblonga.
Cryptocœnia Luciensis, D'Orbigny ; *see* Cyathophora Luciensis.
CYATHINA, *Ehrenberg* . . . . . . Edw. and Haime, pp. xii, 44, 61
Cyathina Bowerbankii, Edw. and Haime, p. 61 ; *see* Caryophyllia Bowerbanki (Duncan, PT. II, p. 31).
Cyathina lævigata, *Edw.* and *Haime* . . . . . . Edw. and Haime, 44
CYATHININÆ, *Edw.* and *Haime* . . . . Edw. and Haime, pp. xii, 44, 61
CYATHOCŒNIA, *Duncan* . . . . . Duncan, PT. IV, pp. 27, 55
Cyathocœnia costata, *Duncan* . . . . . . Duncan, PT. IV, 29
,, dendroidea, *Duncan* . . . . . Duncan, PT. IV, pp. 27, 55
,, globosa, *Duncan* . . . . . . Duncan, PT. IV, 55
,, incrustans, *Duncan* . . . . . . Duncan, PT. IV, 28
CYATHOPORA *Michelin* . . Edw. and Haime, p. 107 ; Duncan, PT. II, p. 21 ; PT. III, 14
Cyathopora *elegans*, Lonsdale ; *see* Holocytis elegans.
,, insignis, *Duncan* . . . . . . Duncan, PT. III, 14
,, Luciensis, *D'Orbigny* . . . . . . Edw. and Haime, 107
,, monticularia, *D'Orb.* . . . . . . Duncan, PT. II, 21
,, Pratti, *Edw.* and *Haime* . . . . . . Edw. and Haime, 108
,, tuberosa, *Duncan* . . . . . . Duncan, PT. III, 15
CYATHOPHYLLIDÆ, *Edw.* and *Haime* . . . . Edw. and Haime, pp. lxv, 143, 145

PAGE

CYATHOPHYLLUM, *Goldfuss* . . . . Edw. and Haime, pp. lxviii, 145
Cyathophyllum novum, *Edw.* and *Haime* . . . Edw. and Haime, 145
CYCLOCYATHUS, *Edw.* and *Haime* . . . Edw. and Haime, pp. xiv, 63
Cyclocyathus Fittoni, *Edw.* and *Haime* . . . Edw. and Haime, pp. xiv, 63
CYCLOLITES, *Lamarck* . . . . Duncan, Pt. ii, p. 24 ; Pt. iii, 23
Cyclolites Beanii, *Duncan* . . . . . . Duncan, Pt. iii, 23
„ *Eudesii*, Michelin ; *see* Discocyathus Eudesi.
„ *lævis*, Blainville ; *see* Anabacia orbulites.
„ Lyceti, *Duncan* . . . . . Duncan, Pt. iii, 23
„ polymorpha, *Goldfuss* . . . . . Duncan, Pt. ii, 24
„ *truncata*, Defrance ; *see* Discocyathus Eudesi.

Decacœnia Michelini, D'Orbigny ; *see* Stylina tubulifera.
Dendrophyllia plicata, M'Coy ; *see* Goniocora sociale.
Dentipora glomerata, M'Coy ; *see* Stylina tubulifera.
DIBLASUS, *Lonsdale* . . . . . . Duncan, Pt. ii, 14
Diblasus Gravensis, *Lonsdale* . . . . . Duncan, Pt. ii, 14
DIMORPHOSERIS, *Duncan* . . . Duncan, Pt. iii, 22
Dimorphoseris oolitica, *Duncan* . . . . Duncan, Pt. iii, 22
DISCOCYATHUS, *Edw.* and *Haime* . . . . Edw. and Haime, pp. xiii, 125
Discocyathus Eudesi, *Michelin* . . Edw. and Haime, 125

ELYSASTRÆA, *Laube* . . . . . Duncan, Pt. iv, 29
Elysastræa Fischeri, *Laube* . . . . . Duncan, Pt. iv, 29
„ Moorei, *Duncan* . . . . . Duncan, Pt. iv, 30
Eunomia Babeana, D'Orbigny ; *see* Cladophyllia Babeana.
„ *radiata*, Lamoureux ; *see* Calamophyllia radiata.
EUPSAMMIDÆ, *Edw.* and *Haime* . . . . Edw. and Haime, pp. li, 54
EUSMILINÆ, *Edw.* and *Haime* . . . Edw. and Haime, pp. xxiii, 47, 57, 68 ; Duncan, Pt. ii, pp. 27, 41
Explanaria flexuosa, Fleming ; *see* Thamnastrea arachnoides.

FAVIA, *Ehrenberg* . . . . . . Duncan, Pt. ii, 21
Favia minutissima, *Duncan* . . . . . . Duncan, Pt. ii, 22
FAVIACEÆ, *Edw.* and *Haime* . . . Duncan, Pt. ii, p. 21 ; Pt. iv, pp. 5, 32, 52
Favosites radiata, Blainville ; *see* Calamophyllia radiata.
FUNGIA, *Lamarck* . . . . . . Edw. and Haime, lxvi
Fungia clathrata, Geinitz ; *see* Micrabacia coronula.
„ *coronula*, Goldfuss ; *see* Microbacia coronula.
„ *lævis*, Goldfuss ; *see* Anabacia orbulites.
„ *orbulites*, Lamoureux ; *see* Anabacia orbulites.
FUNGIDÆ, *Dana* . Edw. and Haime, pp. xlv, 60, 101, 120, 142 ; Duncan, Pt. ii, pp. 24, 37, 42 ; Pt. iii, pp. 16, 19
FUNGINÆ, *Edw.* and *Haime* . . . . . Duncan, Pt. ii, 24

Gemmastrea limbata, M'Coy ; *see* Stylina conifera.
GONIOCORA, *Edw.* and *Haime* . . . . . Edw. and Haime, 92
Goniocora socialis, *Roemer* . . . . . Edw. and Haime, 92

PAGE

GONIOSERIS, *Duncan* . . . . Duncan, Pt. III, 21
Gonioseris angulata, *Duncan* . . . . . . Duncan, Pt. III, 21
„ Leckenbyi, *Duncan* . . . . . . Duncan, Pt. III, 22

HOLOCYSTIS, *Lonsdale* . . . . . . Edw. and Haime, pp. lxiv, 70
Holocystis elegans, *Fitton* . . . . Edw. and Haime, 70
Hydnophora Frieslebenii? Fischer; *see* Stylina tubulifera.

ISASTRÆA, *Edw.* and *Haime* . Edw. and Haime, pp. 74, 94, 113, 138; Duncan, Pt. II, p. 30; Pt. III, p. 15; Pt. IV, pp. 30, 41, 46, 53, 65
Isastræa Conybearii, *Edw.* and *Haime* . . . . . Edw. and Haime, 113
„ endothecata, *Duncan* . . . . . . Duncan, Pt. IV, 53
„ explanata, *Goldfuss* . . . . . . Edw. and Haime, 94
„ explanulata, *M'Coy* . . . . . . Edw. and Haime, 115
„ gibbosa, *Duncan* . . . . Duncan, Pt. III, 15
„ globosa, *Duncan* . . . . Duncan, Pt. IV, 31
„ Greenoughi, *Edw.* and *Haime* . . . . . Edw. and Haime, 96
„ Haldonensis, *Duncan* . . . . . . Duncan, Pt. II, 30
„ insignis, *Duncan* . . . . Duncan, Pt. IV, 54
„ latimæandroidea, *Duncan* . . . . . Duncan, Pt. IV, 65
„ limitata, *Lamouroux* . . . . . . Edw. and Haime, 114
„ Lonsdalii, *Edw.* and *Haime* . . . . . . Edw. and Haime, 139
„ Morrisii, *Duncan* . . . . Duncan, Pt. II, 42
„ Murchisoni, *Wright* . . . . . . Duncan, Pt. IV, 41
„ oblonga, *Fleming* . . . . Edw. and Haime, 73
„ tenuistriata, *M'Coy* . . . . . . Edw. and Haime, 138
„ Tomesii, *Duncan* . . . . Duncan, Pt. IV, 46
„ Richardsoni, *Edw.* and *Haime* . . . . . Edw. and Haime, 138
„ serialis, *Edw.* and *Haime* . . . . . . Edw. and Haime, 116
„ Sinemuriensis, *E. de From.* . . . . . . Duncan, Pt. IV, 30
„ Stricklandi, *Duncan* . . . . . . Duncan, Pt. IV, 54

Lasmophyllia radisensis, D'Orbigny; *see* Montlivaltia dispar.
LATIMÆANDRA, *D'Orbigny* . Edw. and Haime, pp. xxxiv, 136; Duncan, Pt. III, p. 18; Pt. IV, 32
Latimæandra Davidsoni, *Edw.* and *Haime* . . . . Edw. and Haime, 137
„ denticulata, *Duncan* . . . . . . Duncan, Pt. IV, 32
„ Flemingi, *Edw.* and *Haime* . Edw. and Haime, p. 136; Duncan, Pt. III, 18
LEPIDOPHYLLIA, *Duncan* . . . . Duncan, Pt. IV, pp. 53, 62
Lepidophyllia Hebridensis, *Duncan* . . . . . . Duncan, Pt. IV, 62
„ Stricklandi, *Duncan* . . . . . Duncan, Pt. IV, 53
LEPTOCYATHUS, *Edw.* and *Haime* . . . . . . Duncan, Pt. II, 34
Leptocyathus gracilis, *Duncan* . . . . . . Duncan, Pt. II, 34
LITHODENDRON, *Phillips* . . . . . . Edw. and Haime, lxxi
Lithodendron annulare, Keferstein; *see* Thecosmilia annularis.
„ astreatum, *M'Coy* . . . . . . Edw. and Haime, 143
„ *centrale*, Keferstein; *see* Parasmilia centralis.

PAGE

Lithodendron dichotomum, M'Coy; *see* Cladophyllia Conybearii.
„ *dispar*, Goldfuss; *see* Montlivaltia dispar.
„ *Edwardsii*, M'Coy; *see* Rhabdophyllia Phillipsi.
„ *eunomia*, Michelin; *see* Calamophyllia radiata.
„ *oblongum*, Fleming; *see* Isastræa oblonga.
„ *sociale*, Roemer; *see* Goniocora sociale.
„ *trichotomum*, Morris; *see* Thecosmilia annularis.
Lithostrotion oblongum, Morris; *see* Isastræa oblonga.
Lobophyllia trichotoma, M'Coy; *see* Thecosmilia annularis.
LOPHOSERINÆ, *Edw.* and *Haime* . . . . . Duncan, Pt. II, 24, 42

Madrepora arachnoides, Parkinson; *see* Thamnastrea arachnoides.
„ *centralis*, Mant.; *see* Parasmilia centralis.
„ *flexuosa*, Smith; *see* Cladophyllia Babeana.
„ *porpites*, W. Smith; *see* Anabacia orbulites.
„ *turbinata*, Smith; *see* Montlivaltia Smithi.
Meandrina vermicularis, M'Coy; *see* Comoseris vermicularis.
MICRABACIA, *Edw.* and *Haime* Edw. and Haime, pp. xlvii, 60; Duncan, Pt. II, pp. 24, 37
Micrabacia coronula, *Goldfuss* . . . . . Edw. and Haime, 60
„ Fittoni, *Duncan* . . . . . . Duncan, Pt. II, 37
MICROSOLENA, *Lamoureux* . . . . Edw. and Haime, pp. lvi, 122
Microsolena excelsa, *Edw.* and *Haime* . . . . . Edw. and Haime, 124
„ regularis, *Edw.* and *Haime* . . . . Edw. and Haime, 122
Monocarya centralis (pars), Lonsdale; *see* Cyathina lævigata.
„ *centralis* (pars), Lons.; *see* Parasmilia centralis.
MONTLIVALTIA, *Lamoureux* . Edw. and Haime, pp. xxv, 80, 110, 129; Duncan, Pt. III, p. 16; Pt. IV, pp. 7, 35, 39, 46, 51, 56, 58, 63, 68
Montlivaltia brevis, *Duncan* . . . . . . Duncan, Pt. IV, 10
„ *caryophyllata*, Bronn; *see* Montlivaltia trochoides.
„ cupuliformis, *Edw.* and *Haime* . . . . . Edw. and Haime, 132
„ *decipiens*, M'Coy; *see* Montlivaltia Delabechii.
„ Delabechii, *Edw.* and *Haime* . . . . Edw. and Haime, 132
„ depressa, *Edw.* and *Haime* . . . . Edw. and Haime, 134
„ *dilatata*, M'Coy; *see* Montlivaltia dispar.
„ dispar, *Phillips* . . . . . . Edw. and Haime, 80
„ *gregaria*, M'Coy; *see* Thecosmilia gregaria.
„ Guettardi, *Blainville* . . . . . . Duncan, Pt. IV, 51
„ Haimei, *Chapuis* and *Dewalque* . . . . Duncan, Pt. IV, 35
„ Hibernica, *Duncan* . . . . . . Duncan, Pt. IV, 39
„ Holli, *Duncan* . . . . . . Duncan, Pt. III, 16
„ lens, *Edw.* and *Haime* . . . . . Edw. and Haime, 133
„ *Moreausiaca*; *see* Montlivaltia dispar.
„ Morrisi, *Duncan* . . . . . Duncan, Pt. III, 17
„ mucronata, *Duncan* . . . . . . Duncan, Pt. IV, 59
„ Murchisoniæ, *Duncan* . . . . . . Duncan, Pt. IV, 8
„ nummiformis, *Duncan* . . . . . . Duncan, Pt. IV, 60
„ *obconica*, M'Coy; *see* Montlivaltia dispar.

PAGE

Montlivatia Painswicki, *Duncan* . . . . . Duncan, Pt. iii, 17
„ patula, *Duncan* . . . . . Duncan, Pt. iv, 56
„ parasitica, *Duncan* . . . . Duncan, Pt. iv, 9
„ papillata, *Duncan* . . . . . Duncan, Pt. iv, 36
„ pedunculata, *Duncan* . . . . . Duncan, Pt. iv, 10
„ polymorpha, *Terquem* and *Piette* . . . . Duncan, Pt. iv, 8
„ radiata, *Duncan* . . . . . Duncan, Pt. iv, 61
„ rugosa, *Duncan* and *Wright* . . . . Duncan, Pt. iv, 58
„ Ruperti, *Duncan* . . . . . Duncan, Pt. iv, 46
„ simplex, *Duncan* . . . . . Duncan, Pt. iv, 9
„ Smithi, *Edw.* and *Haime* . . . . . Edw. and Haime, 110
„ Stutchburyi, *Edw.* and *Haime* . . . Edw. and Haime, 131
„ tenuilamellosa, *Edw.* and *Haime* . . . . Edw. and Haime, 130
„ trochoides, *Edw.* and *Haime* . . . . Edw. and Haime, 129
„ Victoriæ, *Duncan* . . . . . Duncan, Pt. iv, 63
„ Walliæ, *Duncan* . . . . Duncan, Pt. iv, 7
„ Waterhousei, *Edw.* and *Haime* . . . Edw. and Haime, 111
„ Wrighti, *Edw.* and *Haime* . . . . Edw. and Haime, 131

OCULINIDÆ, *Edw.* and *Haime* . . Edw. and Haime, pp. xix, 53; Duncan, Pt. ii, 14
ONCHOTROCHUS, *Duncan* . . . . . Duncan, Pt. ii, pp. 4, 20
Onchotrochus Carteri, *Duncan* . . . . . Duncan, Pt. ii, 20
„ serpentinus, *Duncan* . . . . . . Duncan, Pt. ii, 4
OPPELISMILIA, *Duncan* . . . . . Duncan, Pt. iv, 39
Oppelismilia gemmans, *Duncan* . . . . . . Duncan, Pt. iv, 39

PARASMILIA, *Edw.* and *Haime* . . Edw. and Haime, pp. xxiv, 47; Duncan, Pt. 11
Parasmilia centralis, *Mantell* . . . Edw. and Haime, p. 47; Duncan, Pt. ii, 12
„ cylindrica, *Edw.* and *Haime* . . . . Edw. and Haime, 50
„ Fittoni, *Edw.* and *Haime* . . . . . Edw. and Haime, 50
„ granulata, *Duncan* . . . . . . Duncan, Pt. ii, 13
„ *Mantelli*, Edw. and Haime; *see* Parasmilia centralis (Edw. and Haime, p. 49; Duncan, Pt. ii, pp. 12, 46)
„ monilis, *Duncan* . . . . . . Duncan, Pt. ii, 12
„ serpentina, *Edw.* and *Haime* . . . . Edw. and Haime, 51
PARASTREA, *Edw.* and *Haime* . . . . Edw. and Haime, pp. xliii, 59
Parastrea stricta, *Edw.* and *Haime* . . . . Edw. and Haime, 59
PEPLOSMILIA, *Edw.* and *Haime* . . Edw. and Haime, pp. xxv, 57; Duncan, Pt. ii, 29
Peplosmilia Austeni, *Edw.* and *Haime* . . . . Edw. and Haime, 57
„ depressa, *E. de Fromentel* . . . . Duncan, Pt. ii, 29
PLACOSMILIA, *Edw.* and *Haime* . . . . . Duncan, Pt. ii, 27
Placosmilia consobrina, Reuss; *see* Placosmilia Parkinsoni.
„ cuneiformis, *Edw.* and *Haime* . . . . Duncan, Pt. ii. 27
„ magnifica, *Duncan* . . . . . Duncan, Pt. ii, 28
„ Parkinsoni, *Edw.* and *Haime* . . . . Duncan, Pt. ii, 28
PODOSERIS, *Duncan* . . . . Duncan, Pt. ii, p. 25; Pt. iii, 24

PAGE

Podoseris constricta, *Duncan* . . . . . . Duncan, Pt. iii, 24
„ elongata, *Duncan* . . . . . . Duncan, Pt. ii, 26
„ mammiliformis, *Duncan* . . . . . . Duncan, Pt. ii, 25
PORITIDÆ, *Edw.* and *Haime* . . . . . Edw. and Haime, pp. lv, 122
Prionastrea alimena, D'Orbigny ; *see* Isastræa limitata.
„ *explanata*, Edw. and Haime ; *see* Isastræa explanata.
„ *limitata*, Edw. and Haime ; *see* Isastræa limitata.
„ *Luciensis*, D'Orbigny ; *see* Isastræa limitata.
PROTOSERIS, *Edw.* and *Haime* . . . . . . Edw. and Haime, 103
Protoseris Waltoni, *Edw.* and *Haime* . . . . . Edw. and Haime, 103

RHABDOPHYLLIA, *Edw.* and *Haime* . Edw. and Haime, p. 87 ; Duncan, Pt. iv, 17
Rhabdophyllia Phillipis, *Edw.* and *Haime* . . . . Edw. and Haime, 87
„ recondita, *Laube* . . . . . . Duncan, Pt. iv, 17

SEPTASTRÆA, *D'Orbigny* . . . . Duncan, Pt. iv, pp. 5, 32, 37, 52
Septastræa Eveshami, *Duncan* . . . . . . Duncan, Pt. iv, 52
„ excavata, *E. de From* . . . . . . Duncan, Pt. iv, 32
„ Fromenteli, *Terquem* and *Piette* . . . . Duncan, Pt. iv, 37
„ Haimei, *Wright* . . . . . . Duncan, Pt. iv, 5
Siderastræa agariciaformis, M'Coy ; *see* Thamnastrea arachnoides.
„ cadomensis, *M'Coy* . . . . . . Edw. and Haime, 143
„ *explanata*, Blainville ; *see* Isastræa explanata.
„ *incrustata*, M'Coy ; *see* Microsolena excelsa.
„ *Lamourouxi*, M'Coy ; *see* Thamnastrea Lyelli.
„ *meandrinoides*, M'Coy ; *see* Comoseris irradians.
SMILOTROCHUS, *Edw.* and *Haime* . . . . . Duncan, Pt. ii, pp. 18, 35, 39
Smilotrochus angulatus, *Duncan* . . . . . . Duncan, Pt. ii, 20
„ Austeni, *Edw.* and *Haime* . . . . Duncan, Pt. ii, pp. 19, 39
„ cylindricus, *Duncan* . . . . . . Duncan, Pt. ii, 36
„ elongatus, *Duncan* . . . . . Duncan, Pt. ii, pp. 19, 36
„ granulatus, *Duncan* . . . . . . Duncan, Pt. ii, 36
„ insignis, *Duncan* . . . . . . Duncan, Pt. ii, 37
„ tuberosus, *Edw.* and *Haime* . . . . Duncan, Pt. ii, 19
STAURIDÆ, *Edw.* and *Haime* . . . . . Edw. and Haime, pp. lxiv, 70
Stephanocœnia concinna, D'Orbigny ; *see* Thamnastrea concinna.
STEPHANOPHYLLIA, *Michelin* . . . . . Edw. and Haime, pp. liii, 54
Stephanophyllia Bowerbankii, *Edw.* and *Haime* . . . . Edw. and Haime, 54
STYLINA, *Lamarck* . . . . . Edw. and Haime, pp. xxix, 76, 105, 128
Stylina Babeana, D'Orbigny ; *see* Stylina solida.
„ conifera, *Edw.* and *Haime* . . . . . . Edw. and Haime, 105
„ Delabechii, *Edw.* and *Haime* . . . . . . Edw. and Haime, 79
„ *Luciensis*, Edw. and Haime ; *see* Cyathophora Luciensis.
„ Ploti, *Edw.* and *Haime* . . . . . . Edw. and Haime, 106
„ solida, *M'Coy* . . . . . . Edw. and Haime, pp. 105, 128
„ tubulifera, *Edw.* and *Haime* . . . . . . Edw. and Haime, 76
„ *tubulosa*, Michelin ; *see* Stylina tubulifera.
STYLINACEÆ, *Edw.* and *Haime* . . . . . Duncan, Pt. ii, 21

PAGE

Stylopora solida, M'Coy . . . . . . Edw. and Haime, 105
SYMPHYLLIA, *Edw.* and *Haime* . . . . . Duncan, Pt. III, 19
Symphyllia Etheridgei, *Duncan* . . . . . Duncan, Pt. III, 19
Synastrea concinna, Edw. and Haime; *see* Thamnastrea concinna.
„ *Defranciana*, Edw. and Haime; *see* Thamnastrea Defranciana.
SYNHELIA, *Edw.* and *Haime* . . . . Edw. and Haime, pp. xx, 53
Synhelia Sharpeana, *Edw.* and *Haime* . . . . Edw. and Haime, 53

THAMNASTREA, *Edw.* and *Haime* Edw. and Haime, pp. xlii, 97, 118, 139; Duncan, Pt. II, p. 22; Pt. III, pp. 16, 19
Thamnastrea arachnoides, *Parkinson* . . . . . Edw. and Haime, 97
Thamnastrea Browni, *Duncan* . . . . . Duncan, Pt. III, 16
„ concinna, *Goldfuss* . . . . . Edw. and Haime, 100
„ Defranciana, *Michelin* . . . . Edw. and Haime, 139
„ fungiformis, *Edw.* and *Haime* . . . . Edw. and Haime, 141
„ Lyelli, *Edw.* and *Haime* . . . . Edw. and Haime, 118
„ M'Coyi, *Edw.* and *Haime* . . . Edw. and Haime, 141
„ mammosa, *Edw.* and *Haime* . . . Edw. and Haime, 119
„ Manseli, *Duncan* . . . . Duncan, Pt. III, 20
„ Mettensis, *Edw.* and *Haime* . . . Edw. and Haime, 141
„ scita, *Edw.* and *Haime* . . . . . Edw. and Haime, 119
„ superposita, *Michelin* . . . . . Duncan, Pt. II, 22
„ Terquemi, *Edw.* and *Haime* . . . . Edw. and Haime, 140
Walcotti, *Duncan* . . . . . Duncan, Pt. III, 19
„ Waltoni, *Edw.* and *Haime* . . . Edw. and Haime, 120
THECOCYATHUS, *Edw.* and *Haime* . . . Edw. and Haime, pp. xiv, 144
Thecocyathus Moorii, *Edw.* and *Haime* . . . . Edw. and Haime, 144
„ *rugosus*, Wright; *see* Montlivaltia rugosa.
Thecophyllia Arduennensis, D'Orbigny; *see* Montlivaltia dispar.
THECOSMILIA, *Edw.* and *Haime* . Edw. and Haime, pp. xxvi, 84, 135; Duncan, Pt. III, pp. 14, 17; Pt. IV. 11. 65
Thecosmilia affinis, *Duncan* . . . . . . Duncan, Pt. IV, 16
„ annularis, *Fleming* . . . . . Edw. and Haime, 84
„ Brodiei, *Duncan* . . . . . . Duncan, Pt. IV, 13
„ *cylindrica*, Edw. and Haime; *see* Thecosmilia annularis.
„ dentata, *Duncan* . . . . . Duncan, Pt. IV, 16
„ gregaria, *M'Coy* . . Edw. and Haime, p. 135; Duncan, Pt III, 18
„ irregularis, *Duncan* . . . . . Duncan, Pt. IV, 15
„ Martini, *E. de From.* . . . . . Duncan, Pt. IV, 14
„ Michelini, *Terq.* and *Piette* . . . . Duncan, Pt. IV, 14
„ mirabilis, *Duncan* . . . . Duncan, Pt. IV, 12
„ obtusa, *D'Orbigny* . . . . . Duncan, Pt. III, 14
„ plana, *Duncan* . . . . . Duncan, Pt. IV, 17
„ rugosa, *Laube* . . . . . Duncan, Pt. IV, 13
„ serialis, *Duncan* . . . . . Duncan, Pt. IV, 12
„ Suttonensis, *Duncan* . . . . . Duncan Pt. IV, 11
„ Terquemi, *Duncan* . . . . . Duncan, Pt. IV, 16

PAGE

Thecosmilia Wrighti, *Duncan* . . . Duncan, Pt. III, 17
Tremocœnia varians, D'Orbigny; *see* Thamnastrea concinna.
TROCHOCYATHACEÆ, *Edw.* and *Haime* . . . Duncan, Pt. II, 32
TROCHOCYATHUS, *Edw.* and *Haime* Edw. and Haime, pp. xiv, 63, 126, 145; Duncan, Pt. II, p. 32
Trochocyathus conulus, *Phillips* . . . Edw. and Haime, 63
„ Harveyanus, *Edw.* and *Haime* . . Edw. and Haime, p. 65; Duncan, Pt. II, 32
„ *Koenigi*, Mantell; *see* Trochocyathus Harveyanus (Edw. and Haime, p. 66; Duncan, Pt. II, pp. 32, 46)
„ Magnevillianus, *Michelin* . . . . . Edw. and Haime, 126
„ primus, *Edw.* and *Haime* . . . . . Edw. and Haime, 145
„ *Warburtoni*, Edw. and Haime; *see* Trochocyathus Harveyanus (Edw. and Haime, p. 67; Duncan, Pt. II, pp. 32, 46)
Trochocyathus Wiltshirei, *Duncan* . . . . . Duncan, Pt. II, 34
TROCHOSMILIA, *Edw.* and *Haime* . Edw. and Haime, pp. xxiv, 58, 68; Duncan, Pt. II, pp. 5, 8, 41, 50
Trochosmilia (Cœlosmilia) cornucopiæ, *Duncan* . . . . Duncan, Pt. II, 8
„ „ cylindrica, *Duncan* . . . . Duncan, Pt. II, 10
„ „ granulata, *Duncan* . . . . Duncan, Pt. II, 10
„ „ laxa, *Duncan* . . . . Duncan, Pt. II, 8
„ Meyeri, *Duncan* . . . . . Duncan, Pt. II, 41
„ sulcata, *Edw.* and *Haime* . . . . Edw. and Haime, 68
„ tuberosa, *Edw.* and *Haime* . . . Edw. and Haime, 58
„ (Cœlosmilia) Woodwardi, *Duncan* . . . . Duncan, Pt. II, 9
„ „ Wiltshirei, *Duncan* . . . . Duncan, Pt. II, 9
TROCHOSMILIACEÆ, *Edw.* and *Haime* . . . Duncan, Pt. II, pp. 11, 41
TURBINOLIA, *Lamarck* . . . . . . Edw. and Haime, xvi
Turbinolia centralis, Roemer; *see* Parasmilia centralis.
„ *compressa*, Lam.; *see* Trochosmilia tuberosa.
„ *conulus*, Michelin; *see* Trochocyathus conulus.
„ didyma? *Goldfuss* . . . . . . Edw. and Haime, 104
„ *dispar*, Phillips; *see* Montlivaltia dispar.
„ *excavata*, Hag.; *see* Parasmilia centralis.
„ *Konigi*, Mantell; *see* Trochocyathus Konigi.
„ *Magnevilliana*, Michelin; *see* Trochocyathus Magnevillianus.
TURBINOLIACEÆ, *Edw.* and *Haime* . . . . Duncan, Pt. II, pp. 4, 18, 35, 39
TURBINOLIDÆ, *Edw.* and *Haime* Edw. and Haime, pp. xi, 44, 61, 125, 144; Duncan, Pt. II, pp. 2, 4, 18, 31, 35, 39
TURBINOLINÆ, *Edw.* and *Haime* . . Edw. and Haime, p. 2; Duncan, Pt. II, 35
TURBINOSERIS . . . . . Duncan, Pt. II, 42
Turbinoseris De-Fromenteli, *Duncan* . Duncan, Pt. II. 43

ZAPHRENTIS, *Rafinesque* and *Clifford* . . . . Edw. and Haime, pp. lxv, 143
Zaphrentis Waltoni, *Edw.* and *Haime* . . . Edw. and Haime, 143
ZOANTHARIA, *Gray* . . . . Edw. and Haime, pp. ix, 44

PLATE I.

TO ILLUSTRATE THE STRUCTURE OF CORALS.

(See *Introduction.*)

FIG.

1. The calice of *Bathycyathus Sowerbyi*[1] (after Milne-Edwards and Jules Haime). The projection of the *costæ* externally and of the *septa* internally is shown; the existence of a *wall* between the junctions of the septa and costæ is evident. There is no *columella.*

2. The *costæ* running down the outside of the corallum of *Trochosmilia tuberosa* (after Milne-Edwards and Jules Haime).

3. A section of a corallite of *Lophohelia anthophyllites*, Ellis, showing the dense *wall*, with the projection inwards of the *septa*. There are no costæ. From nature, magnified.

4. A corallite of *Cœnocyathus Adamsi*, Duncan,[2] showing the *base*, the *body*, and the *calicular termination.* The base is rough, and was formerly strongly attached to a foreign substance; the body has a few aborted buds on it, and the upper extremity shows faint costæ terminating in septa.

5. A longitudinal section of *Sphenotrochus intermedius*[3] (after Milne-Edwards and Jules Haime). The central styliform process is the *columella;* it arises from the base internally, and is joined to the septa by lateral processes. It is an "*essential*" columella. The septa are shown as broad plates, granulated and arched; they are attached externally to the wall. Outside the faint shading of the wall is the slight projection of one of the costæ. This corallite is open from the *calicular margin* to the *base.*

6. The calice of *Placotrochus costatus,* Duncan.[4] The upper and free surface of a long *columella* is shown, also the same structures as in fig. 1. Magnified.

7. The external surface of the same coral, showing the irregular *calicular margin,* the strong costæ, and the delicate *peduncle* of the base.

8. Part of a calice of *Placocyathus Moorei,* Duncan,[5] showing the costæ, septa, and part of a long columella, as in fig. 6; but there are *pali* on the ends of four of the septa. Magnified.

9. The calice of *Trochocyathus obesus*[6] (after Milne-Edwards and Jules Haime), magnified. The larger septa are separated by three smaller, of which the middle one is the longest. There are twelve large septa, and every other one is a primary septum. The pali are before the primary, the secondary, and the tertiary septa. There are four cycles of septa.

10. The calice of *Discocyathus Eudesii*[7] (after Milne-Edwards and Jules Haime), magnified. The columella is lamellar, and the large pali are before the antepenultimate cycle (or the third). There are five cycles.

11. Two corallites of *Heliastræa endothecata,*[8] Duncan, magnified. The costæ seem to be united by transverse exothecal dissepiments, and the tooth of a small costa projects in the space formed by the dissepiments and the costæ. Some cœnenchyma exists between the corallites.

12. A longitudinal section of *Conosmilia anomala,*[9] Duncan, magnified. The twisted processes forming the essential columella are seen, and one side of the lamina of a septum. This is granular, and is marked by a broken ridge, which once was continued to the next septum as a dissepiment. The wall is seen externally.

13. A section of a corallite of *Calamophyllia Stokesi*[10] (after Milne-Edwards and Jules Haime), magnified. The formation of a rudimentary columella is shown, and the sections of oblique dissepiments between the septa and crossing the interseptal loculi are seen.

14. A longitudinal section of the upper part of a corallum of *Caryophyllia cyathus*[11] (after Milne-Edwards and Jules Haime), magnified. The wall is the external and structureless part, and it has no costæ projecting from it. The lateral view of the septa shows them to be granular, arched above, and slightly exsert. The pali are attached to the inner margin of the septa and to the outer part of the columella, which is formed by many twisted processes. A line drawn from the top of opposite septa forms the upper limit of the calicular fossa, and whose base is the top of the columella centrally, and the top of the pali. There are no dissepiments.

15. A longitudinal section of part of the corallum of *Antillia Lonsdaleia,*[12] Duncan, magnified. The thin wall gives off internally many dissepiments, which are joined by their side to the septum. Externally, it is in contact with a few oblique exothecal dissepiments. The granulated structure crossed by the exotheca, and external to the wall, is a costa, and is seen to emerge into a septum superiorly. The septum is very exsert, is bilobate, dentate, and is marked by radiating ornamental ridges. The columella is dense. The endotheca is vesicular.

16. A corallum of the genus *Montlivaltia,* showing the epitheca with circular rings.

17. A diagram of the relation of the hard and soft parts of a coral. The parts shaded are the wall, the part of the sclerenchyma below the newest dissepiment, and the columella. All the rest is in contact with soft tissues. The mouth and tentacles are shown.

18. A diagram of the hard parts of a coral. The living tissues only cover the portion above the topmost exothecal and endothecal dissepiments. The base is pedunculate, and embraces a foreign substance; the columella springs from the inside of the base, and is in contact laterally with the pali. The septa, wall, costæ, endothecal and exothecal dissepiments, are shown, and the trace of an epitheca quite externally and inferiorly also.

19. Corallites of a *Sarcinula*[13] (after Milne-Edwards and Jules Haime), united by peritheca; the costæ are rudimentary.

---

[1] 'Brit. Foss. Corals,' tab. ii.
[2] "Corals of Maltese Miocene," 'Ann. Mag. Nat. Hist.,' s. 3, vol. xv, pl. xi.
[3] 'Brit. Foss. Corals,' tab. i, fig. 5.
[4] Duncan and Wall, Jamaica, 'Quart. Jour. Geol. Soc.,' Feb , 1865.
[5] Duncan and Wall, op. cit.
[6] 'Ann. des Sciences Nat.,' 3me série, "Zool.," tom. ix, pl. x, fig. 2.
[7] Ibid., pl. ix, fig. 7.
[8] Duncan, "Foss. Corals of West Indies," 'Quart. Jour. Geol. Soc.,' Nov., 1863, vol. xix, p. xv.
[9] Duncan, "Foss. Corals of Australian Tertiaries," 'Ann. Mag. Nat. Hist.,' Sept., 1865.
[10] 'Brit. Foss. Corals,' part iii.
[11] 'Ann. des Sci. Nat.,' ut supra, tom. ix, p. 85.
[12] 'Foss. Corals of West Indies,' part ii, pl. iii.
[13] 'Ann. des Sci. Nat.,' ut supra, tom. x, pl. vi.

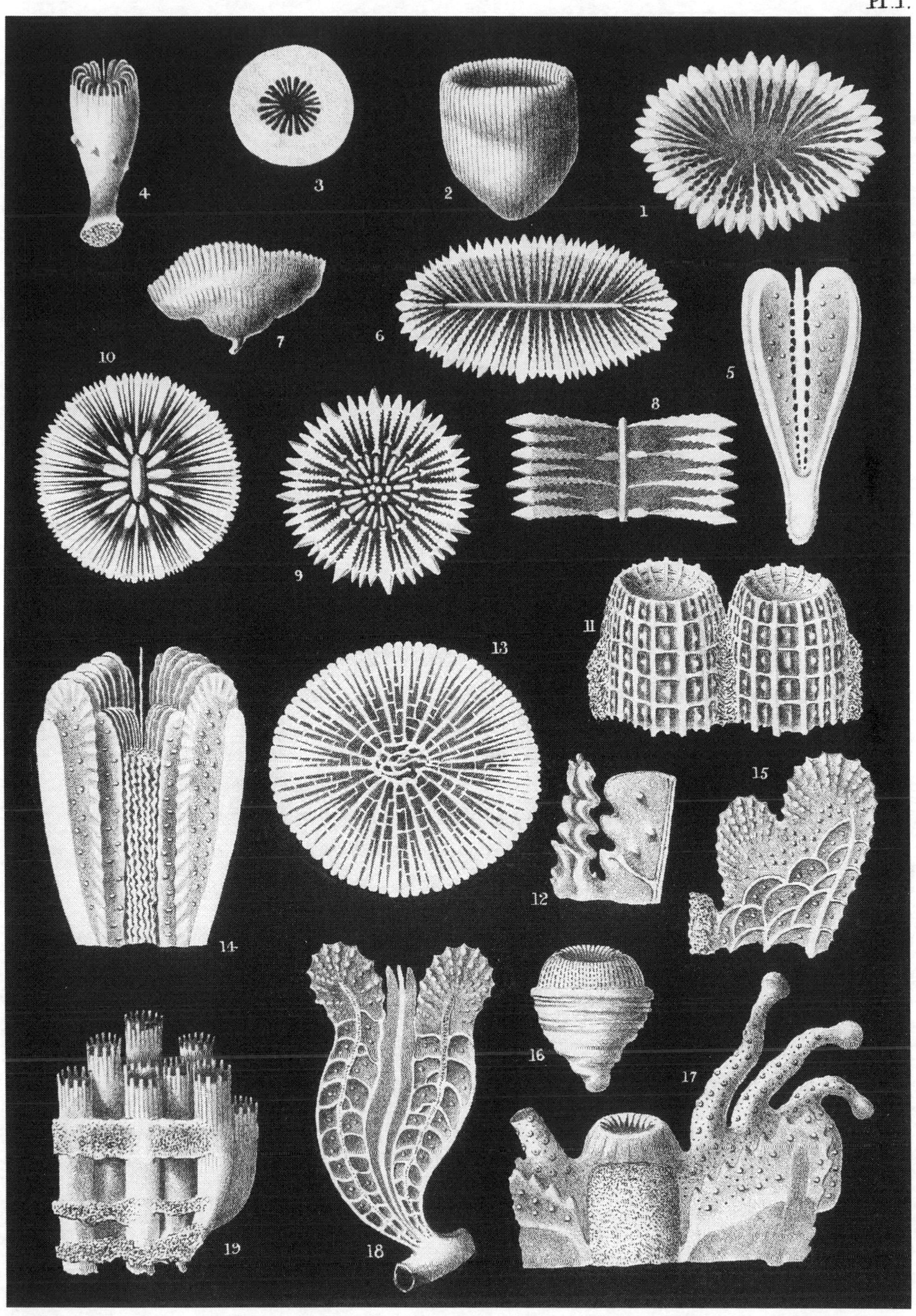

De Wilde lith.

M&N.Hanhart imp.

CORALS

PLATE II.

TO ILLUSTRATE THE STRUCTURE OF CORALS.

FIG.

1 to 8, and fig. 19. The soft parts of *Cladocora cæspitosa*[1] (after Jules Haime). Fig. 4. The tentacules, tentacular disc, mouth, and radiating lines on the lips. Fig. 3. Magnified view of a section of part of a tentacule; the arrangement and nature of the nematocysts and of the large transparent vesicles of the verrucose prominences are shown; the structure of the internal layer, with its colour-bearing cells, is also shown. Fig. 1. A portion of the terminal swelling of a tentacule; the two kinds of nematocysts are very well seen. Figs. 5, 6, 7, 8. Nematocysts of the terminal swelling. Fig. 2. The tubular processes attached to mesenteric folds; they are covered with cilia, and contain nematocysts. Fig. 19. A portion of a tentacule, magnified, showing the terminal swelling and the verrucose swellings.

9, 11, 12, 15, 18, 20.[2] The soft tissues of *Caryophyllia clavus* (*borealis*). Fig. 12. The polype attached to a Ditrupa by a fine peduncle; the costæ are seen to be covered with a transparent tissue, which gives them a rounded outline; the tentacules overlap the calicular margin, and are fully expanded (slightly magnified). Fig. 9. The tentacules of various orders fully expanded, the central mouth, the lips, and the disc immediately around them, with the radiating lines, are shown. The hard parts of the calice are completely covered and hidden. Fig. 11. A magnified view of the tentacular disc, the tentacules not being fully expanded. The septa are seen, but are covered with soft tissue. The mouth, lips, and disc, with the radiating lines, are shown.

15. The top of a tentacule, magnified, showing scutiform processes analogous to the verrucose projections of *Cladocora*. Figs. 18 and 20. The same processes, highly magnified.

10. The tentacular discs[3] of the corallites of *Heliastræa cavernosa*, magnified. The mouth is projected on a truncated process, and the tentacular development is small.

13. *Lithophyllia Cubensis*,[4] in the living state. The costæ are quite hidden by the soft parts, and the large disc, with its central mouth and radiating lines, is seen. The base is very broad.

14. *Colpophyllia gyrosa*,[5] from a living specimen. The three mouths to a part of a serial calice.

16. *Manicina areolata*,[6] showing the relation of the tentacules to the mouths in the serial calice.

17. A coral of the same species,[7] with the prehensile cirrhi fully expanded. The tentacules are small, and there are two mouths to the serial calice.

---

[1] 'Hist. Nat. des Corall.,' vol. ii, plate, A, IV.

[2] These beautiful illustrations were drawn for me, from nature, by Mr. Peach, who also gave me his notes on the anatomy of the *C. borealis*, Fleming.

[3] [4] [5] [6] [7] The figures are after Michelotti et Duchassaing, op. cit., pl. V.

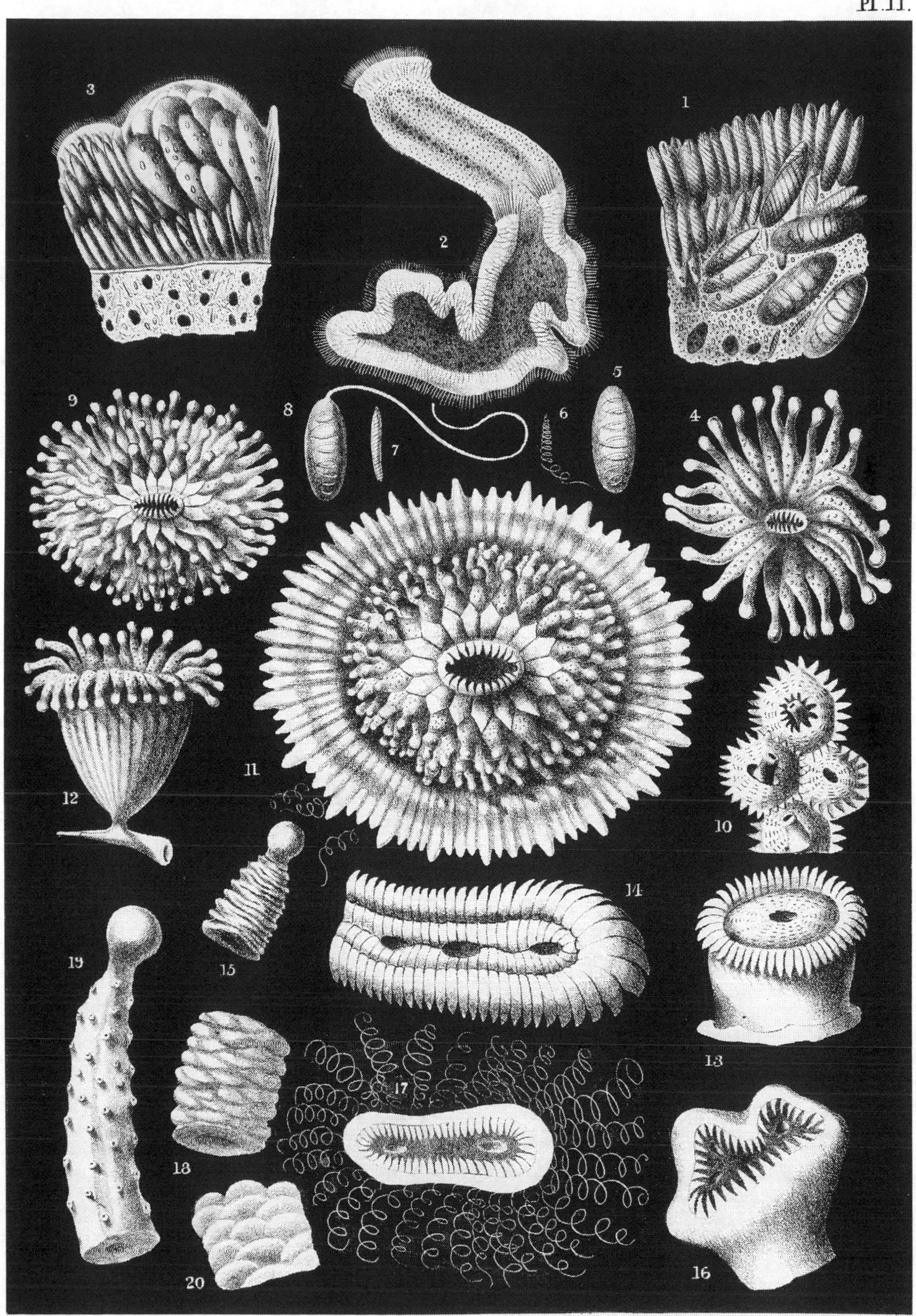

De Wilde lith.

CORALS

M&N Hanhart imp

PLATE III.

TO ILLUSTRATE THE STRUCTURE OF CORALS.

FIG.

1, 2, and 7. These illustrate the nature of synapticulæ, from species of *Micrabacia* and *Mycedium*. The cross bars are not in the nature of dissepiments, and must not be considered to be the upper surfaces of very oblique or nearly vertical dissepiments.

3. Corallites of *Alveopora dædalæa*,[1] showing the regular perforations in the wall, constituting the species a "perforate" or porose coral.

4. The perforate septa and walls of *Litharæa Websteri*. Compare these cribriform septa with those of *Sphenotrochus intermedius* in Plate I, fig. 5.

5. The wall, septa, and false columella of *Alveopora fenestrata*.[2]

6. The structure of the septa of the same coral seen longitudinally. They consist of a simple series of projections, and do not form a continuous plate or lamina.

8. The calice of *Pocillopora crassoramosa*,[3] showing a horizontal dissepiment (a tabula) closing the calice below. It is marked by faint septa near the calicular margin; the cœnenchyma external to the calice is very dense and granular.

9. A diagram of a longitudinal section of the same species. The tabulæ with arched superior surfaces and the dense cœnenchyma with its granules are shown.

10. The tubuliform structures marked across by lines are corallites of *Heliolites Murchisoni*; the tabulæ represented by the lines are close; the wall of the corallites is very slender, and there is much cellular cœnenchyma between the corallites.

11. A longitudinal section of a tabulate coral, a *Favosites*. There is no cœnenchyma, but the walls are fused.

12. Calices of *Heliolites interstincta*, magnified. The cœnenchyma is cellular.

13. Calices of an *Alveolites*.

14. Perforate walls of a *Favosites*.

15. A calice of *Stauria astrææformis*,[4] with three calicular buds. The quadriseptate arrangement is very evident. Magnified.

16. Longitudinal section of a corallite of the same species.[5] The dense walls, the endotheca forming cellular dissepiments externally and horizontal tabulæ internally, and the septa, are shown. Magnified.

17. The calice of *Anisophyllum Agassizi*,[6] magnified, showing three large septa.

18. The calice of *Cyathaxonia cornu*,[7] magnified.

19. The calice of *Aulacophyllum mitratum*,[8] magnified.

20. The calice of *Ptychophyllum expansum*.[9]

---

[1] 'Foss. Corals of West Indies,' pl. xiv.

[2] 'Ann. des Sc. Nat.,' t. ix, pl. v.

[3] 'Fossil Corals of West Indies,' pt. ii, pl. 5.

[4–9] Selected from 'Polyp. Foss. des Terr. Paléo.,' MM. Milne-Edwards et Jules Haime; they are intended to illustrate the Introduction which will appear when the palæozoic species are described.

Pl. III.

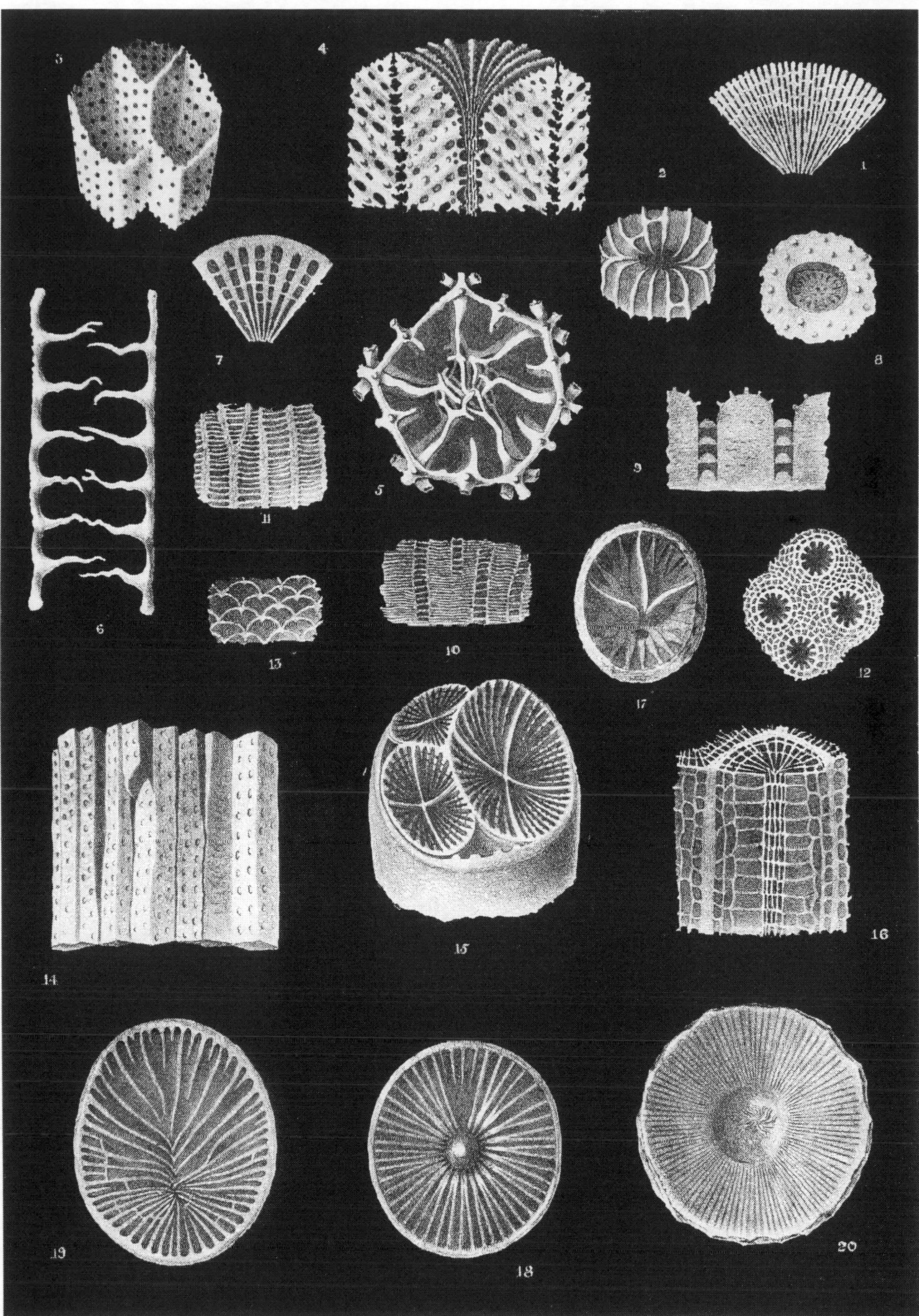

De Wilde lith

M&N Hanhart imp

CORALS

PLATE IV.

TO ILLUSTRATE THE STRUCTURE OF CORALS.

FIG.

1. Magnified view of part of a transverse section of the corallum of *Antillia Walli.*[1] The upright plates are septa, and the lowest structure at right angles to the septa, and which has its lower margin somewhat wavy, is part of the epitheca. The structure parallel with the epitheca, and separated from it by the short costæ and intercostal spaces, is the true wall. Higher up are two transverse dense layers of sclerenchyma; they spread from septum to septum across the interseptal loculi and simulate secondary walls. They are highly developed masses of dissepiments, whose intercellular spaces have been filled up with carbonate of lime.

2. A longitudinal section of part of a corallite of *Lonsdaleia Bronni,*[2] magnified. The columella has been removed. The tabulæ are seen stretching across, but not interfering with the growth of the septa; externally, the vesicular endotheca partly produces a false wall. The dense wall is shown.

3 and 4. Examples of inner and outer walls in Rugose corals.

5. The septa and the cut edges of oblique dissepiments in a large species of *Zaphrentis,* from nature.

6. Part of a corallite of *Zaphrentis gigantea,*[3] showing the granular epitheca, the slight true wall, the septa, and the interseptal loculi, with dissepiments.

7. Calices and cœnenchyma of *Lyellia Americana,*[4] magnified.

8. Calicinal gemmation in a *Caryophyllia;* it is fatal to the parent, and is accidental. From nature.

9 and 10. Calicinal gemmation in a *Cyathophyllum.* The normal and the budding corallites are shown.

11. Calicinal gemmation close to the margin, in the genus *Isastræa,* magnified.

12. Fissiparous division of calices in *Dichocœnia.*

13. Fissiparous division of calices in *Leptastræa Roissyana,*[5] magnified.

14. A serial calice of the genus *Thysanus.*

15. Calices (serial) of a *Mæandrina.*

16. An example of extracalicular gemmation, from nature.

17. A corallum of *Oculina Halensis.* The centre is occupied by the parent stem, and the buds radiate from it.[6]

18. A section of a branch of a species of *Madrepora,* magnified. The parent corallite occupies the centre, and the younger arise from it more or less at right angles. The peculiar septal arrangement of the genus and the porose condition of the sclerenchyma are shown. From nature.

---

[1] Duncan and Wall, op. cit., pl. ii.

[2] [3] [4] From 'Polypiers Fossiles des Terr. Pal.,' MM. Milne-Edwards et Jules Haime.

[5] 'Ann. des. Sc. Nat.,' t. x, pl. ix.

[6] "Foss. Corals from Sinde," 'Ann. and Mag. Nat. Hist.,' &c., April, 1864.

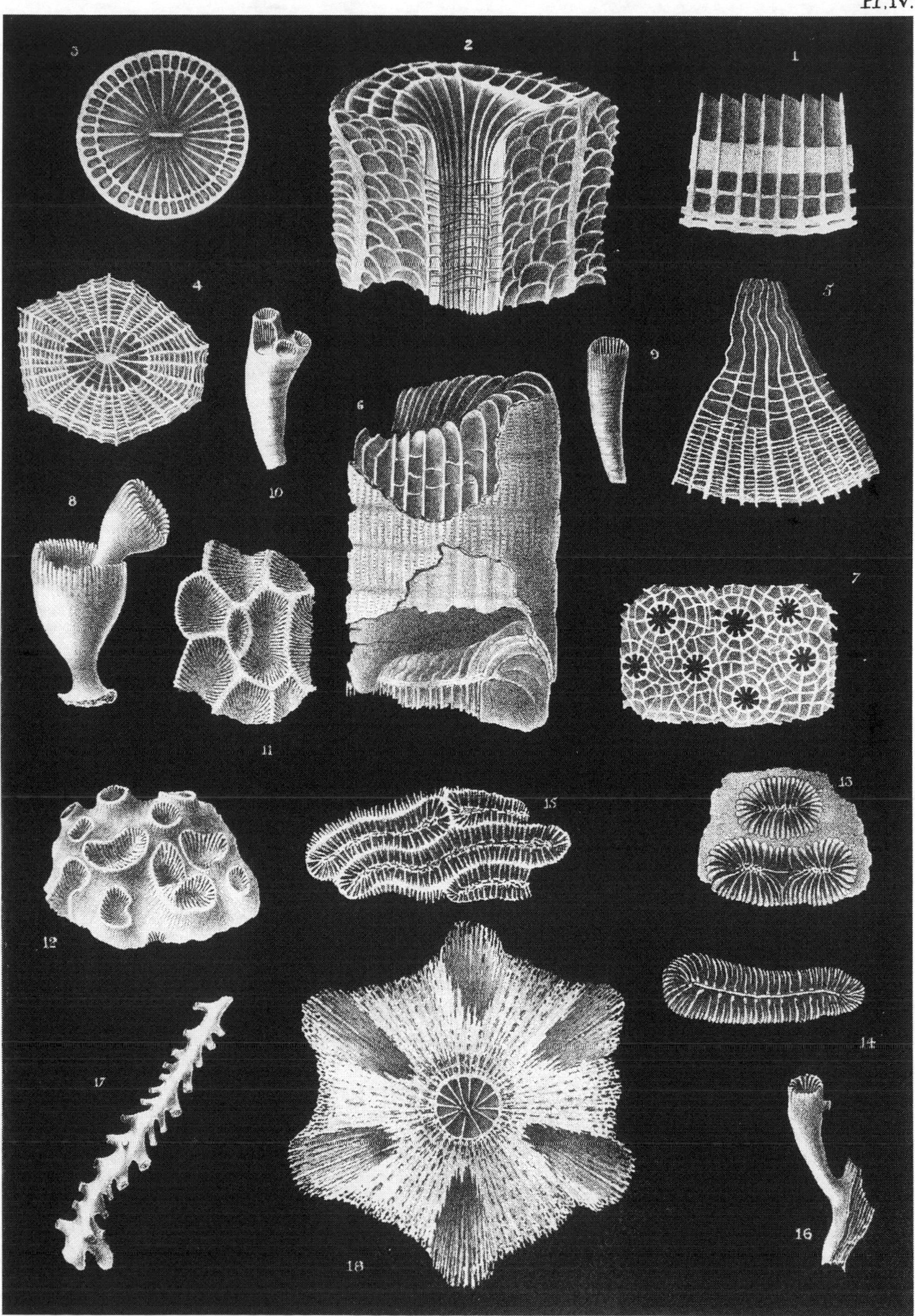

De Wilde lith

M&N Hanhart imp

CORALS

PLATE V.

CORALS FROM BROCKENHURST.

Solenastræa cellulosa, Duncan. (P. 41.)

Fig.

1. View of the upper surface of the corallum.
2. Lateral view, showing the cellular exotheca.
3. One system, with its four cycles; the abundant endotheca is shown, also the rudimentary costæ. Much magnified.
4. A transverse section of a corallite close to the calice, magnified.
5. A lateral view of a corallite covered with exotheca, magnified.
6. Part of a corallite above the exotheca, and showing the costæ; magnified.
7. View of the upper surface, magnified.

Solenastræa Koeneni, Duncan. (P. 42.)

8. The corallum.
9. A calice, highly magnified.

Solenastræa Reussi, Duncan. (P. 43.)

10. The corallum, showing the banded exotheca.
11. Costæ; there is exotheca above and below them.
12. Upper surface of corallum, highly magnified, showing the granular surface of the upper layer of the exotheca, the banded structure of part of the exotheca, and the calices.
13. Upper surface of corallum, worn.
14. Exotheca, cellular and banded.
15. Side view of one of the septa, magnified.
16. One system of septa, showing five cycles.

Pl. V.

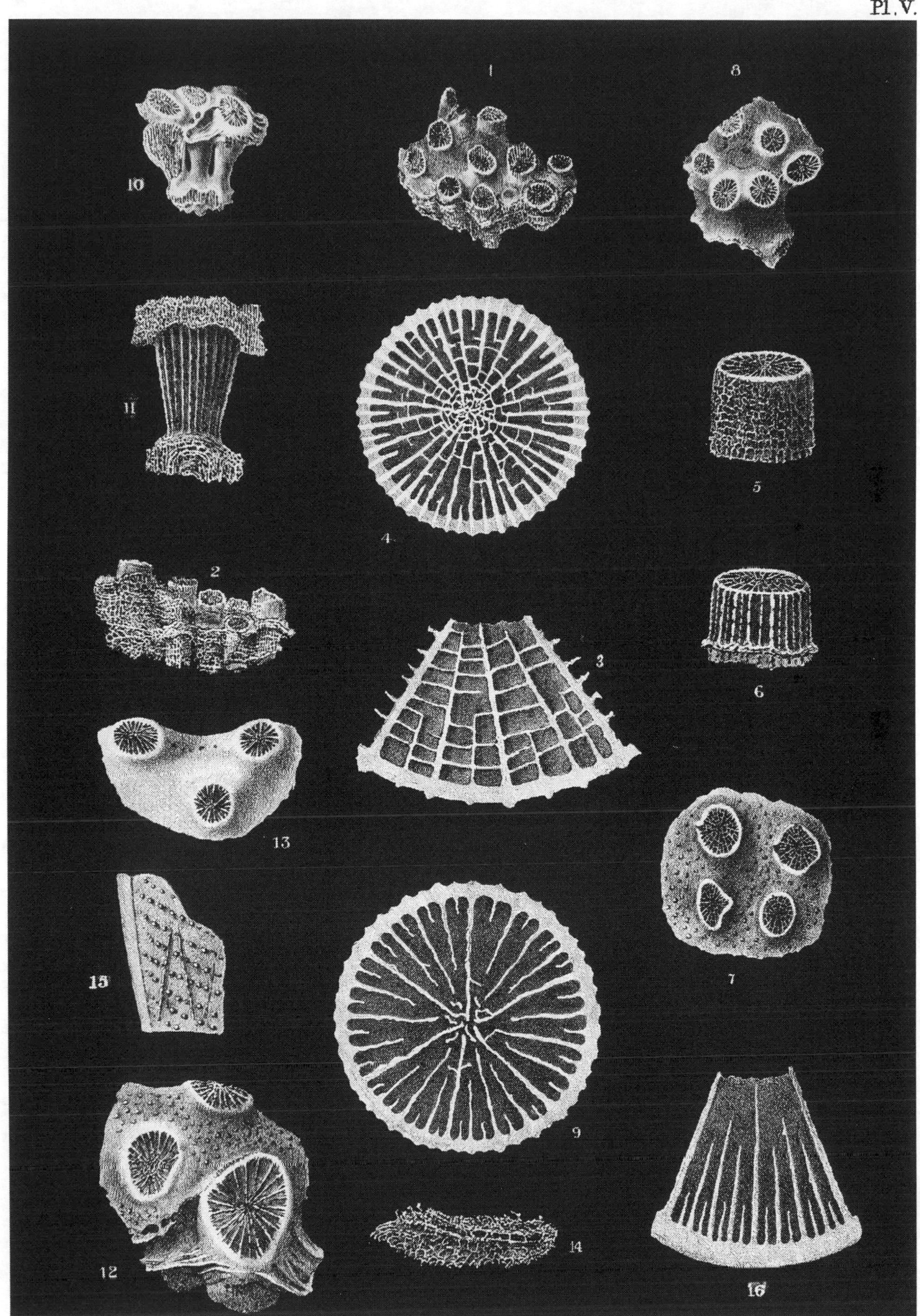

De Wilde lith.

M & N Hanhart imp.

TERTIARY CORALS

PLATE VI.

CORALS FROM BROCKENHURST.

Solenastræa gemmans, Duncan. (P. 44.)

FIG.

1. The corallum, a side view.
2. Corallites showing the method of gemmation, slightly magnified.
3. Costæ, magnified.
4. Two corallites, united by exotheca, magnified.
5. A view a little below the calice, magnified.
6. One of the septa; the lateral processes join endothecal dissepiments.
7. Granular and endothecal markings on the side of one of the septa, magnified.

Solenastræa Beyrichi, Duncan. (P. 44.)

8. The corallum, its upper surface.
9. Lateral view of corallites and exotheca, slightly magnified.
10. One of the septa, showing the thick wall and inclined endotheca, magnified.
11. Costæ, thick wall, and septa, magnified.
12. Transverse section, close to a calice, magnified.
13. A deformed calice, magnified.

Solenastræa granulata, Duncan. (P. 45.)

14. Upper surface of a worn corallum.
15. Cellular and banded exotheca uniting corallites, magnified.
16. Corallite wall without exotheca; exotheca in cells and bands; the costæ are also shown. Magnified.
17. Transverse section of a corallite, magnified.
18. The septa at the calicular margin, showing the paliform lobe, magnified.

Pl. VI.

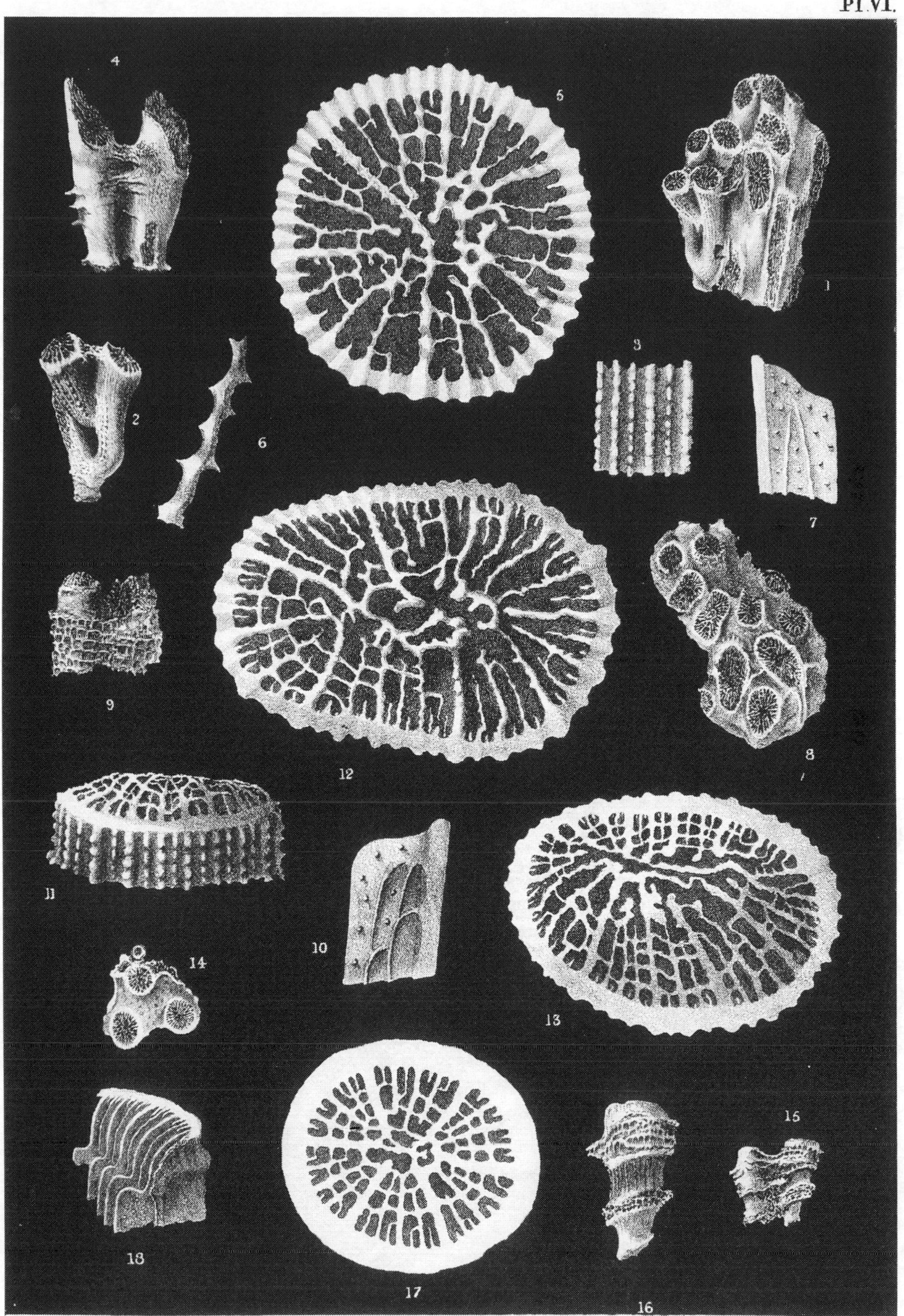

De Wilde lith. M & N Hanhart imp.

TERTIARY CORALS

PLATE VII.

CORALS FORM BROCKENHURST.

Balanophyllia granulata, Duncan. (P. 47.)

Fig.

1. The corallum fixed to a shell.
2. General view of the costæ, magnified.
3. Larger or inferior end of the costæ, magnified.
4. Costæ higher up, magnified, to show their granules.
5. The rough and elevated granular surface of the smaller costæ, magnified.

Lobopsammia cariosa, Goldf., sp. (P. 48.)

6. Lateral view of a corallum.
7. Costæ, magnified.
8. A corallum with fissiparous calices.
9. A fissiparous calice, magnified.
10. The base of a corallum.

Axopora Michelini, Duncan. (P. 50.)

11. Corallum.
12. Magnified view of calices, with the columella and cœnenchyma.
13. Magnified view of corallites in longitudinal section.
14. Columella, tabulæ, and cœnenchyma, highly magnified.
15. Longitudinal view of corallites in longitudinal section, magnified. (Figs. 13 and 15 are from a variety.)

Litharæa Brockenhursti, Duncan. (P. 49.)

16. Calices, magnified.
17. A calice, highly magnified.

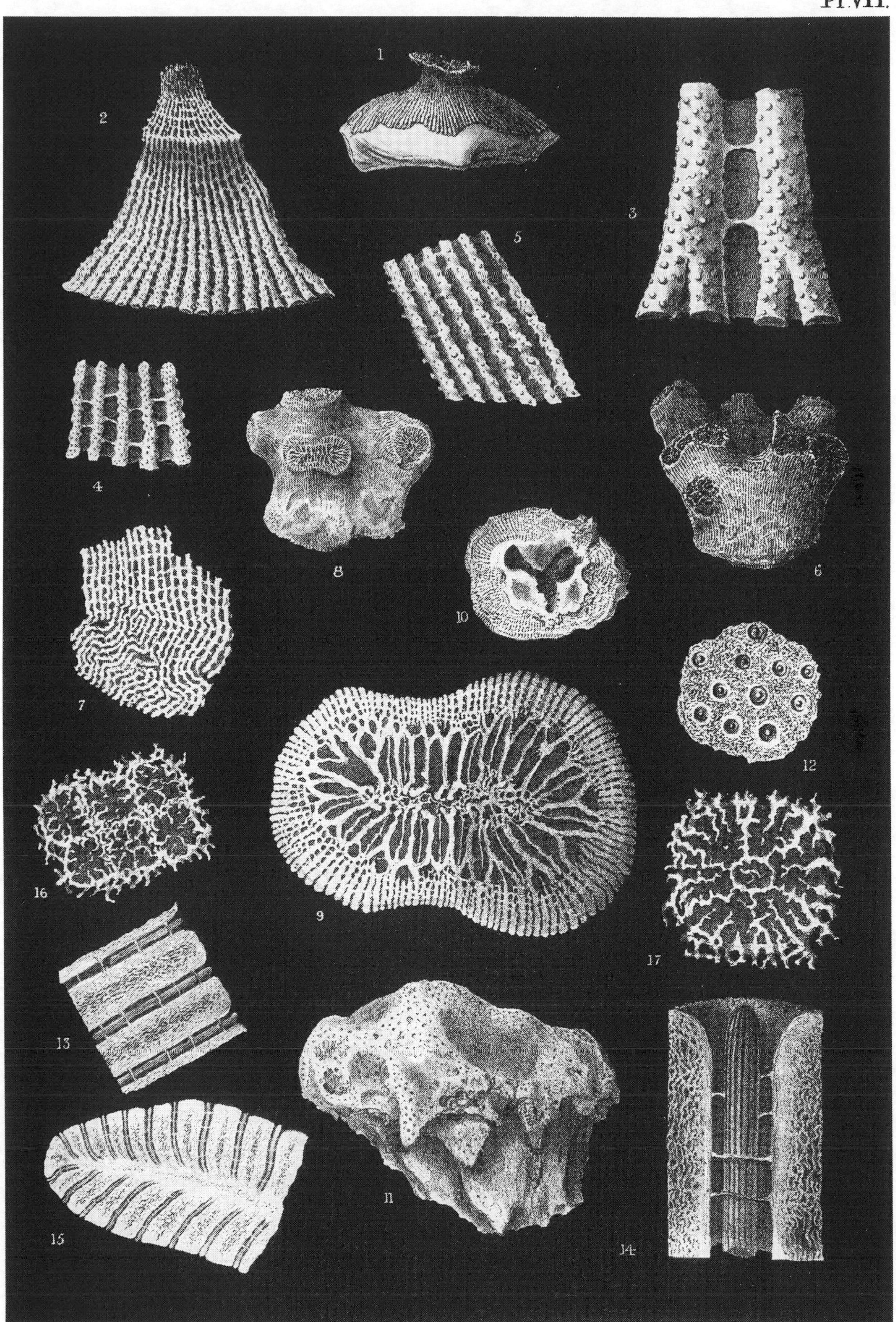

De Wilde lith

M. & N. Hanhart imp

TERTIARY CORALS

PLATE VIII.

CORALS FROM BROCKENHURST.

Madrepora Anglica, Duncan. (P. 51.)

Fig.

1. The corallum.
2. Group of calices from the end of an aborted branch; the union of opposite primary septa is well seen. Magnified.
3. A diagram of the septal arrangement, the wall, and the faint costæ.
4. Slightly projecting calices, separated by much papillate cœnenchyma, magnified.
5. One of the calices, magnified, showing the papillate cœnenchyma also.
6. Longitudinal section of two corallites and the intervening cœnenchymal cells; the papillæ on the surface are shown. Magnified.
7. Magnified view of a projecting tubuliform calice, with costæ ending inferiorly in the cœnenchymal papillæ.

Madrepora Roemeri, Duncan. (P. 51.)

8. The coalesced branches of part of the corallum.
9. Diagram of the septal arrangement.
10. A tubuliform calice, with projecting costæ, magnified.
11. A branch (worn), magnified.

Madrepora Solanderi, Defrance. (P. 51.)

12. Part of a corallum.
13. Group of calices, magnified.
14. Group of calices and surrounding granular cœnenchyma, magnified.

Pl. VIII.

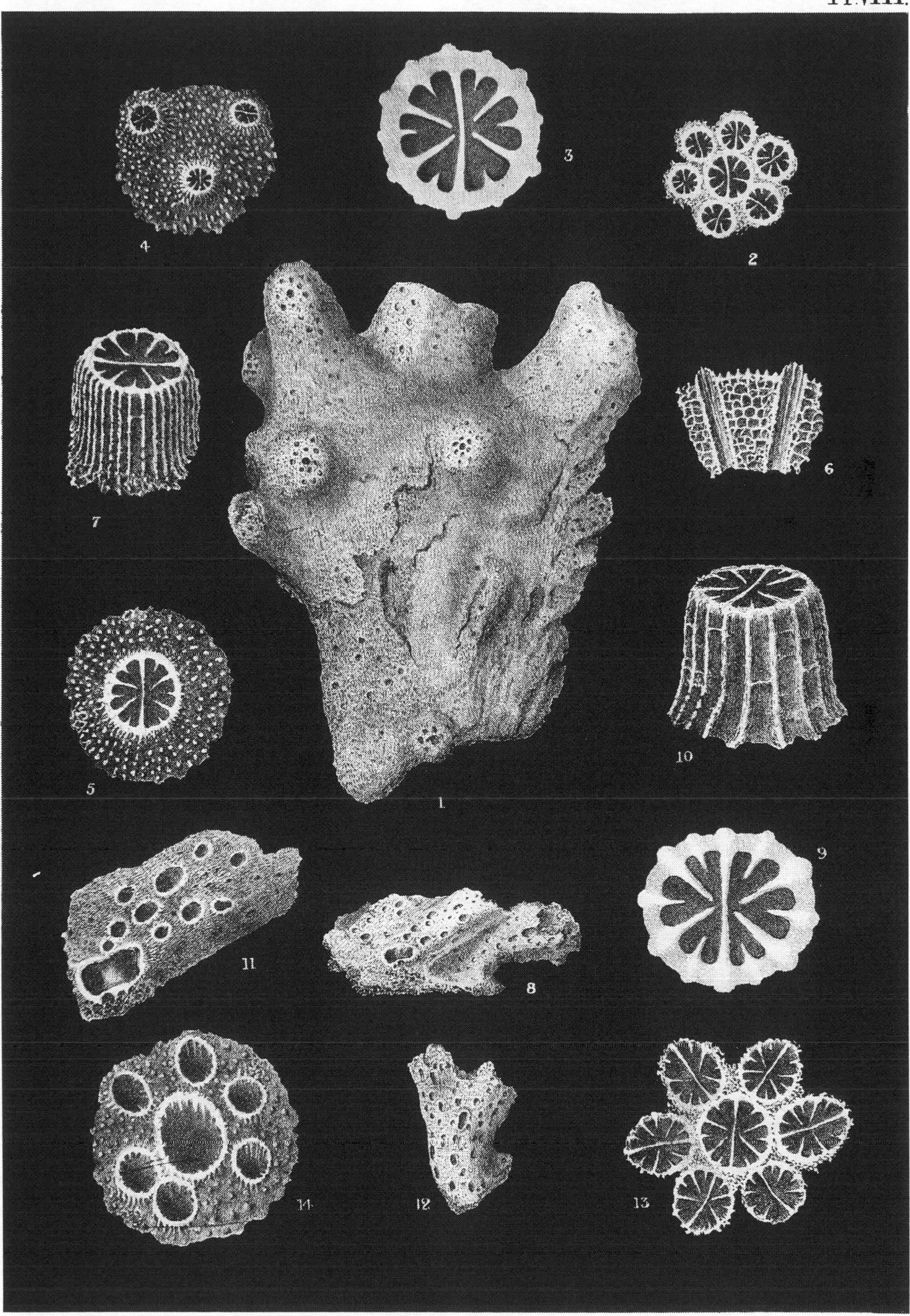

De Wilde lith.

M & N. Hanhart imp.

TERTIARY CORALS

PLATE IX.

CORALS FROM HIGH-CLIFF, BROOK, BRACKLESHAM, BRAMSHAW, AND BARTON.

Turbinolia affinis, Duncan. (P. 54.)

FIG.

1. Corallum, natural size.
2. Corallum, highly magnified.
3. Calice, magnified.

The costal markings are shown in fig. 2.

Turbinolia exarata, Duncan. (P. 55.)

4. Corallum, natural size.
5. The same, highly magnified.
6. The calice, highly magnified, showing the projecting costæ, the thin wall, and the small columella.
7. The rudimentary exotheca on the side of one of the costæ, and its attachment to the thin wall. The portion of the wall is at the bottom of an intercostal space. Magnified.

Turbinolia Forbesi, Duncan. (P. 55.)

8. Corallum, natural size.
9. The same, magnified; the rudimentary costæ are seen between those well developed, close to the calicular margin.
10. Calice, magnified, showing the irregular septal arrangement, the rudimentary costæ, and the angular shape of the columella.
11. A part of a calice, highly magnified, to show the rudimentary and the perfect costæ; the rudimentary costæ are sharp, and have no septa.

Paracyathus Haimei, Duncan. (P. 59.)

12. Corallum, natural size.
13. The calice, magnified. It is worn.
14. Costæ, magnified. The exotheca is shown.

Trochocyathus Austeni, Duncan. (P. 57.)

15. Corallum.
16. Calice, magnified.
17. One of the septa joined to a costa, showing the spinules; magnified.

Paracyathus cylindricus, Duncan. (P. 58.)

18. Corallum, natural size; adult.
19. Young corallum.
20. Calice, magnified.
21. Side view of a magnified septum, showing the large granules.

Oculina incrustans, Duncan. (P. 60.)

22. Part of a corallum, slightly magnified.
23. Part of a corallum, showing the faint costal striæ and the absence of granules; slightly magnified.
24. Calice, much magnified.

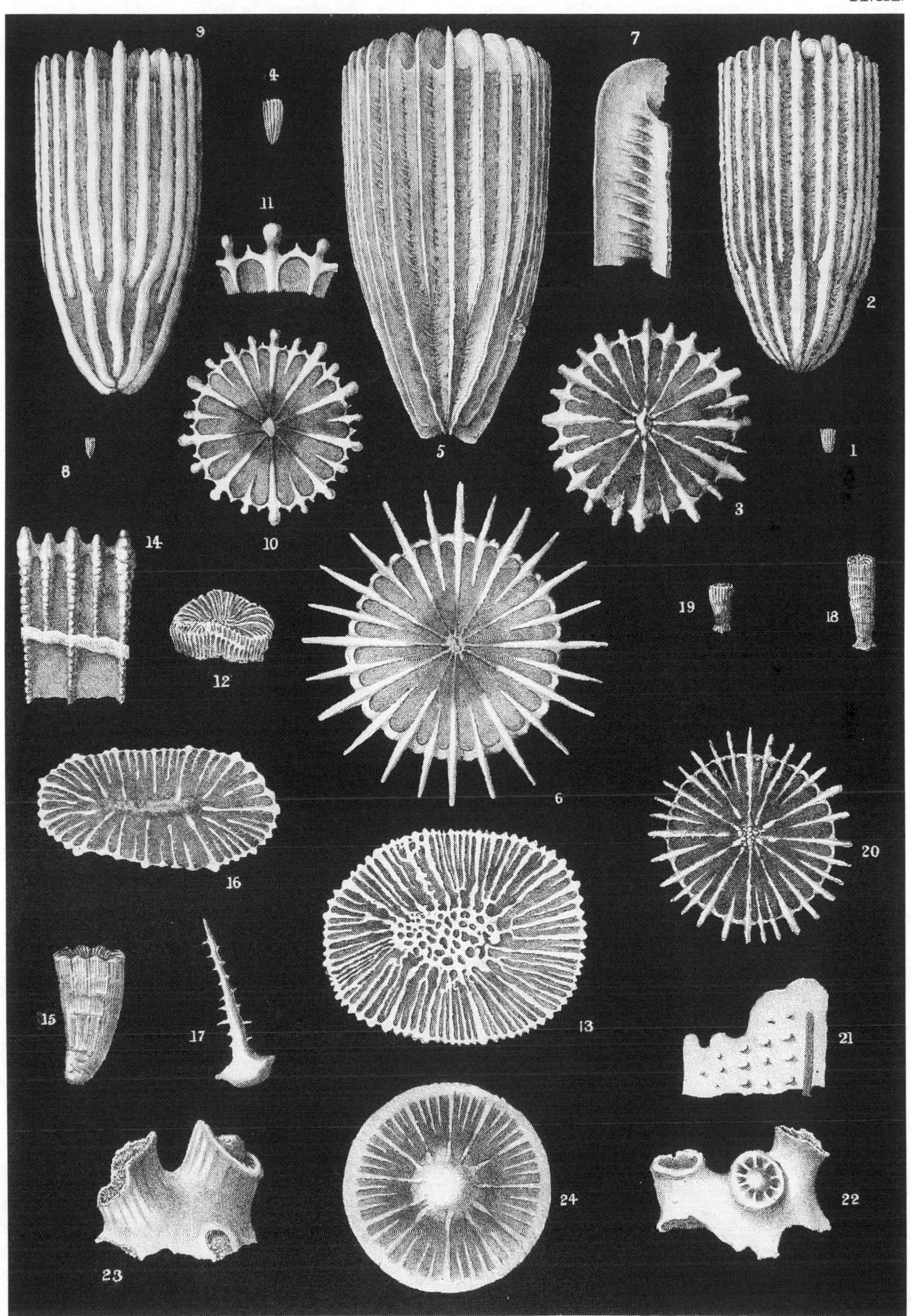

De Wilde lith.

M & N Hanhart imp.

TERTIARY CORALS

PLATE X.

CORALS FROM BRACKLESHAM, WHETSTONE, AND FINCHLEY.

Trochocyathus insignis, Duncan. (P. 57.)

FIG.

1. Corallum.
2. The calice, magnified.
3. One of the septa, magnified, to show the lateral spinules and the wavy shape.
4. Costæ, magnified (at the calicular margin).

Oculina Wetherelli, Duncan. (P. 60.)

5.\
6.} Corallites, showing the broad base.

7. The calice, magnified.

Porites panicea, Lonsdale. (P. 63.)

8. Corallum.
9. Calices and intercalicular tissue, magnified, showing the columella, pali, and granules.
10. The inter-corallite tissue, magnified.

Dendracis Lonsdalei, Duncan. (P. 62.)

11. Corallum, natural size.
12. Calice, highly magnified, showing the granules around the calice.
13. Transverse section of a branch of the corallum, showing its reticulate appearance.
14. Intercalicular or cœnenchymal granules, highly magnified.

Dendrophyllia elegans, Duncan. (P. 61.)

15. Corallum.
16. A calice, highly magnified.
17. The method of gemmation.
18. Costæ, near the calices, magnified.
19. Costæ, near the base, magnified.

Axopora Fisheri, Duncan. (P. 64.)

20. Corallum.
21. Calices and intercalicular tissue, magnified.
22. A calice, columella, and cœnenchyma, highly magnified.

De Wilde lith

M&N Hanhart imp

TERTIARY CORALS

PLATE I.

CORALS FROM THE CHALK.

Fig.

1. The corallum of *Caryophyllia Lonsdalei*, Duncan. (P. 3.)
2. The calice and columella, magnified.
3. The costæ, magnified.
4. The corallum of *Caryophyllia Tennanti*, Duncan. (P. 4.)
5. The calice, magnified.
6. The costæ, magnified.
7. *Caryophyllia cylindracea*, Reuss, sp. (P. 3.) On a Belemnite.

8., 9., 10. Unusual shapes of this species.

11. A calice, magnified, showing the small pali noticed in many specimens.
12. A septum, its dentation, and a portion of one of the pali, magnified.

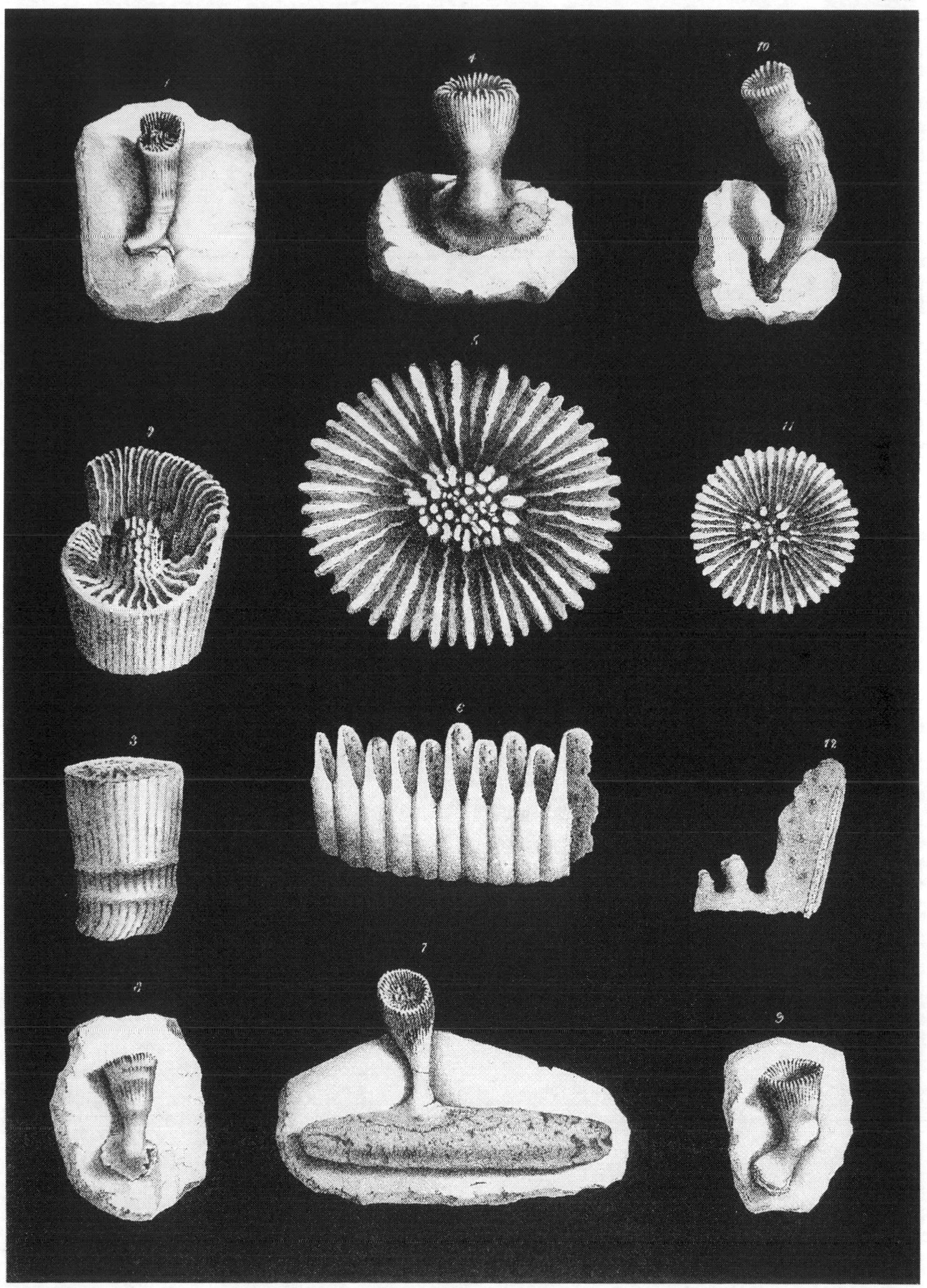

De Wilde lith.

M & N Hanhart imp

CORALS FROM THE CHALK

PLATE II.

CORALS FROM THE CHALK.

Fig.

Fig.	
1. 2. 3. 7. 10. 11.	Various shapes of the corallum of *Diblasus Gravensis*, Lonsdale. (P. 14.)
4. 6. 9.	The costæ, magnified.
5.	The peculiar appearance of tolerably well preserved calices, induced by fossilization, magnified.
8.	The method of gemmation, and the appearance of a large calice with the septa worn out of it, magnified.

De Wilde lith

M & N Hanhart imp

CORALS FROM THE CHALK

PLATE III.

CORALS FROM THE CHALK.

FIG.

1. The corallum of *Trochosmilia* (*Cœlosmilia*) *Wiltshiri*, Duncan. (P. 9.)
2. A portion of the calice, magnified.
3. A side view of one of the septa, magnified.

4., 5. Magnified views of the costæ.

6. The corallum of *Trochosmilia* (*Cœlosmilia*) *cornucopiæ*, Duncan. (P. 8.)
7. The calice, magnified.
8. The costæ near the calicular margin, magnified.
9. The arrangement of the septa as regards their size (a diagram).
10. The peduncle, magnified.

11., 14. Specimens of *Trochosmilia* (*Cœlosmilia*) *laxa*, Ed. and H., varieties. (P. 8.)

12., 13. Magnified portions.

15., 16., 18. Costæ, magnified.

17. A diagram of the septal arrangement.

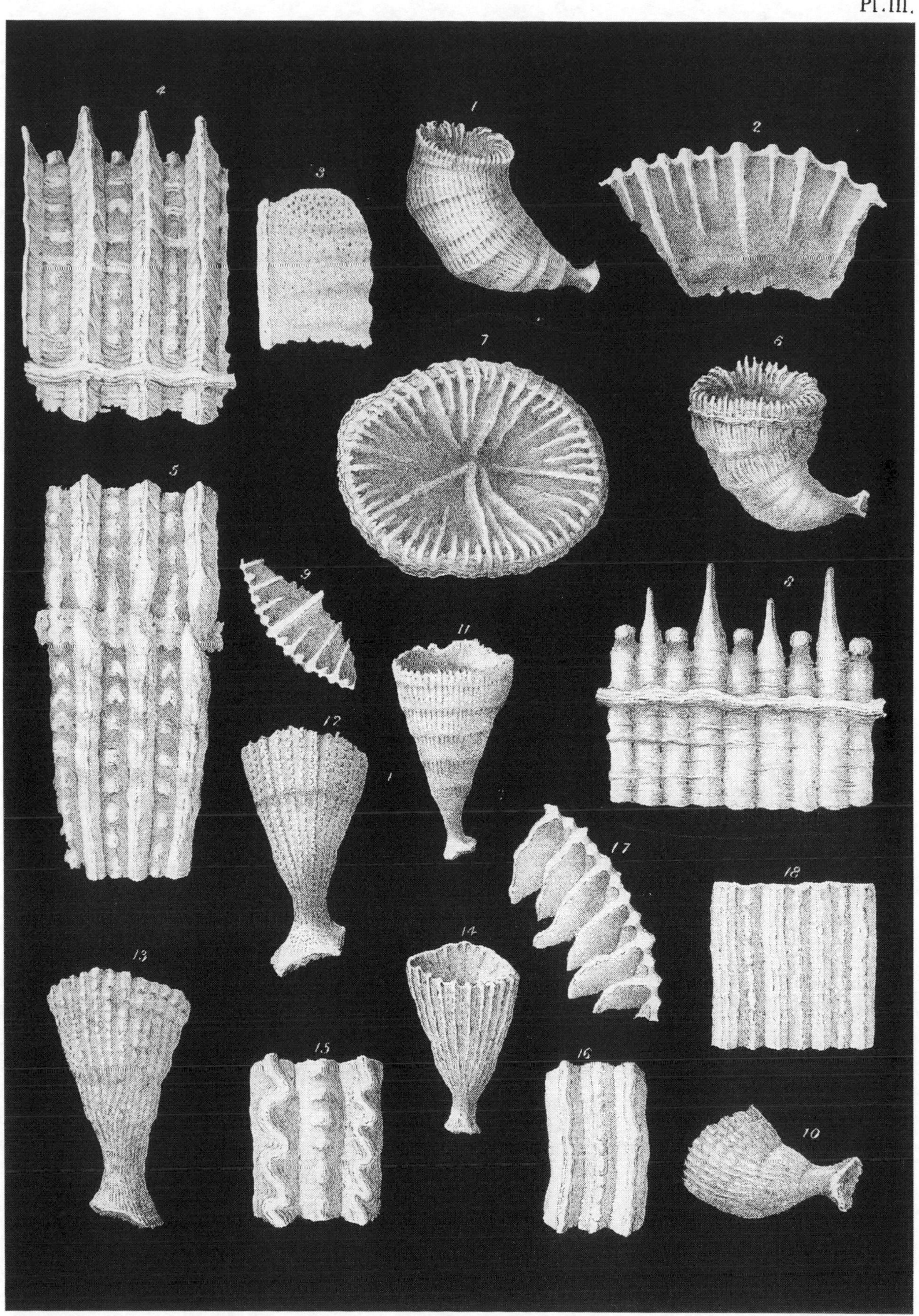

De Wilde lith

M&N Hanhart imp

CORALS FROM THE CHALK

PLATE IV.

CORALS FROM THE CHALK.

Fig.

1. The corallum of *Trochosmilia* (*Cœlosmilia*) *granulata*, Duncan. (P. 10.)
2. The costæ, magnified.
3. The cellular margin, magnified.
4. The peduncle, magnified.
5. The corallum of *Trochosmilia* (*Cœlosmilia*) *Woodwardi*, Duncan. (P. 9.)
6. The costæ, magnified.
7. The septa, magnified.
8. The peduncle, magnified.

9., 10. The corallum (nat. size and enlarged) of a variety of *Trochosmilia* (*Cœlosmilia*) *laxa*, Ed. and H. (P. 8.)

11. The costæ, magnified.
12. The calice, magnified, showing the septa of the fourth cycle.

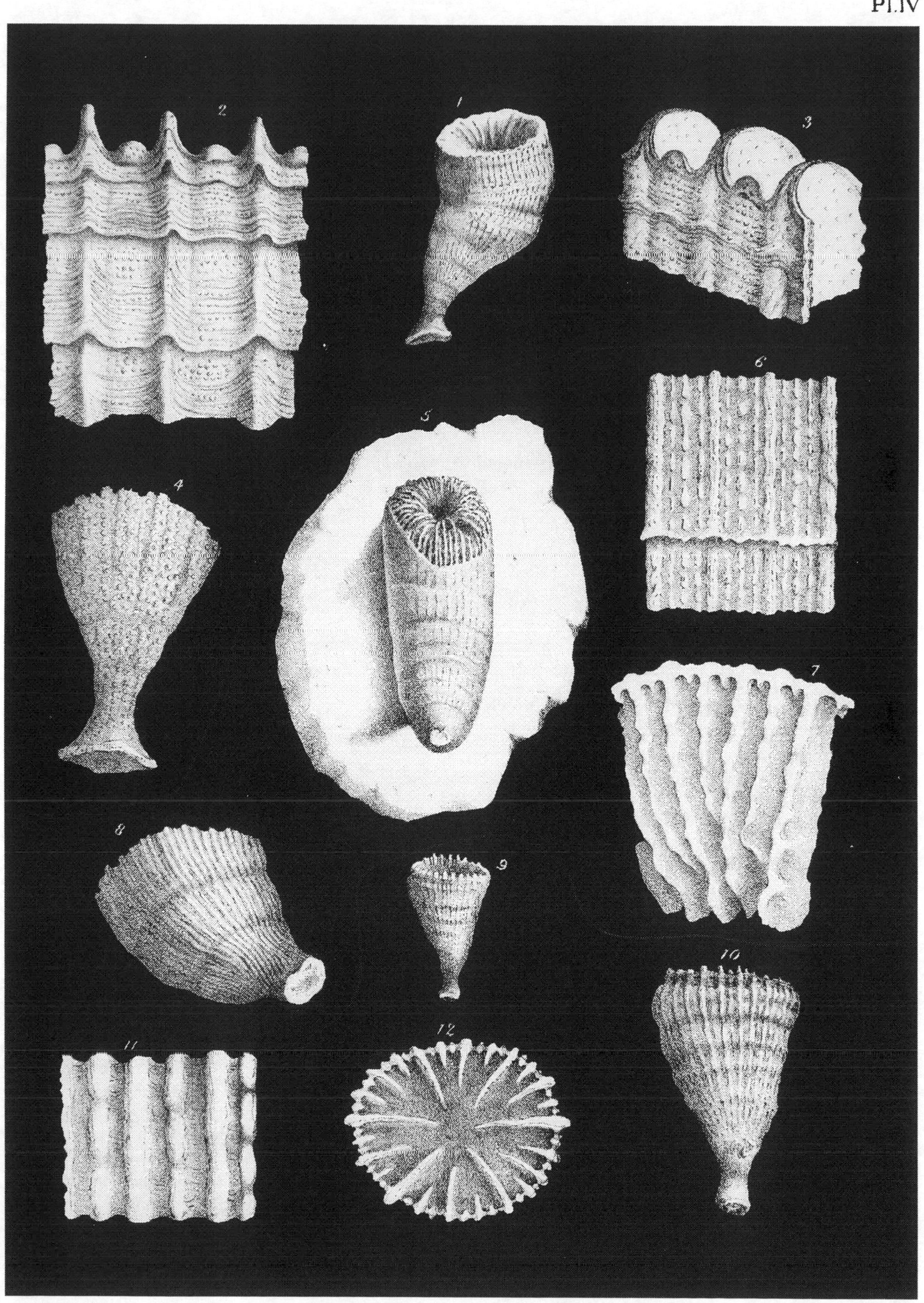

De Wilde lith.

M.&N. Hanhart imp.

CORALS FROM THE CHALK

PLATE V.

CORALS FROM THE CHALK.

Fig.

1. The corallum of *Trochosmilia* (*Cœlosmilia*) *cylindrica*, Duncan. (P. 10.)
2. A fractured portion of the corallum, showing the endothecal dissepiments and the septa.
3. Costæ, magnified.
4. The corallum of *Parasmilia monilis*, Duncan. (P. 12.)
5. A magnified view of the costæ on the peduncle.
6. A magnified view of the costæ high up.
7. The costæ on the body of the corallum, magnified.

8., 9. The corallum of *Parasmilia centralis*, Ed. and H., sub-species *Gravesana*. (P. 12.)

10., 13. The corallum of *Parasmilia centralis*, showing the typical costal arrangement.

11., 12., 14., 15. The costæ of the corallum, magnified.

Pl.V.

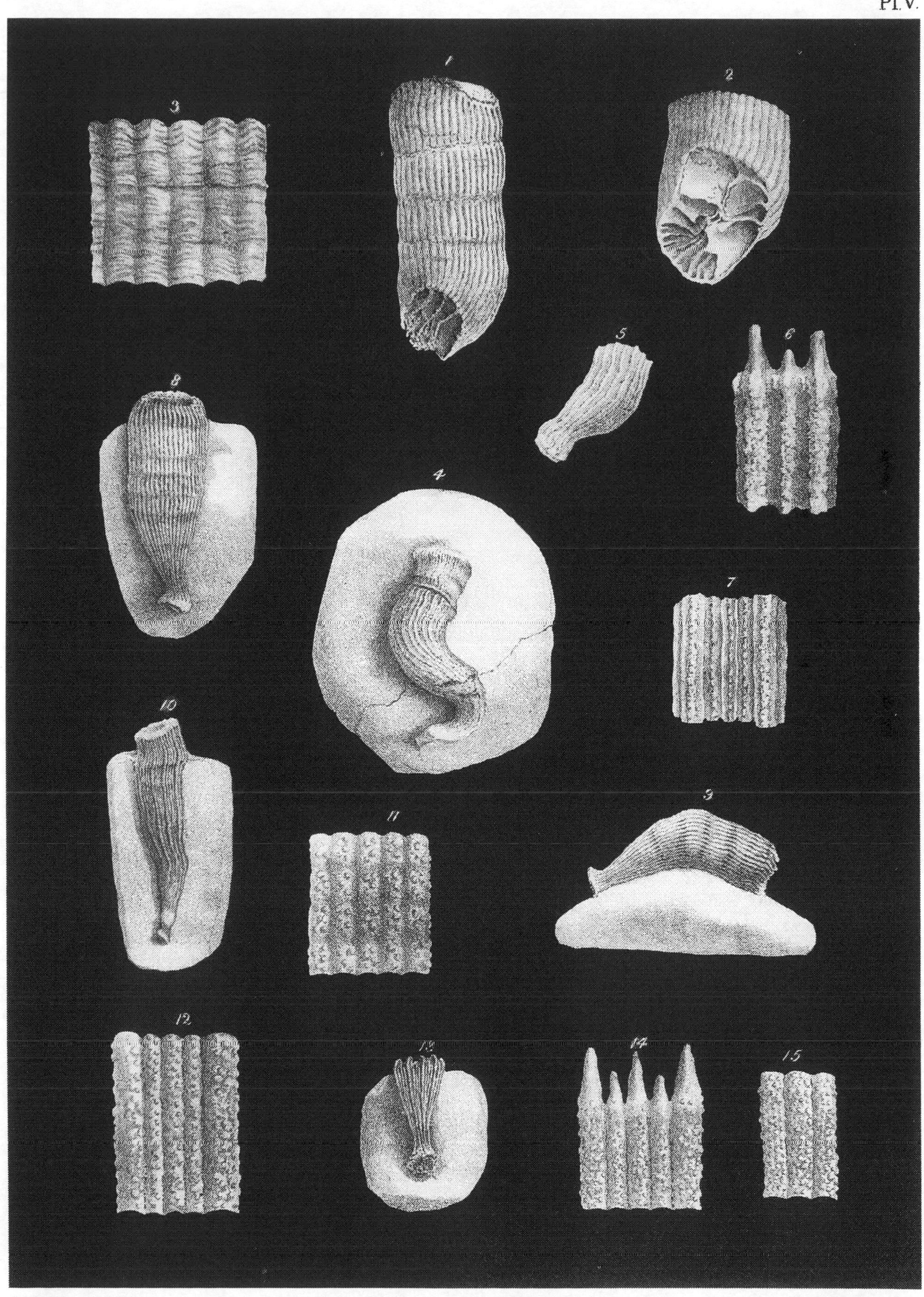

De Wilde lith. M.&N.Hanhart imp.

CORALS FROM THE CHALK

PLATE VI.

CORALS FROM THE CHALK.

Fig.

1. The corallum of *Onchotrochus serpentinus*, Duncan. (P. 4.)
2. The calicular end of the corallum, magnified.
3. The corallum of a small specimen.
4. The costæ, magnified.
5. The corallum of *Parasmilia granulata*, Duncan. (P. 13.)
6. The costæ, magnified.
7. The calice, magnified.
8. The peduncle and its costæ, magnified.
9. A longitudinal section of the corallum of *Trochosmilia* (*Cœlosmilia*) *granulata*, Duncan, showing the wavy inner ends of the septa, and the scanty endotheca.
10. The corallum of a *Caryophyllia*, showing irregular growth.
11. The calice, magnified, showing a distorted arrangement of the septa.

12., 13. Longitudinal sections of *Parasmilia centralis*, showing the large columella and the scanty endotheca.

14. A corallum of a young *Parasmilia centralis*, variety *Mantelli*. (P. 12.)

16. A younger specimen.

15. The costæ, magnified.

17. A portion of the calice, magnified.

18. A distorted corallum of *Parasmilia centralis*.

19., 20., 21. Its costæ, magnified.

De Wilde lith.

M & N Hanhart imp

CORALS FROM THE CHALK

PLATE VII.

CORALS FROM THE UPPER GREENSAND.

FIG.

1. } Various shapes of the corallum of *Smilotrochus elongatus*, Duncan. (P. 19.) The
2. } specimens are worn, and the corallites are in the form of casts. Small portions
3. } of the original hard parts still remain.

4. A transverse section of a corallum, slightly magnified.

5. }
6. } The casts of the intercostal spaces simulating costæ, slightly magnified.

7. The corallum of *Smilotrochus angulatus*, Duncan. (P. 20.)

8. The transverse section, slightly magnified. (The specimens are in the form of casts.)

9. The corallum of *Favia minutissima*, Duncan. (P. 22.)

10. A portion, magnified.

11. Endothecal structures of the corallum, magnified.

12. The corallum of *Smilotrochus Austeni*, Edwards and Haime. (P. 19.) Copied from the Hist. Nat. des Coralliaires.'

13. The corallum of *Thamnastræa superposita*, Michelin, sp. (P. 22.)

14. }
15. } Specimens from the French Upper Greensand.

16. A corallum, magnified.

17. A calice, close to the edge of the corallum, magnified and drawn with the camera lucida. The continuous costæ are to be observed inferiorly.

Pl. VII.

De Wilde lith

M&N Hanhart imp

CORALS FROM THE UPPER GREENSAND

PLATE VIII.

CORALS FROM THE UPPER GREENSAND.

FIG.

1., 2., 3., 4., 5., 11., 13. Corallites of *Onchotrocus Carteri*, Duncan. (P. 20.)

6. A worn calice, magnified. Fossilization has produced a false union of the septa, and a central space.

7., 8. Sections of the same specimen. The central tissue is due to fossilization.

9. A normal calice, magnified.

10., 12., 14. The costal and epithecal structures of three different specimens.

15. The corallum of *Cyathophora monticularia*, D'Orb., sp. (P. 21.)

16. A portion, magnified.

17., 18. Sections magnified, showing the endotheca, and in fig. 18 one of the tablulæ.

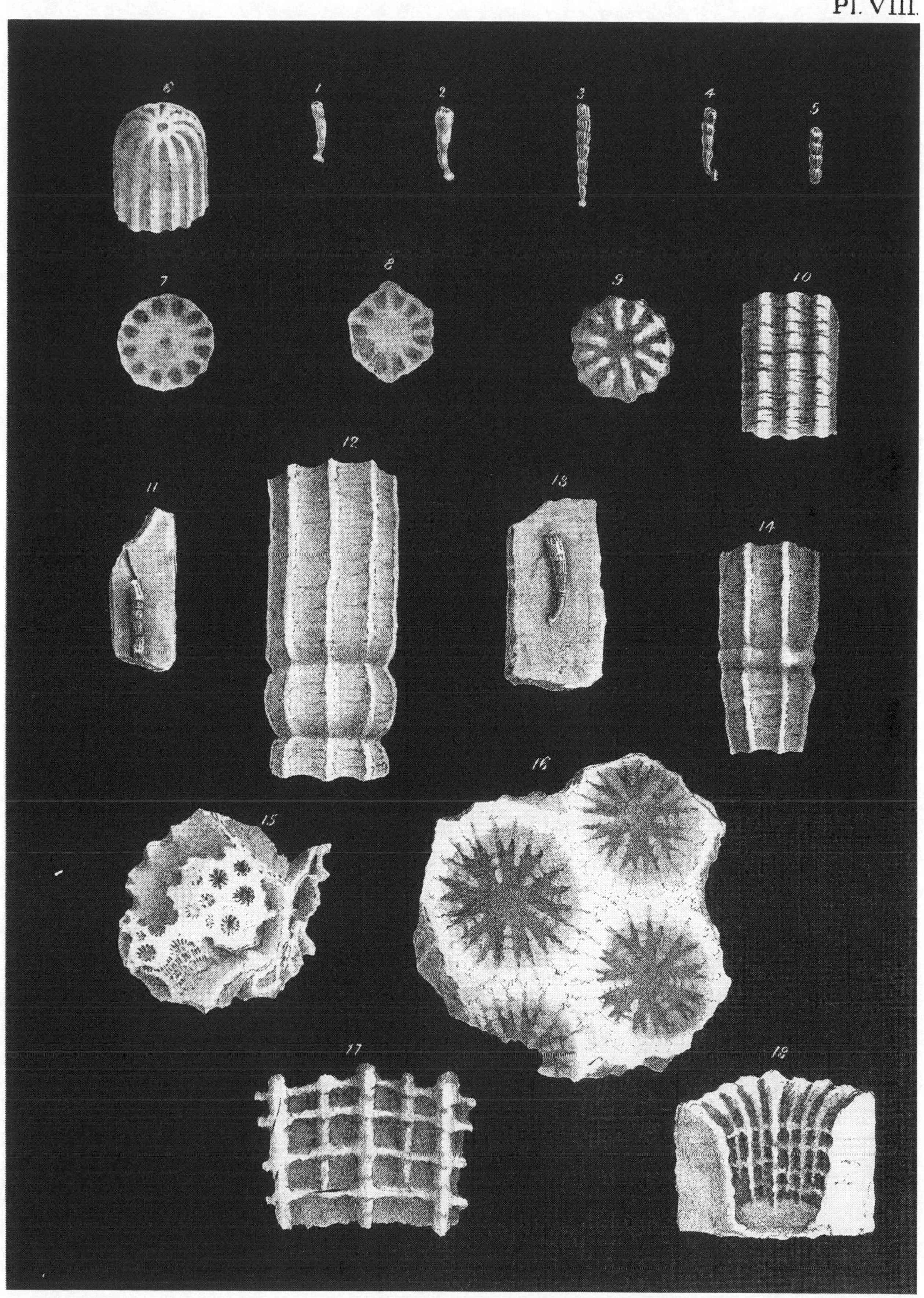

De Wilde lith.

M & N Hanhart imp

CORALS FROM THE UPPER GREENSAND

PLATE IX.

CORALS FROM THE RED CHALK OF HUNSTANTON.

IG.

1. A variety of *Micrabacia coronula*, Goldfuss. Natural size. (P. 24.)

2. The usual appearance presented by the worn specimens of *Podoseris mammiliformis*, Duncan. (P. 25.)

3.
5. } The calice, magnified.

6. A specimen with a large base.

4. A natural section (longitudinal) showing the synapticulæ, magnified.

7. A specimen showing a convex calice, the costæ and synapticulæ, magnified.

10. A specimen with epitheca, magnified.

11.
12. } Natural size.

13. An irregularly shaped corallum.

14. Its base, magnified.

9. The side view, magnified.

15. A short specimen.

8. A magnified view of it, showing the synapticulæ.

16. The corallum of *Podoseris elongata*, Duncan. (P. 26.)

17. Its costæ, magnified.

18. The corallum of *Cyclolites polymorpha*, Goldfuss. (P. 24.)

De Wilde del et lith.

M & N Hanhart imp

CORALS FROM THE RED CHALK.

PLATE X.

CORALS FROM THE UPPER GREENSAND OF HALDON.

FIG.

1. The corallum of *Placosmilia cuneiformis*, Ed. and H. (P. 27.)
2. Part of a septum, magnified.
3. The costæ, magnified.
4. Oblique view of the costæ, magnified.
5. The calice, magnified.
6. The corallum of *Placosmilia Parkinsoni*, Ed. and H. (P. 28.)
7. The calice, magnified.
8. The corallum of *Peplosmilia depressa*, E. de From. (P. 29.)
9. The costæ, magnified.
10. The calice, magnified.

11., 12., 13. The corallum of *Placosmilia magnifica*, Duncan. (P. 28.)

Pl. X.

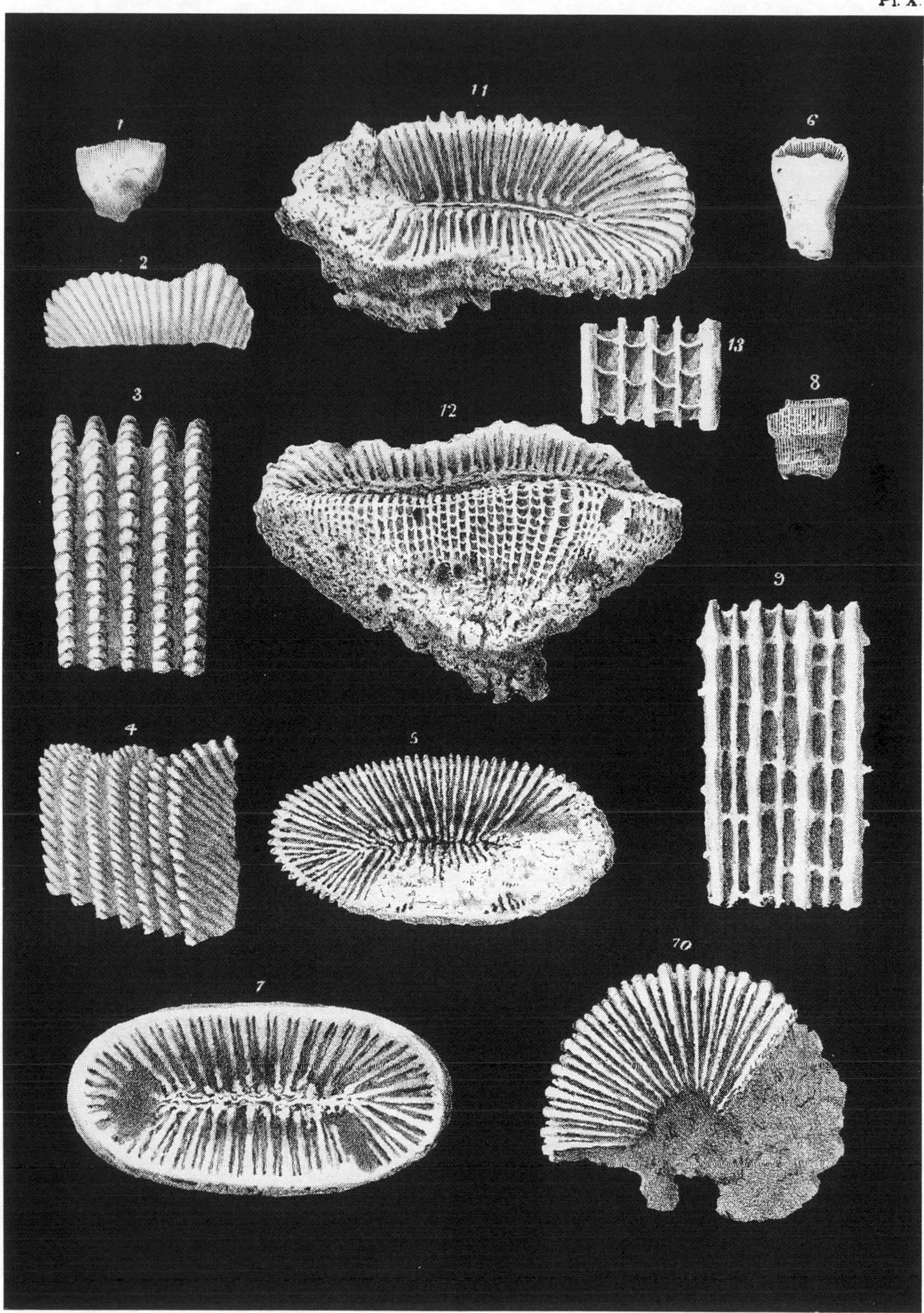

De Wilde del et lith.

M & N Hanhart imp

CORALS FROM THE UPPER GREENSAND.

PLATE XI.

FIG.

1. The corallum of *Astrocœnia decaphylla*, Ed. and H. (P. 29.)
2. The same, magnified.
3. The upper part of a calice, magnified.
4. The corallum of a variety.
5. The upper part of a calice, magnified.
6. The corallum, magnified.

7.
8. } The corallum and calices of *Isastræa Haldonensis*, Duncan. (P. 30.)

De Wilde del. et lith.

M & N Hanhart imp.

CORALS FROM THE UPPER GREENSAND.

PLATE XII.

CORALS FROM THE GAULT.

Fig.

1.
2. } Varieties of *Trochocyathus Harveyanus*, Ed. and H. (P. 33.)

3. Magnified view of the ends of the costæ of one of the varieties.

4. A longitudinal section of a variety, slightly magnified.

5. A variety of *Bathycyathus Sowerbyi*, Ed. and H. (P. 35.)

6. Its costæ, magnified.

7. The corallum of *Bathycyathus Sowerbyi*, Ed. and H. (P. 35.)

8. A variety of *Caryophyllia Bowerbanki*, Ed. and H. (P. 32.)

9. Its costæ, magnified.

10 to 16. Views of *Smilotrochus elongatus*, Duncan. (P. 36.)

12. Costæ, magnified.

14. The calice of a young specimen, magnified.

16. The costæ, magnified.

17. Corallum of *Smilotrochus insignis*, Duncan. (P. 37.)

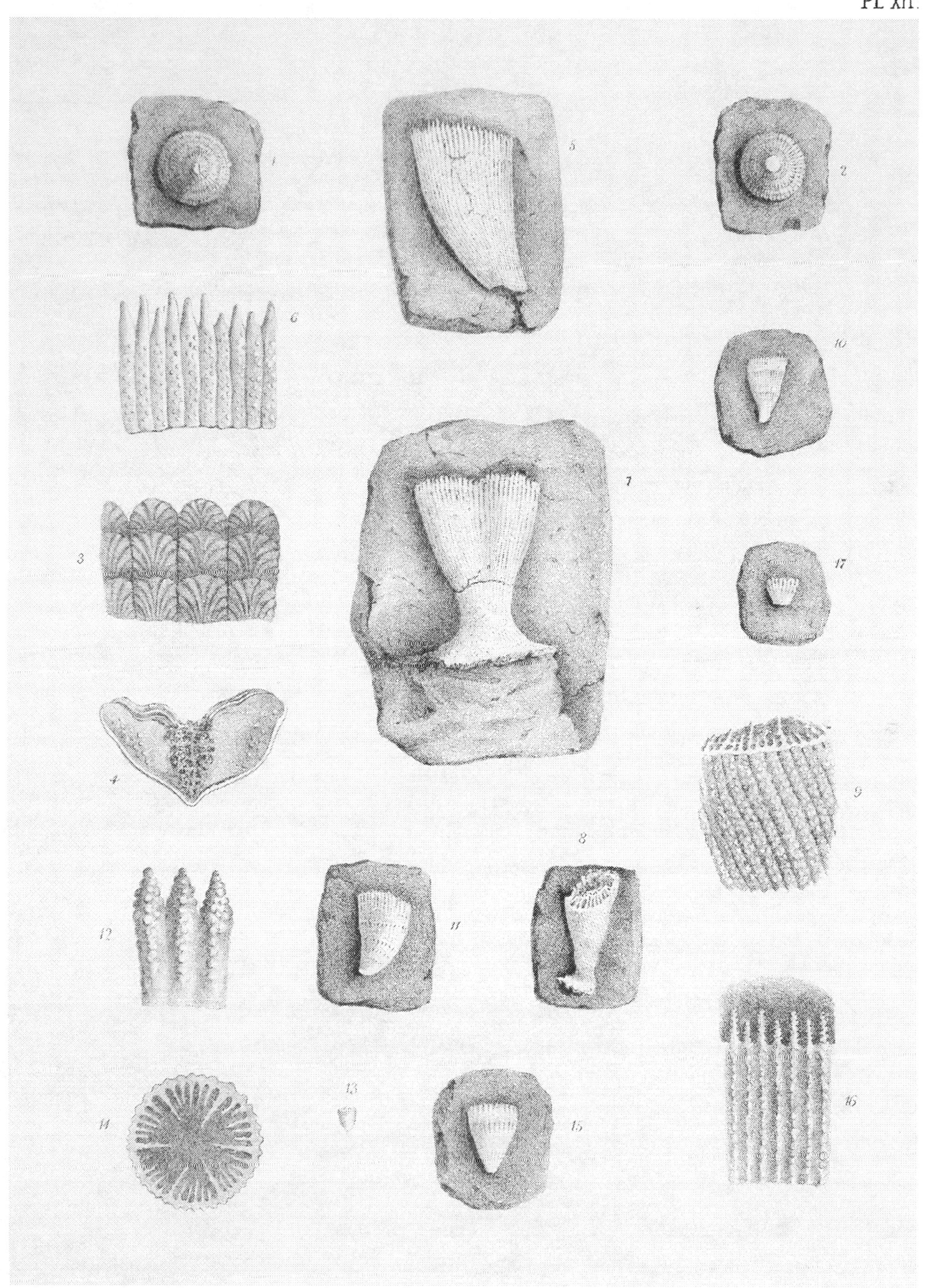

De Wilde del et lith.

M & N Hanhart imp

CORALS FROM THE GAULT.

PLATE XIII.

CORALS FROM THE GAULT.

FIG.

1. A variety of *Trochocyathus Harveyanus*, Ed. and H. The base. (P. 32 and 33.)
2. Costæ and septa, magnified.
3. A variety of the same species.
4. Costæ and septa, magnified.
13. A transverse section, magnified.
5. *Leptocyathus gracilis*, Duncan. Under surface. (P. 34.)
6. The under surface or base, magnified.
7. A transverse section, magnified.
8. A side view, magnified.
9. *Smilotrochus insignis*, Duncan. (P. 37.)

10.\
11. } Young of *Smilotrochus elongatus*, Duncan. (P. 36.)\
12./

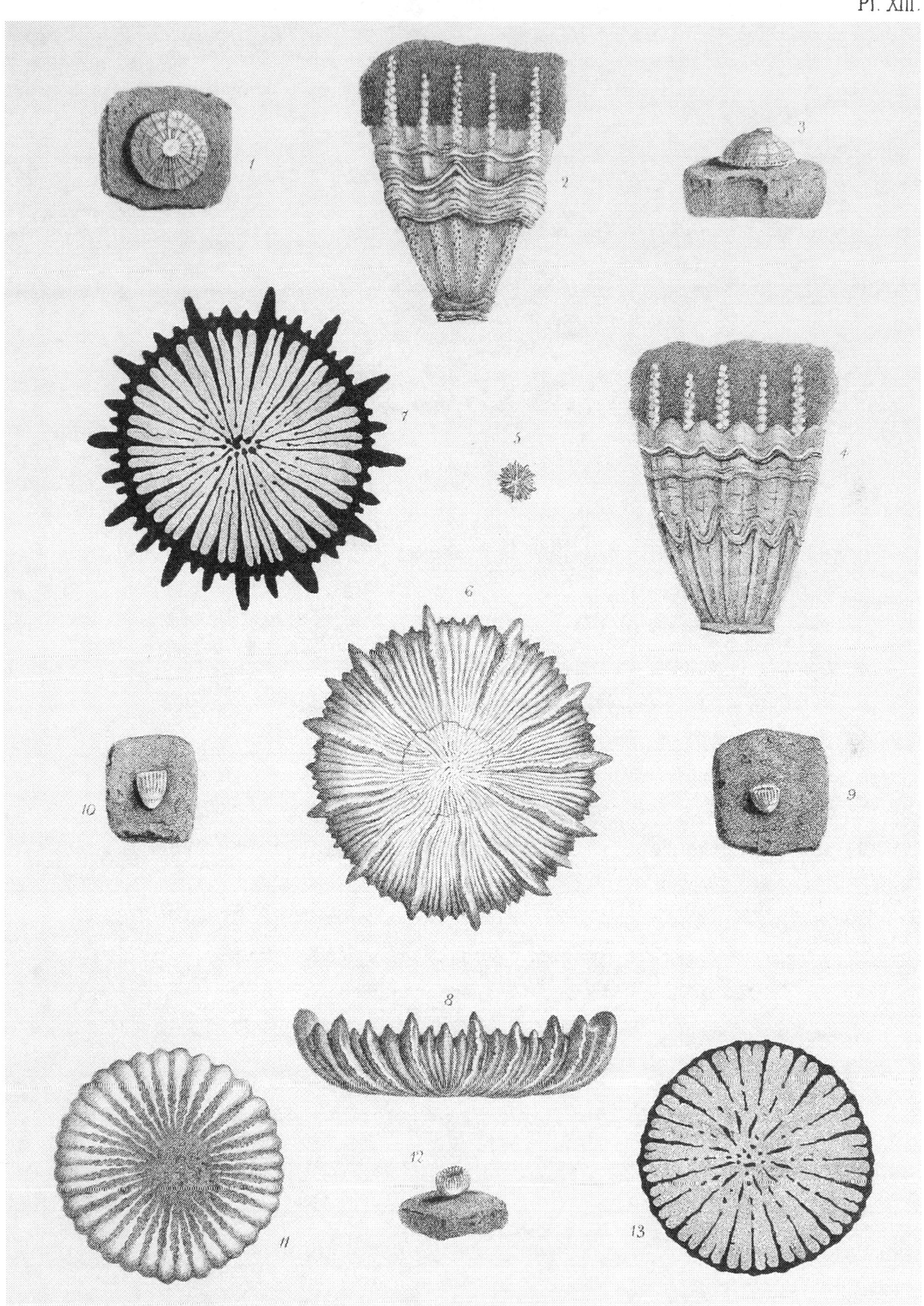

De Wilde del et lith.

M & N Hanhart imp

CORALS FROM THE GAULT.

PLATE XIV.

CORALS FROM THE GAULT.

FIG.

1., 2. Abnormal form of *Trochocyathus Harveyanus*, Ed. and H. (P. 34.)

3., 4., 5. Magnified views.

6. Base of *Micrabacia Fittoni*, Duncan. (P. 37.)
7. The same, magnified.
8. Side view of the corallum, magnified.
9. Junction of septa and costæ, magnified.
10. Corallum of *Trochocyathus Wiltshirei*, Duncan. (P. 34.)
11. Magnified view.
12. The calice, magnified.
13. *Smilotrochus elongatus*, Duncan. Adult form. (P. 36.)
14. The same, magnified.
15. The calice, magnified.
16. *Smilotrochus cylindricus*, Duncan. Corallum, magnified. (P. 36.)
17. *Smilotrochus granulatus*, Duncan. Corallum, magnified. (P. 36.)
18. *Smilotrochus insignis*, Duncan. Corallum, magnified. (P. 37.)

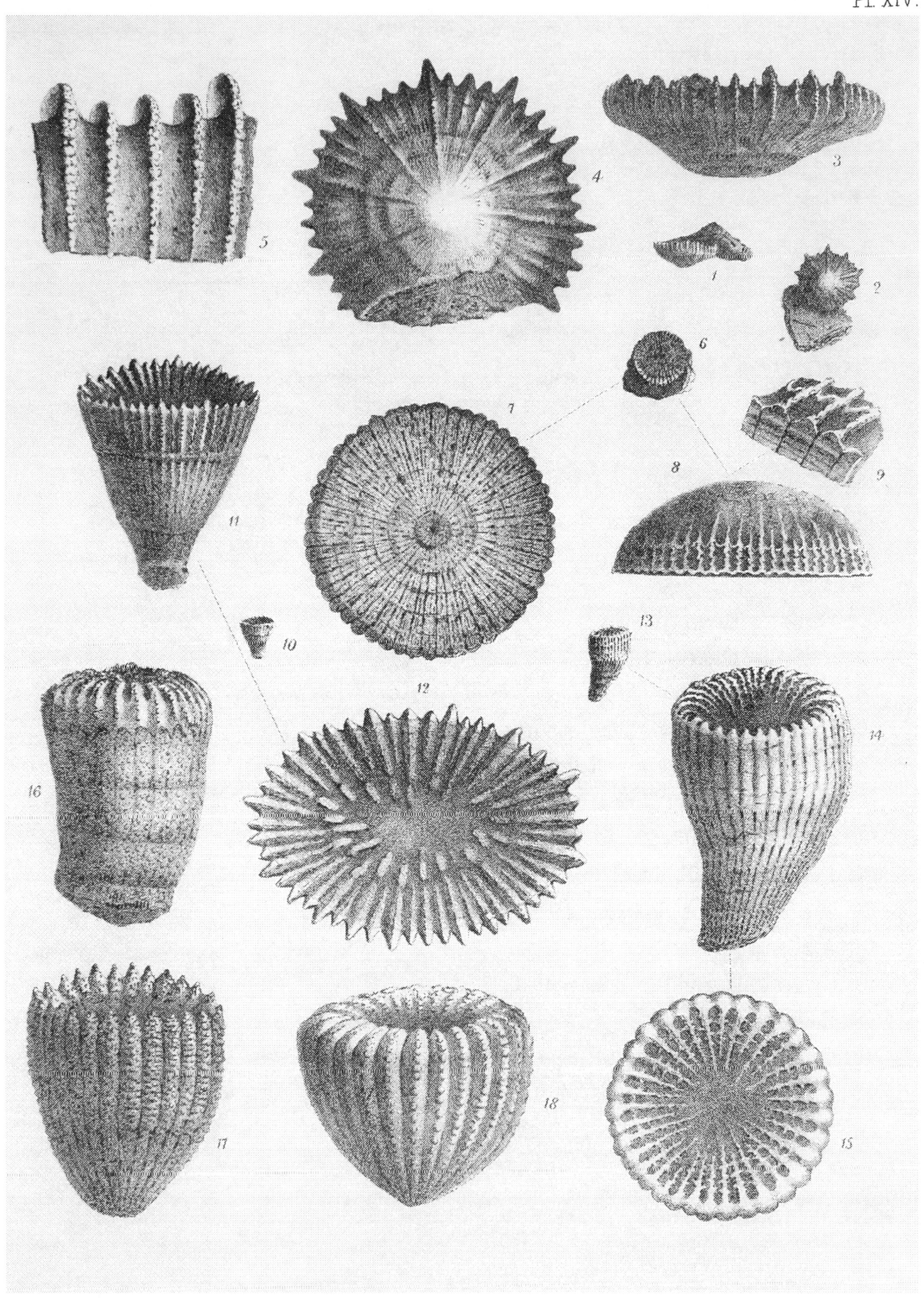

De Wilde del et lith. M & N Hanhart imp

CORALS FROM THE GAULT.

PLATE XV.

CORALS FROM THE LOWER GREENSAND.

FIG.

1.
2. Corallites of *Trochosmilia Meyeri*, Duncan. (P. 41.)
3.

4.
5. Calices, magnified.

6. Variety with broad base.

7. Its calice, magnified.

8. Part of the corallum of *Brachycyathus Orbignyanus*, Ed. and H. (P. 40.)

9. Longitudinal view of the septa and pali, magnified. The notch indicates the commencement of pali attached to tertiary septa.

10. Corallum (cast) of *Isastræa Morrisii*, Duncan. (P. 42.)

11. The cast, magnified.

12. Impression, magnified.

13. The corallum of *Turbinoseris De-Fromenteli*, Duncan. (P. 43.)

14. A variety.

15. Synapticulæ and septa, magnified.

16. Calice, size of life.

17. Costæ, magnified.

18. The unusual appearance of septa ending in intercostal spaces, magnified.

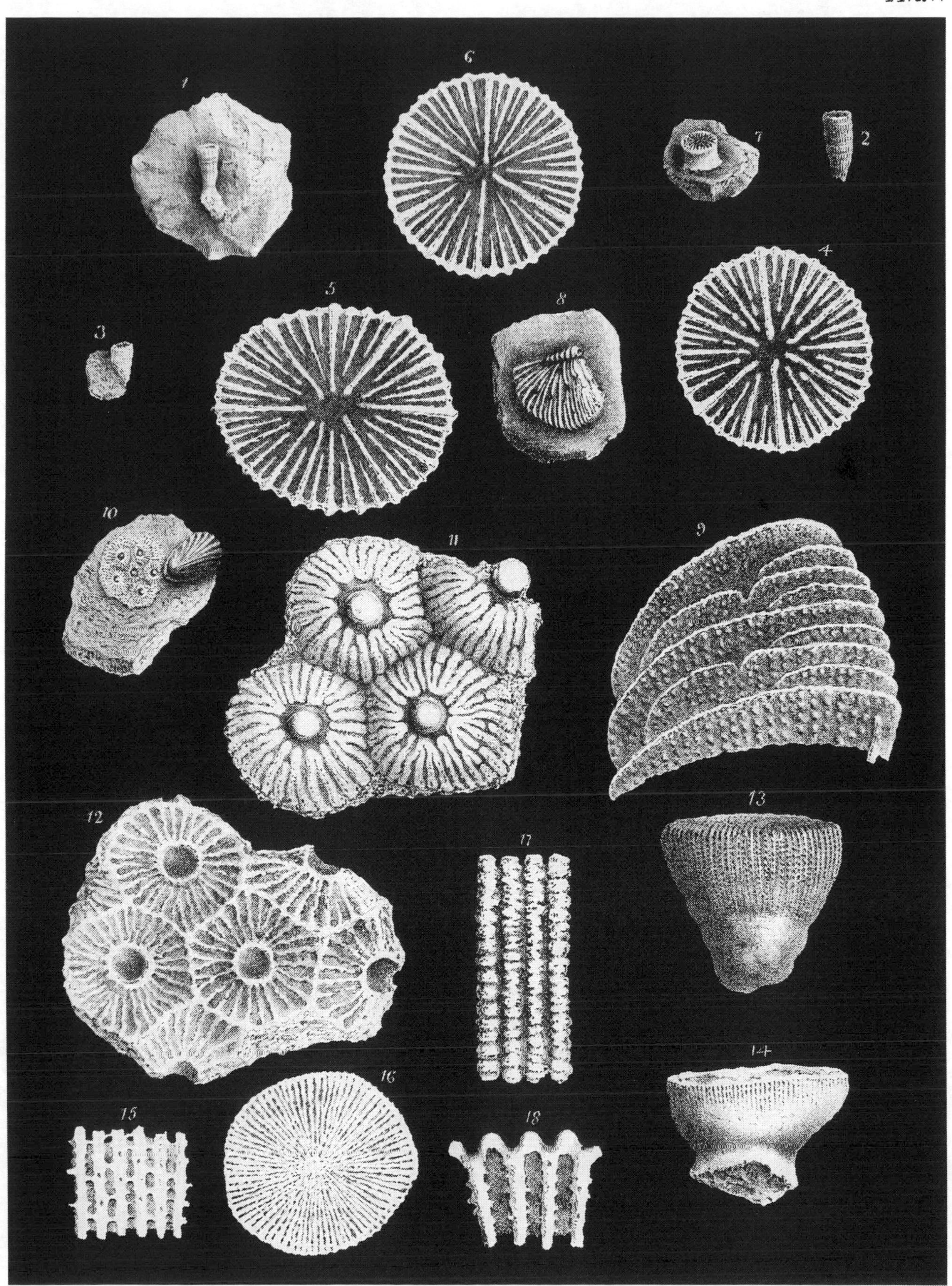

De Wilde del et lith. M & N Hanhart imp

CORALS FROM THE LOWER GREEN SAND.

PLATE I.

CORALS FROM THE GREAT OOLITE AND FROM THE INFERIOR OOLITE.

Fig.

1. *Thecosmilia obtusa*, d'Orb. The corallum. (Great Oolite.) (P. 14.)
2. The upper margin of a septum, magnified.
3. The calicular surface, magnified.
4. Costæ, magnified, and epitheca.
5. *Montlivaltia Holli*, Duncan. (Inferior Oolite.) (Page 16.)
6. The calicular surface, magnified.
7. The calice, magnified.
8. A corallum with calicular gemmation.
9. *Cyathophora insignis*, Duncan. (Great Oolite.) (Page 14.)
10. A calice, magnified.
11. Three calices, slightly magnified.
12. *Montlivaltia Painswicki*, Duncan. (Inferior Oolite.) (Page 17.)

De Wilde del et lith

M & N Hanhart imp

CORALS FROM THE GREAT OOLITE, AND FROM THE INFERIOR OOLITE.

PLATE II.

Fig.

1. *Thamnastræa Browni*, Duncan. (Great Oolite). (Page 16.)
2. The calices, magnified.
3. The costæ and epitheca, magnified.
4. A septum, magnified.
5. A costa, magnified.
6. *Thamnastræa Waltoni*, Ed. & H. (Inferior Oolite.)

7.
8. } Details, magnified.
9.

10.
11. } *Isastræa gibbosa*, Duncan, and magnified view. (Page 15.)

12. *Montlivaltia trochoides*, Ed. & H. (Inferior Oolite.)
13. *Montlivaltia Morrisi*, Duncan. (Inferior Oolite.) (Page 17.)

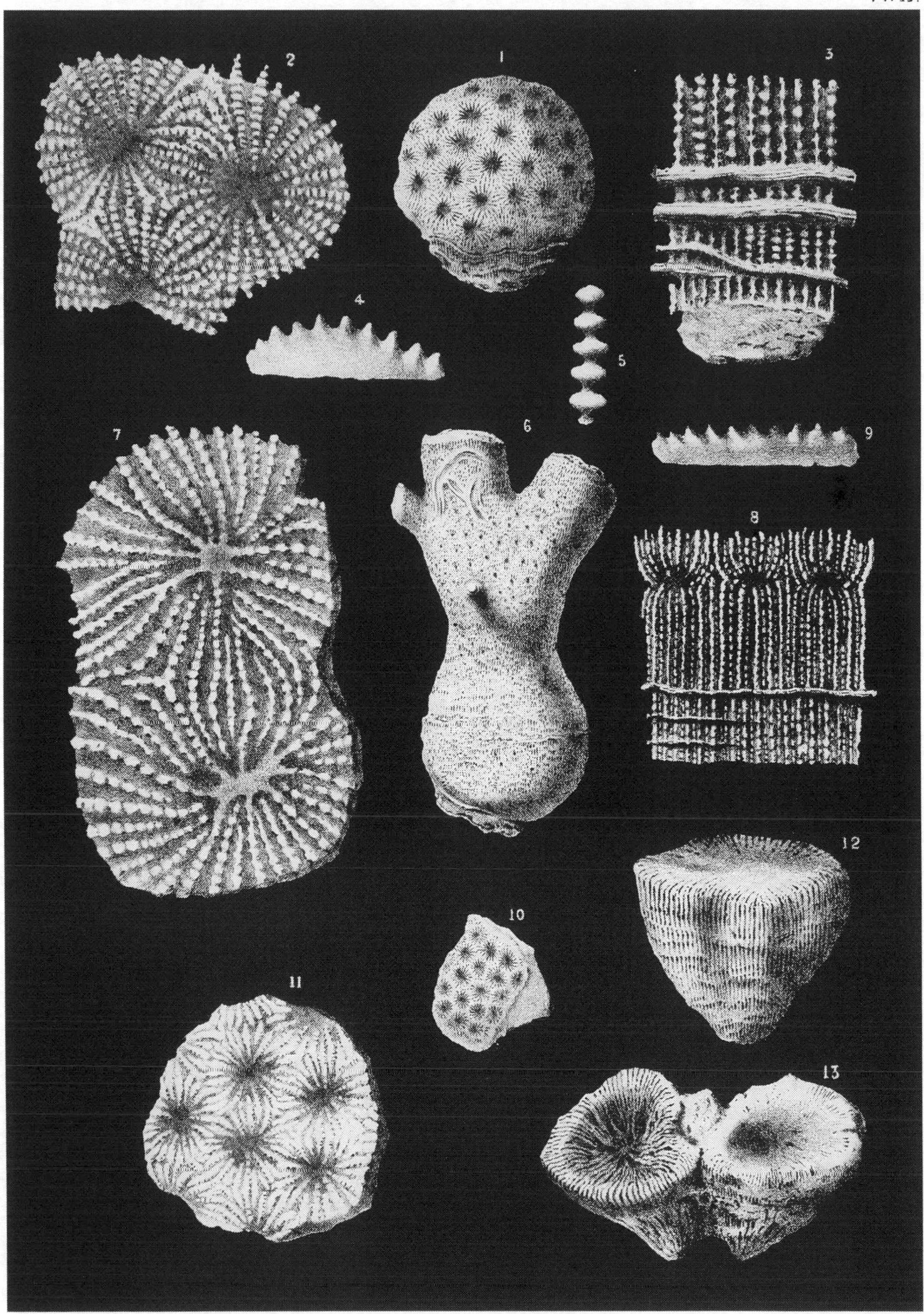

De Wilde del et lith.

M & N Hanhart imp

CORALS FROM THE GREAT OOLITE, AND FROM THE INFERIOR OOLITE.

PLATE III.

FIG.

1., 2., 3., 4. *Cladophyllia Babeanna,* Ed. & H. Corallum and details. (Page 3.)

5. *Podoseris constricta,* Duncan. (Inferior Oolite.) (Page 24.)
6. Corallum, magnified.

7., 8. *Cyclolites Lyceti,* Duncan. (Inferior Oolite.) (Page 23.)

9. The costæ, magnified.
10. *Cyclolites Beanii,* Duncan. (Inferior Oolite.) (Page 23.)
11. The septa and synapticulæ, magnified.

12., 13., 14. A *Montlivaltia* with distinct costæ. (Inferior Oolite.)

15. *Cyathophora tuberosa,* Duncan. (Great Oolite.) (Page 15.)
16. A calice, magnified.
17. Calices, magnified.
18. Side view of calice, magnified.

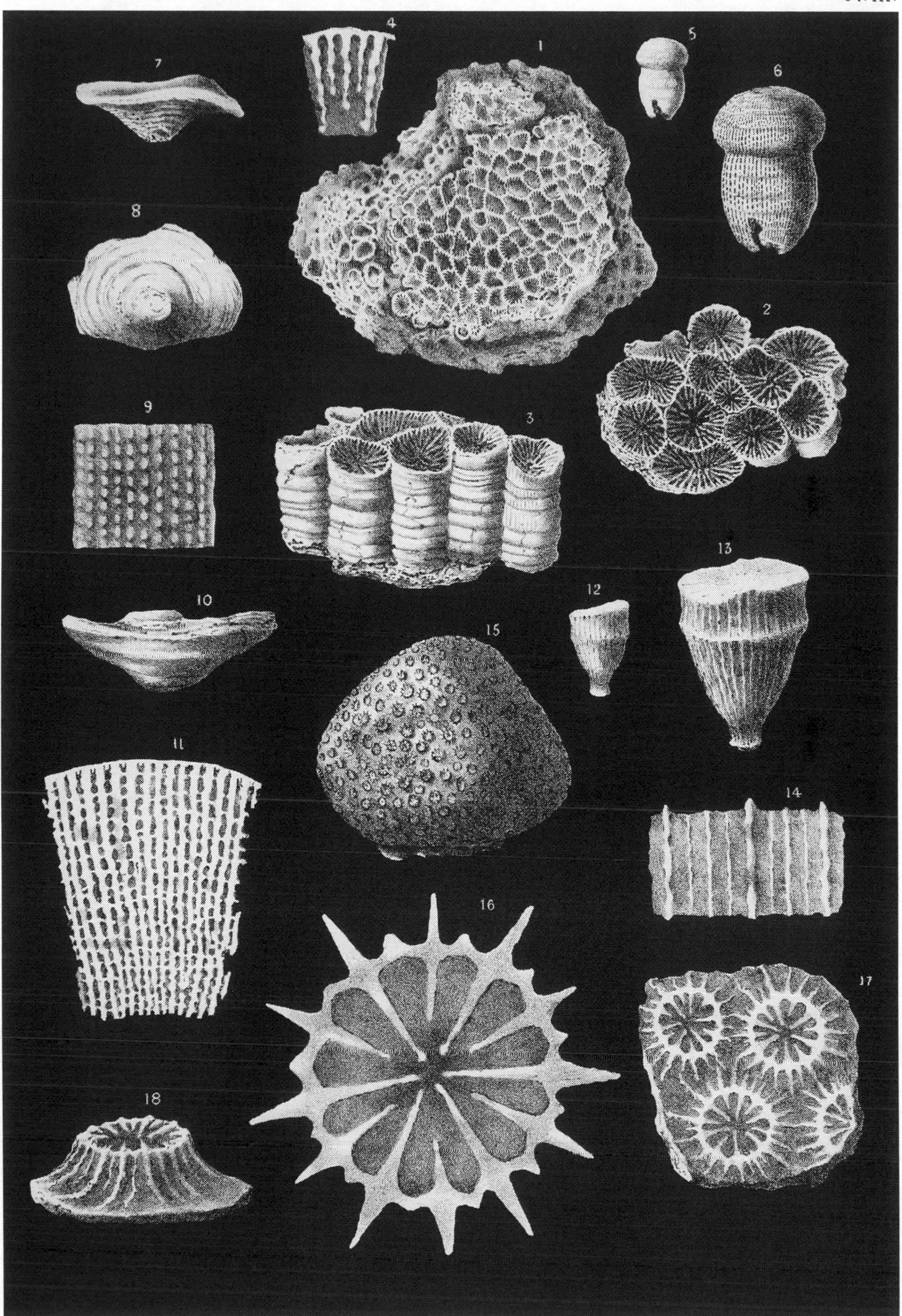

De Wilde del et lith. M & N Hanhart imp

CORALS FROM THE GREAT OOLITE, AND FROM THE INFERIOR OOLITE.

PLATE IV.

Fig.

1. *Dimorphoseris oolitica*, Duncan. (Inferior Oolite.) (Page 22.)
2. Calicular surface.
3. A calice, magnified.
4. Synopticulæ, magnified.

5.–10. Corallum and details of *Thamnastræa Walcotti*, Duncan. (Inferior Oolite.) (P. 19.)

11.–14. *Thamnastræa Manseli*, Duncan. (Inferior Oolite.) (Page 20.)

De Wilde del et lith.

M & N Hanhart imp

CORALS FROM THE INFERIOR OOLITE.

PLATE V.

FIG.

1.–4. The corallum and details of *Thecosmilia Wrighti*, Duncan. (Inferior Oolite.) (Page 17).

5. A calice, magnified.

6.–7. Views of *Latimæandra Flemingi*, Ed. & H. (Page 18.)

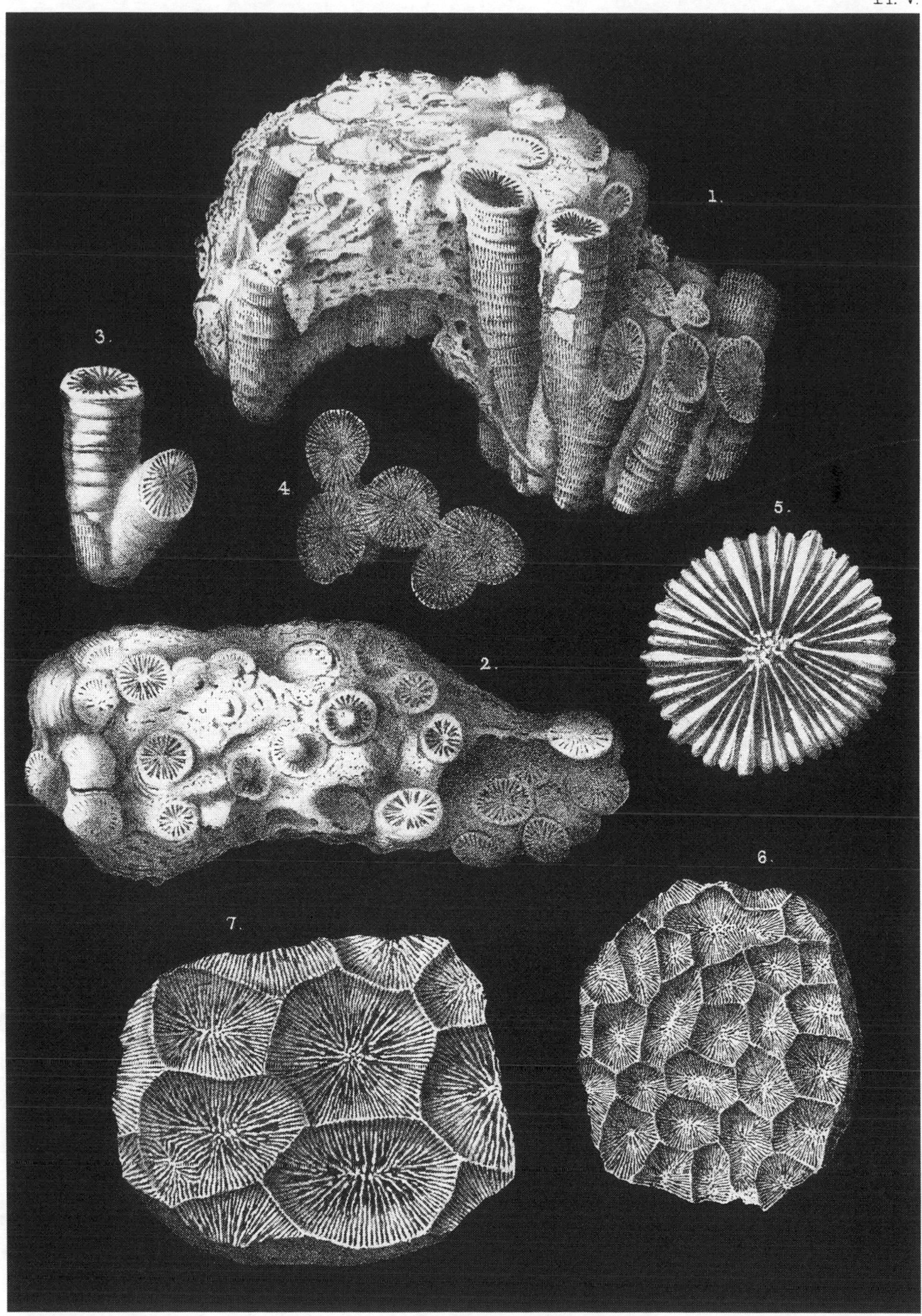

G R DeWilde lith

W West & C° imp

CORALS FROM THE INFERIOR OOLITE.

PLATE VI.

FIG.

1. *Thecosmilia gregaria*, M'Coy. Variety. (Page 18.)

2. }
3. } Parts of corallum, magnified slightly.

4. Costæ and exotheca, magnified.

5. The corallum of *Symphyllia Etheridgei*, Duncan. (Page 19.)

6. }
7. } Calices, magnified.

8. Costæ, magnified.

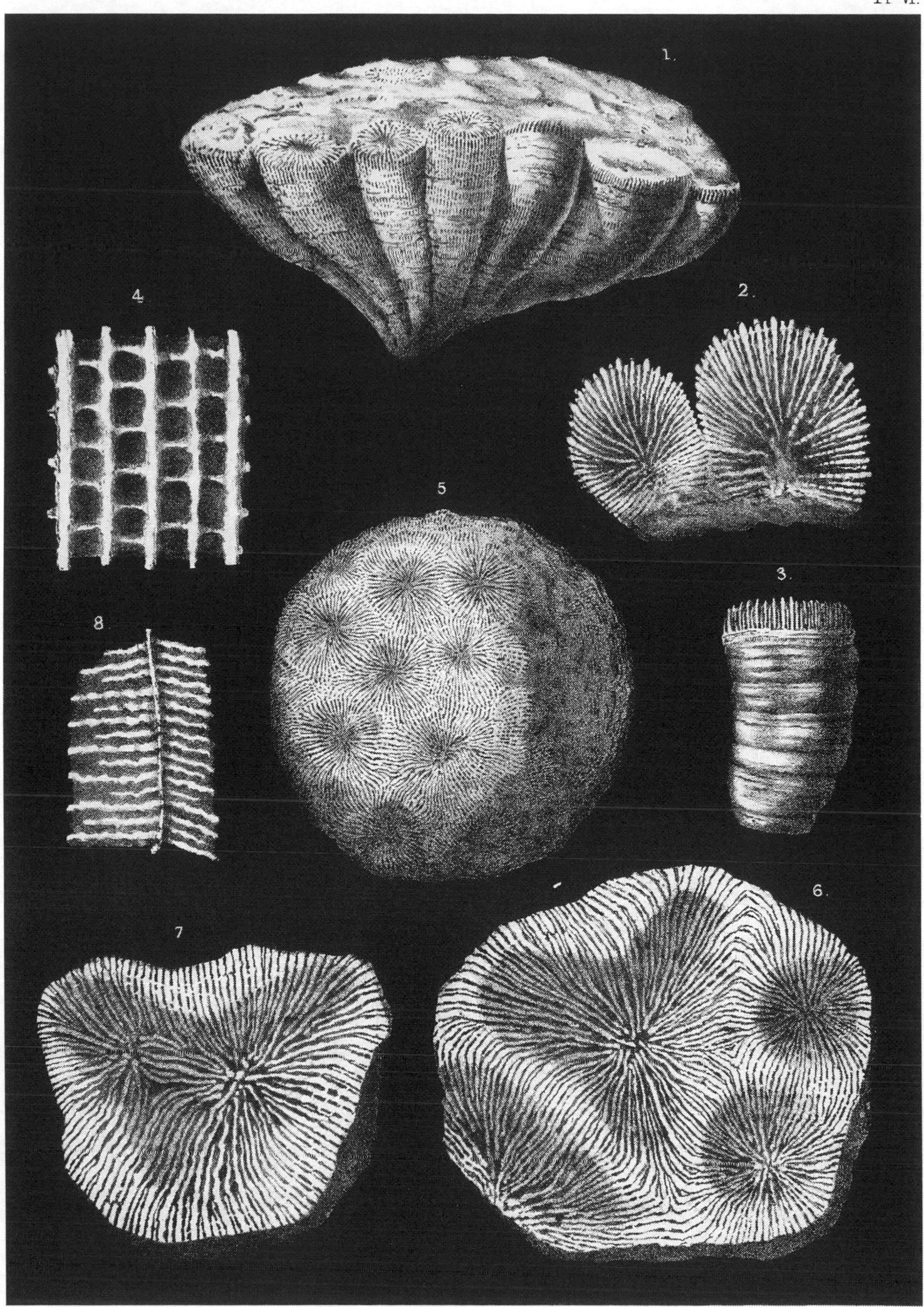

G R DeWilde lith W West & Co imp

CORALS FROM THE INFERIOR OOLITE.

PLATE VII.

Fig.

1. *Gonioseris angulata*, Duncan. (Page 21.)
2. Under side—the base.
3. Side view.
4. Costæ.
5. Synapticulæ.
6. *Gonioseris Leckenbyi*, Duncan. (Page 22.)
7. Base.
8. Side view.
9. Costæ at the angle of the base.

10., 11. } Small specimen of young *Gonioseris*.

12., 13., 14., 15. } Symphyllian form of a *Thecosmilian*. (Page 9.)

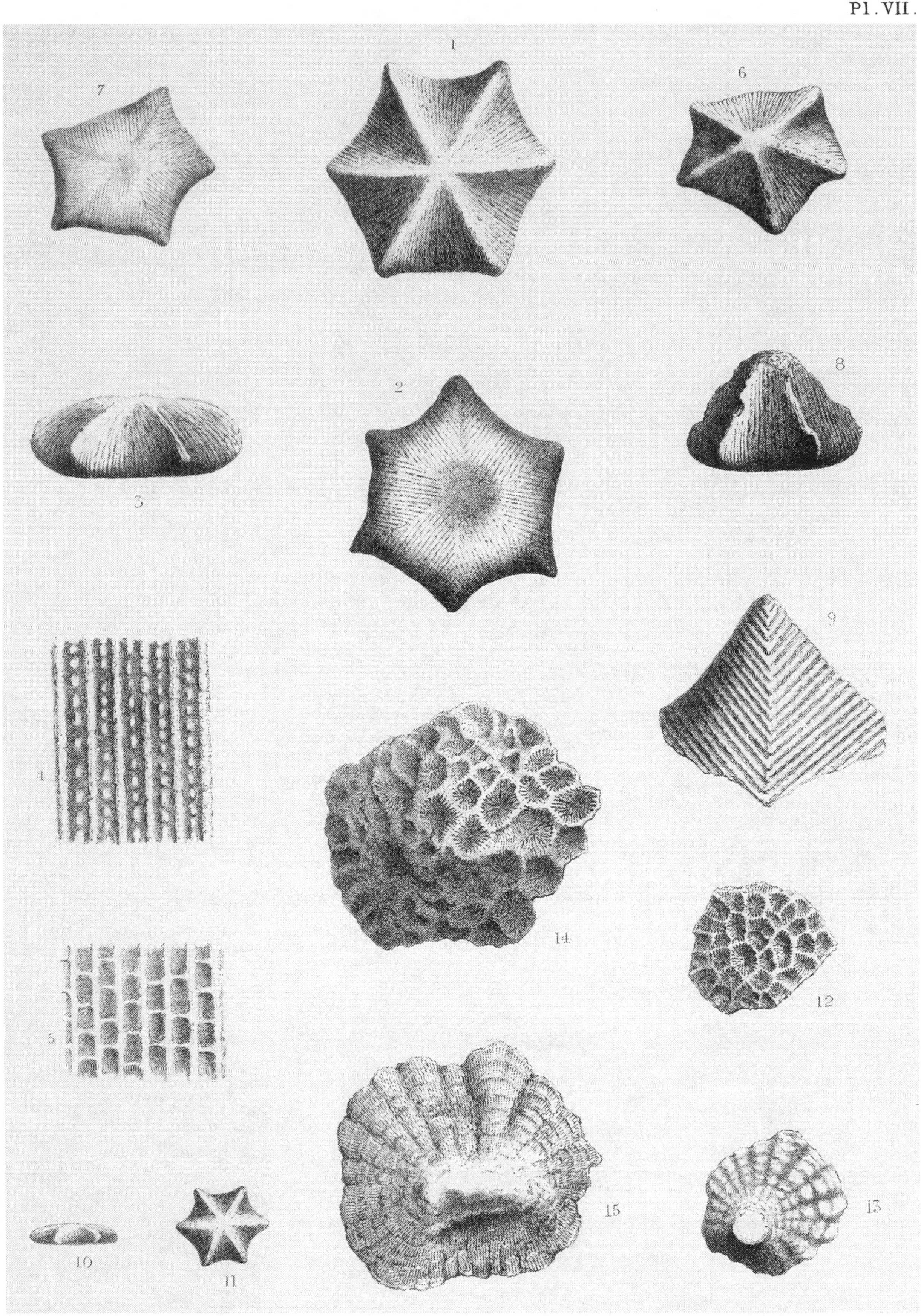

De Wilde lith.

M & N Hanhart imp

CORALS FROM THE SUPERIOR OOLITE

PLATE I.

LIASSIC CORALS FROM STREET AND BROCASTLE.

FIG.

1. *Septastræa Haimei*, Wright, sp. (P. 5.)
2. The base of the corallum.
3. Calices, magnified.
4. Fissiparous calice, magnified.
5. Septa, magnified.
6. *Septastræa excavata*, E. de From. (P. 32.) A calice, magnified.
7. The usual appearance of longitudinal sections in the Brocastle beds, magnified.

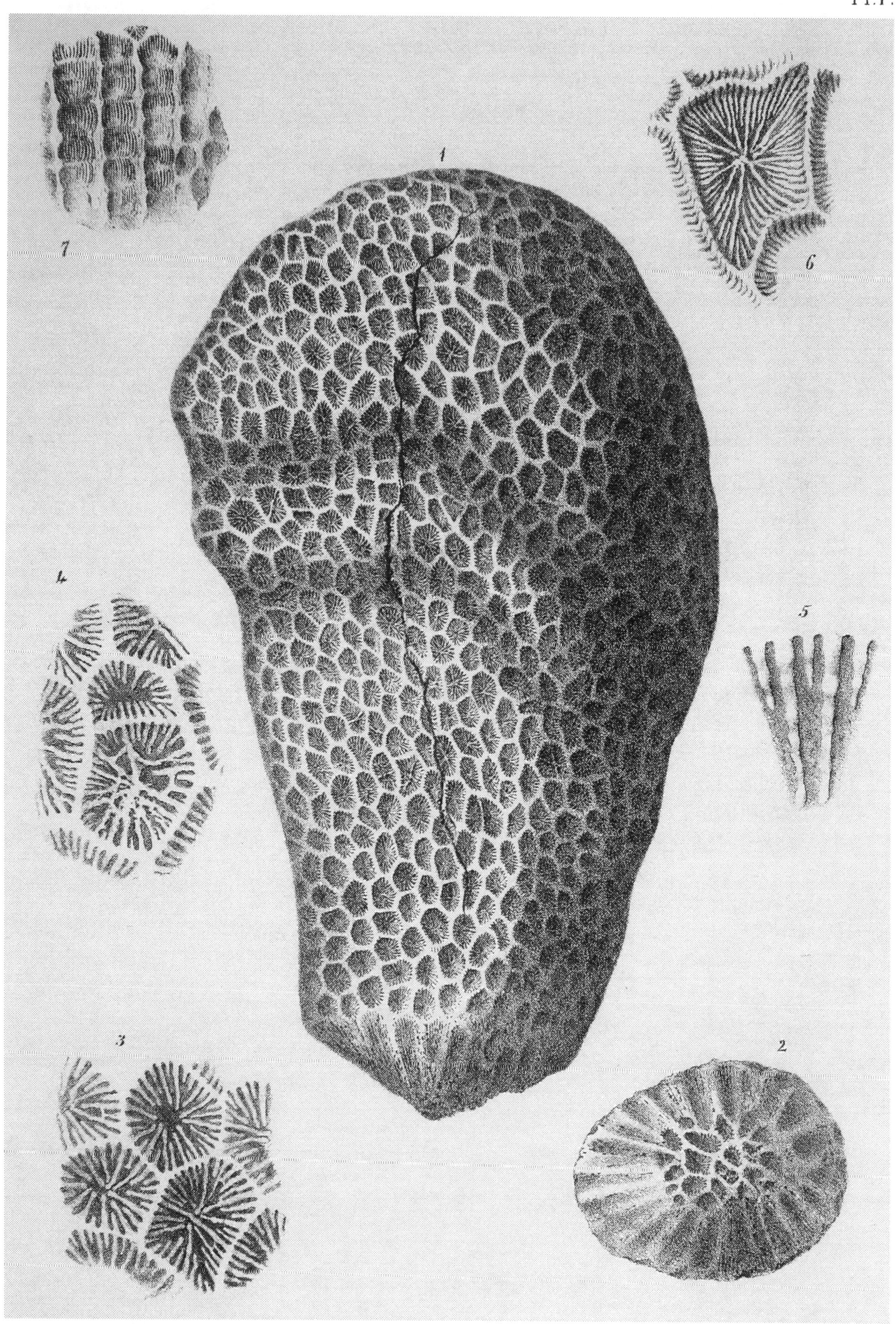

De Wilde lith. M & N Hanhart imp

LIASSIC. CORALS.

PLATE II.

LIASSIC CORALS FROM THE SUTTON STONE.

Fig.

1. *Thecosmilia rugosa*, Laube. (P. 13.)
2. Its calice, magnified.
3. A fissiparous calice, magnified.
4. A corallite, magnified.
5. Part of the wall, some septa, and some dissepiments, magnified.
6. A deformed corallite.
7. *Rhabdophyllia recondita*, Laube. (P. 17.)
8. Part of its transverse section, highly magnified, to show the septal arrangement.
9. The costæ, magnified. The corallite has *Astrocœnia parasitica* upon it.
10. *Thecosmilia mirabilis*, Duncan. (P. 12.)
11. A calice, highly magnified.
12. *Montlivaltia pedunculata*, Duncan. (P. 10.)
13. Its costæ and epitheca, magnified.

14., 15., 16. Peduncles of *Thecosmiliæ*.

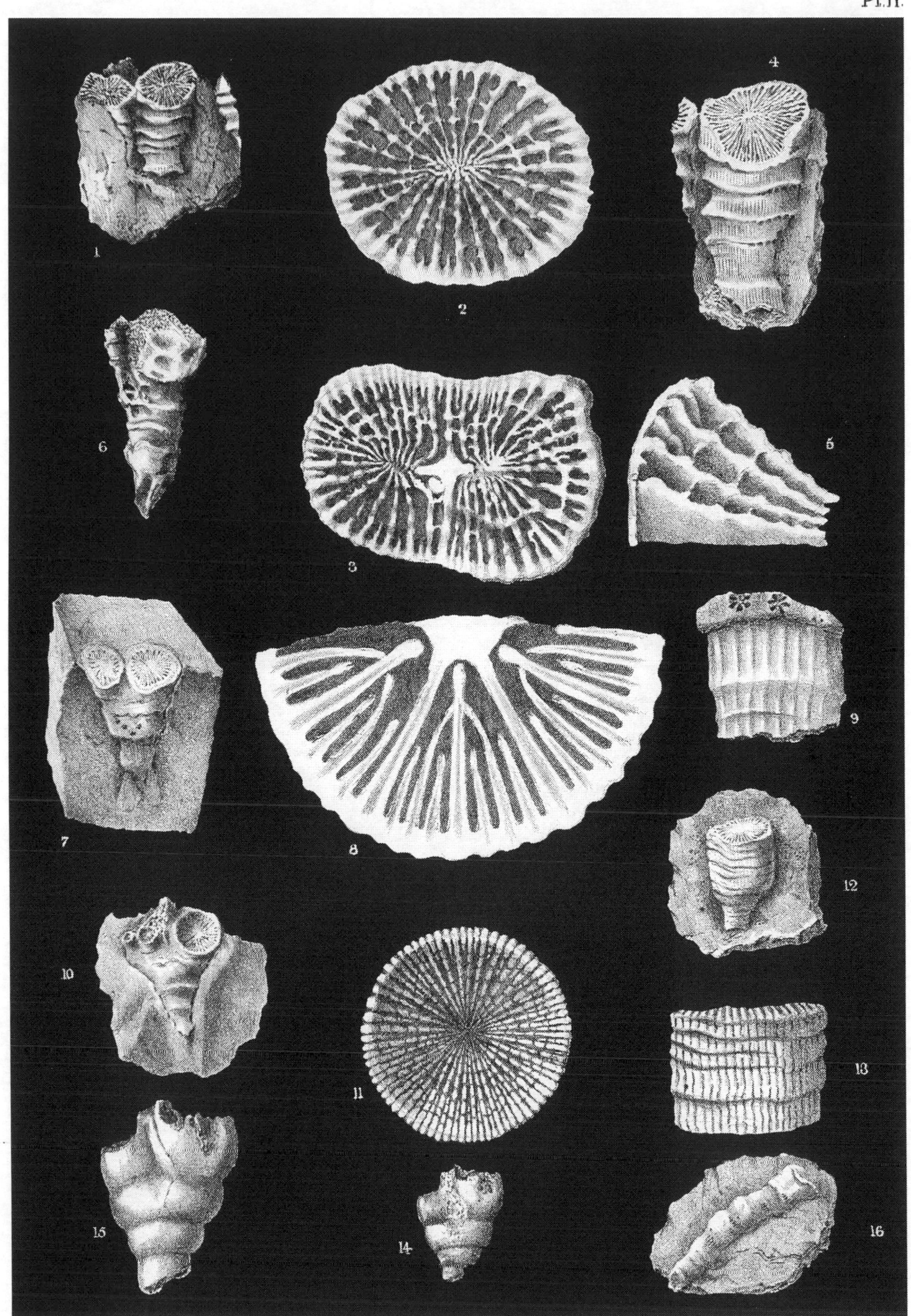

De Wilde lith. M & N Hanhart imp

LIASSIC. CORALS.

PLATE III.

LIASSIC CORALS FROM BROCASTLE.

Fig.

1. *Thecosmilia irregularis*, Duncan. (P. 15.)
2. Its upper surface, magnified.
3. Its epitheca, calices, and base, magnified.
4. A circular calice, magnified.
5. Septa (upper margin), magnified.
6. Side view of a septum, with terminal tooth.
7. *Thecosmilia Terquemi*, Duncan. (P. 16.)
8. Its base, magnified; the epitheca has been worn, and the costæ are seen with dissepiments.
9. Its upper surface, magnified.

10. } Its calices, magnified.
11. }

12. A side view of a septum, magnified.
13. A variety of *Thecosmilia irregularis*, Duncan. (P. 15.)
14. Its calice, magnified.
15. A side view of a septum, magnified.
16. *Montlivaltia simplex*, Duncan. (P. 9.)
17. Its calice, magnified.
18. *Thecosmilia affinis*, Duncan. (P. 16.)
19. A calice, magnified.
20. A side view of a septum, magnified.
21. *Thecosmilia dentata*, Duncan. (P. 16.)
22. Its calice, magnified.
23. A septum, magnified.
24. *Thecosmilia plana*, Duncan. (P. 17.)
25. Part of its calice, magnified.

De Wilde lith

M & N Hanhart imp

LIASSIC. CORALS.

PLATE IV.

LIASSIC CORALS FROM THE SUTTON STONE.

Fig.

1. *Cyathocœnia incrustans*, Duncan. (P. 28.)
2. Some calices, magnified.
3. Casts of *Astrocœnia gibbosa*, Duncan. (P. 18.)
4. *Astrocœnia reptans*, Duncan. (P. 20.)

5. }
15. } Its calices, magnified.

6. Calices altered by fossilization, magnified.
7. *Thecosmilia Suttonensis*, Duncan. (P. 11.)
8. Side view of the corallum.
9. A calice, magnified.
10. *Thecosmilia serialis*, Duncan. (P. 12.)
11. Upper surface of the corallum.
12. A serial calice, magnified.
13. *Montlivaltia parasitica*, Duncan. (P. 9.)
14. Its calice, magnified.

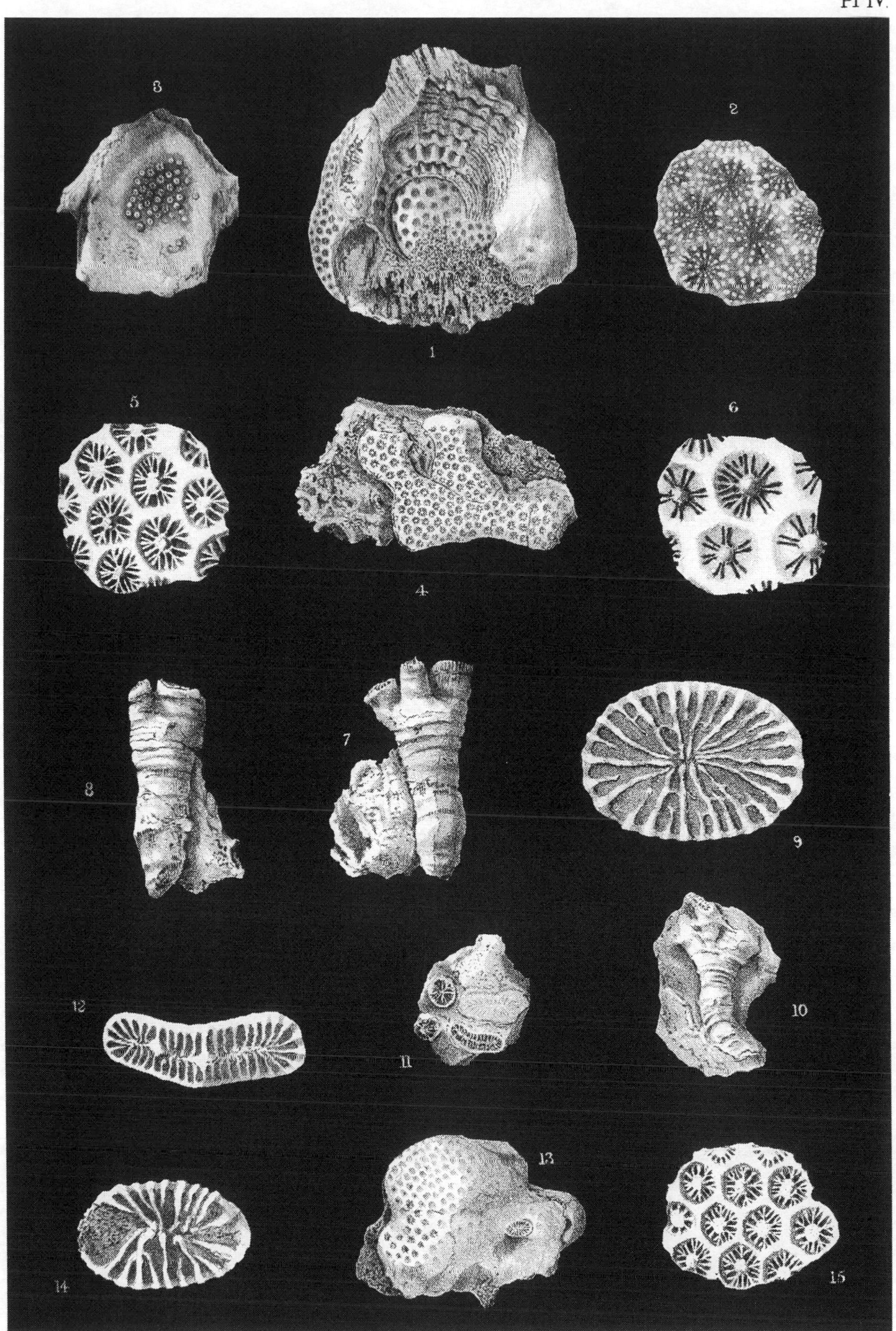

De Wilde lith.

M & N Hanhart imp

LIASSIC. CORALS.

PLATE V.

LIASSIC CORALS FROM THE SUTTON STONE AND BROCASTLE.

Fig.

1. *Astrocœnia plana*, Duncan. (P. 19). The corallum, natural size.
2. *Astrocœnia gibbosa*, Duncan. (P. 18.) A corallum, with much cœnenchyma.
3. A corallum somewhat worn.
4. The usual appearance of the Coral in the Sutton Stone; all the calices have been worn away, and it requires some trouble to distinguish the fossil.
12. A section at right angles to the corallites, highly magnified. The columella, the faint lateral dentations of the septa, and the round ornamentation between the costal ends are shown.
5. *Astrocœnia parasitica*, Duncan. (P. 20.)
6. The same, magnified. The Coral is parasitic on *Rhabdophyllia recondita*.
7. *Astrocœnia pedunculata*, Duncan. (P. 20.) The corallum, magnified.
8. The corallum, natural size.
9. A view of the peduncle and base, magnified.
10. *Cyathocœnia costata*, Duncan. (P. 29.) The corallum, natural size.
11. Some calices, magnified.

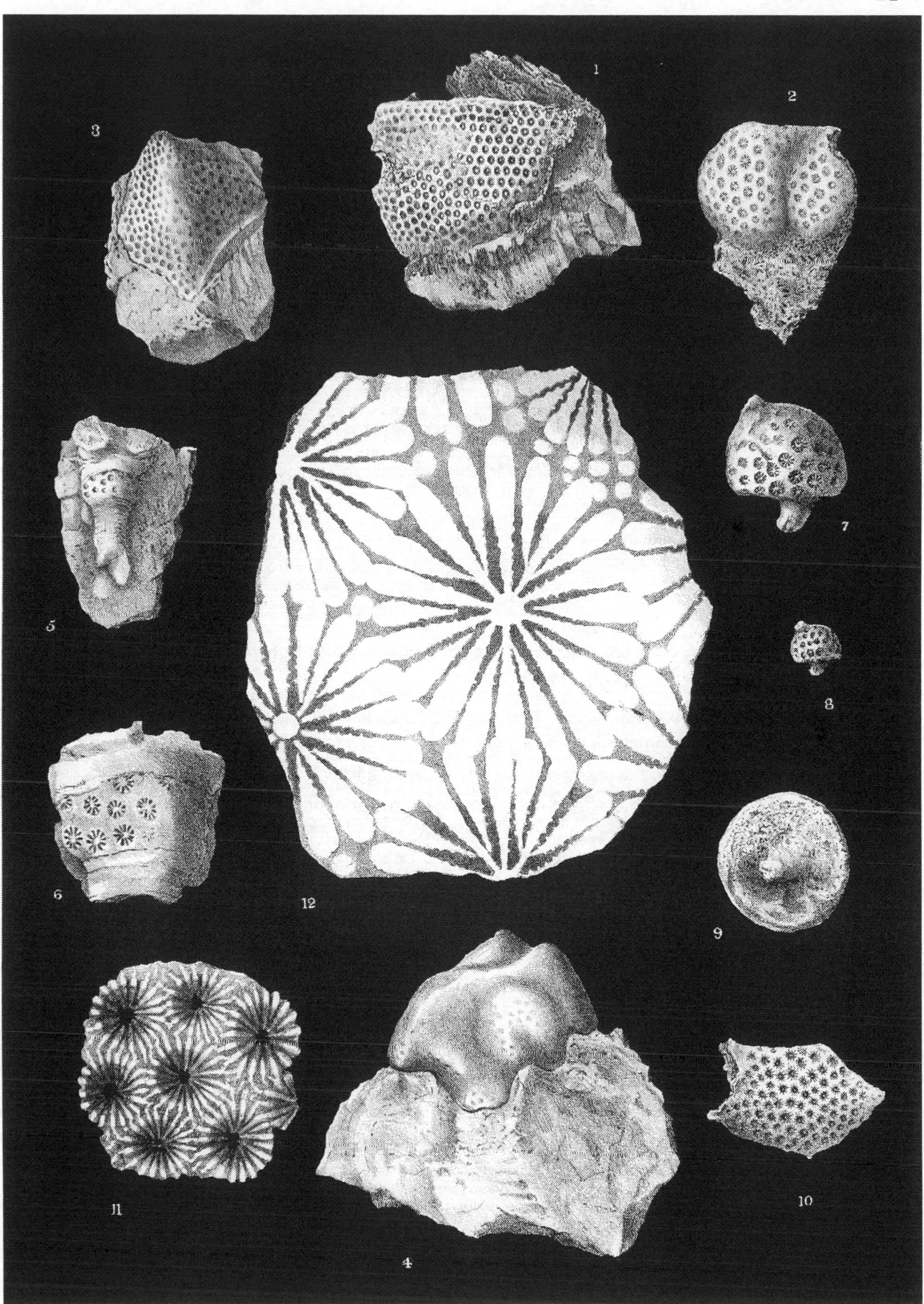

De Wilde lith.

M & N Hanhart imp.

LIASSIC. CORALS.

PLATE VI.

LIASSIC CORALS FROM THE SUTTON STONE AND BROCASTLE.

FIG.

1. *Astrocœnia gibbosa*, Duncan. (P. 18.) Some calices, magnified, showing a very usual state of preservation.
2. A worn calice, magnified.
3. A side view of worn calices, showing the dense intermediate tissue, and faint traces of endotheca, magnified.
4. A side view of a calice, magnified.
5. *Elysastræa Fischeri*, Laube. (P. 29.) A transverse section of part of the corallum, slightly magnified.
6. A transverse section showing some corallites not united by their walls.
7. The septa of neighbouring calices, the walls being united, magnified.
8. Corallites which are separate, and covered with epitheca, magnified.
9. This is a diagram, and shows the plan of the genus.
10. *Elysastræa Moorei*, Duncan. (P. 30.) Upper surface of corallum.
11. A calice, magnified.

12.} 14.} " "

13. United calices, magnified.
15. A corallite, showing costæ, the epitheca having been worn off.

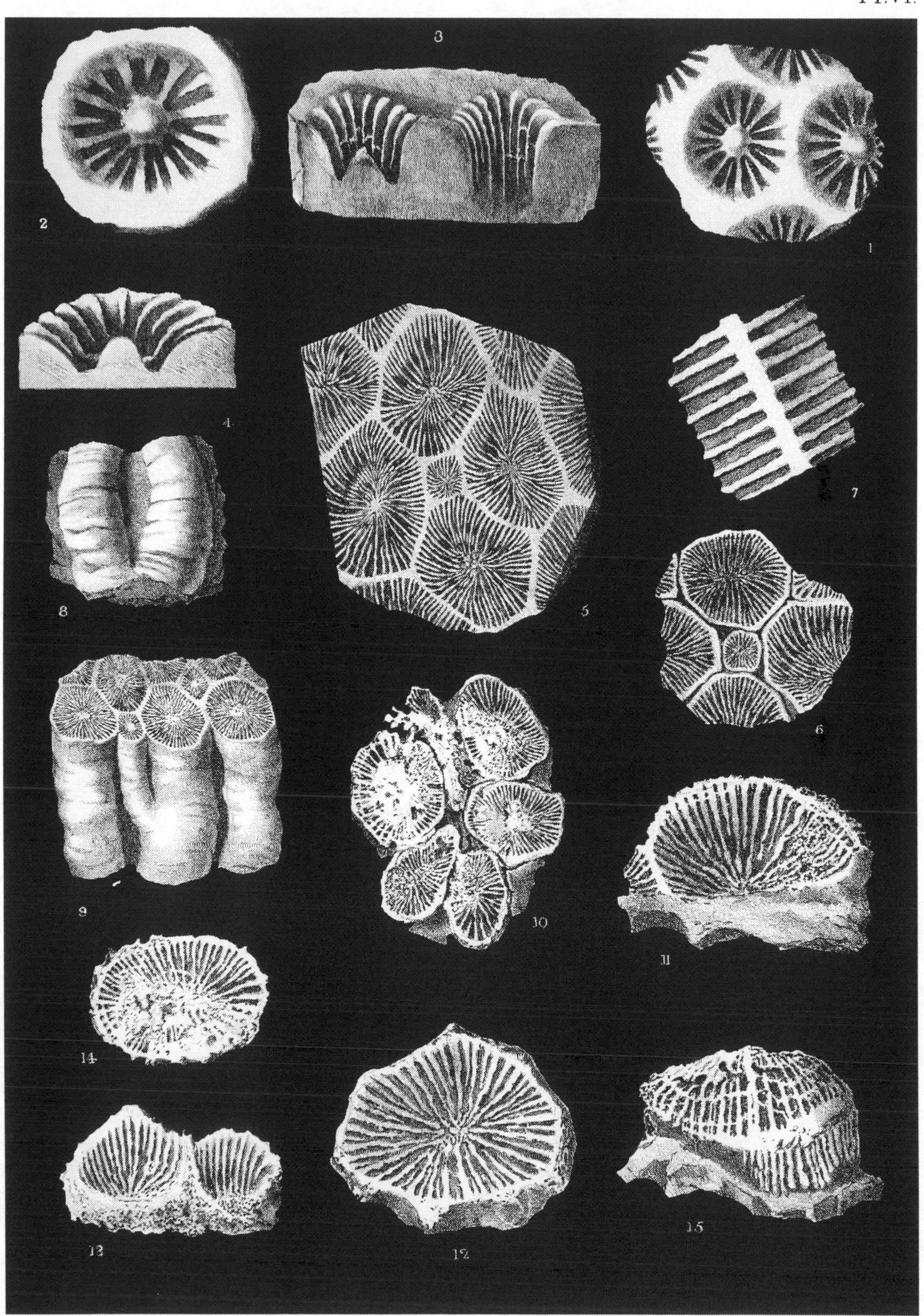

De Wilde lith.

M & N Hanhart imp

LIASSIC. CORALS.

PLATE VII.

LIASSIC CORALS FROM BROCASTLE.

FIG.

1. *Isastræa Sinemuriensis*, E. de From. (P. 30.)
2. The upper part of its corallum.
3. The calices slightly magnified to show the marginal gemmation.
4. Another view.

5. }
6. } Calices, magnified.

7. Septa, magnified.
8. A corallum with larger calices than is usual.
9. Calices, magnified.
10. *Thecosmilia Michelini*, Terquem et Piette. (P. 14.) A large variety.
11. Its calice.
12. A corallum bifurcating.
13. Its calice, magnified.
14. *Montlivaltia polymorpha*, Terquem et Piette. (P. 8.) A fractured corallum.
15. A transverse section, magnified.

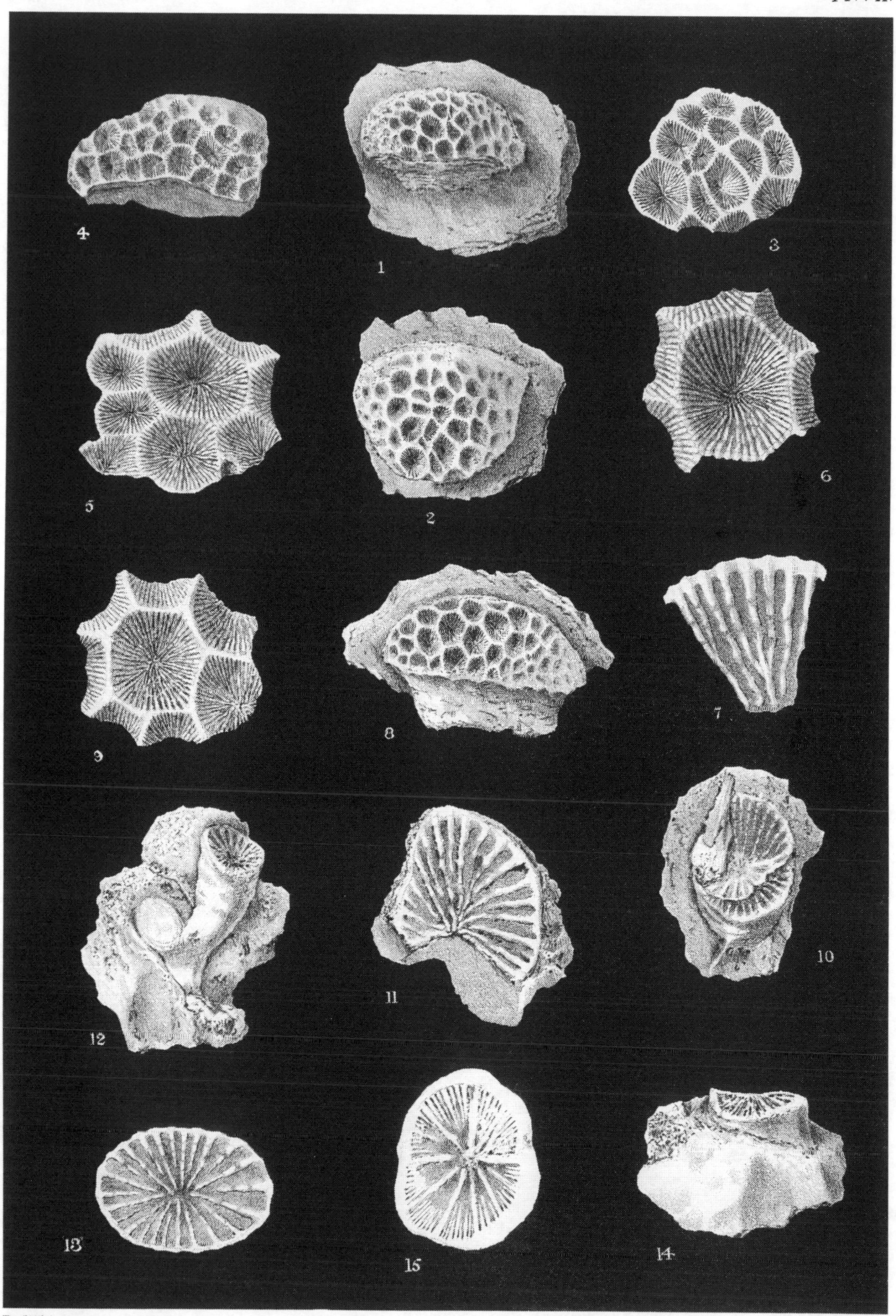

De Wilde lith

M & N Hanhart imp

LIASSIC. CORALS.

PLATE VIII.

LIASSIC CORALS FROM THE SUTTON STONE AND FROM BROCASTLE.

Fig.

1. *Montlivaltia polymorpha*, Terquem et Piette. (P. 8.) A long and large specimen.
2. A part of its transverse section, magnified.
3. Exothecal and endothecal dissepiments, costæ, and septa, magnified.
4. A smaller corallum.
13. Two corallites springing from a common base.
14. Septa of a young corallite, magnified.
15. Costæ and exotheca of a young corallite, magnified. (See also Pl. VII, figs. 14 and 15.)
5. *Montlivaltia Walliæ*, Duncan. (P. 7.) A corallum in the rock.
6. A calice, slightly magnified.
7. A side view of a septum, magnified.
8. *Montlivaltia brevis*, Duncan. (P. 10.) A corallum on the rock.
9. A calice, magnified.
10. *Montlivaltia Murchisoniæ*. (P. 8.) A corallum.
11. A part of the calice, magnified.
12. The peculiar costal arrangement and septa, magnified.
16. *Montlivaltia pedunculata*, Duncan. (P. 10.) A corallum.
17. *Isastræa globosa*, Duncan. (P. 31.) A corallum, the calices are worn.
18. Calices, magnified.

De Wilde lith.

M & N Hanhart imp.

LIASSIC. CORALS.

PLATE IX.

LIASSIC CORALS FROM BROCASTLE.

Fig.

1. *Astrocœnia insignis,* Duncan. (P. 19.) A corallum.
2. Calices, magnified.
3. *Astrocœnia superba,* Duncan. (P. 21.) Part of a corallum.
4. A calice, magnified.
5. A side view of a calice, magnified.
6. *Cyathocœnia dendroidea,* Duncan. (P. 27.) A corallum.
7. A calice, magnified.
8. A calice, magnified ; a side view.
9. A transverse section of a stem, showing the concavities produced by the calices and the intermediate cœnenchyma.
10. *Astrocœnia dendroidea,* Duncan. (P. 22.) A part of a corallum.
11. A calice, magnified.
12. *Astrocœnia favoidea,* Duncan. (P. 21.) A corallum.
13. Calices, magnified.
14. A side view of a calice, magnified.
15. *Astrocœnia costata,* Duncan. (P. 21.) A corallum.
16. Calices, magnified.
17. A corallum.
18. *Astrocœnia minuta,* Duncan. (P. 22.) A corallum.
19. A calice, magnified.
20. A side view of a calice, magnified.

Pl. IX.

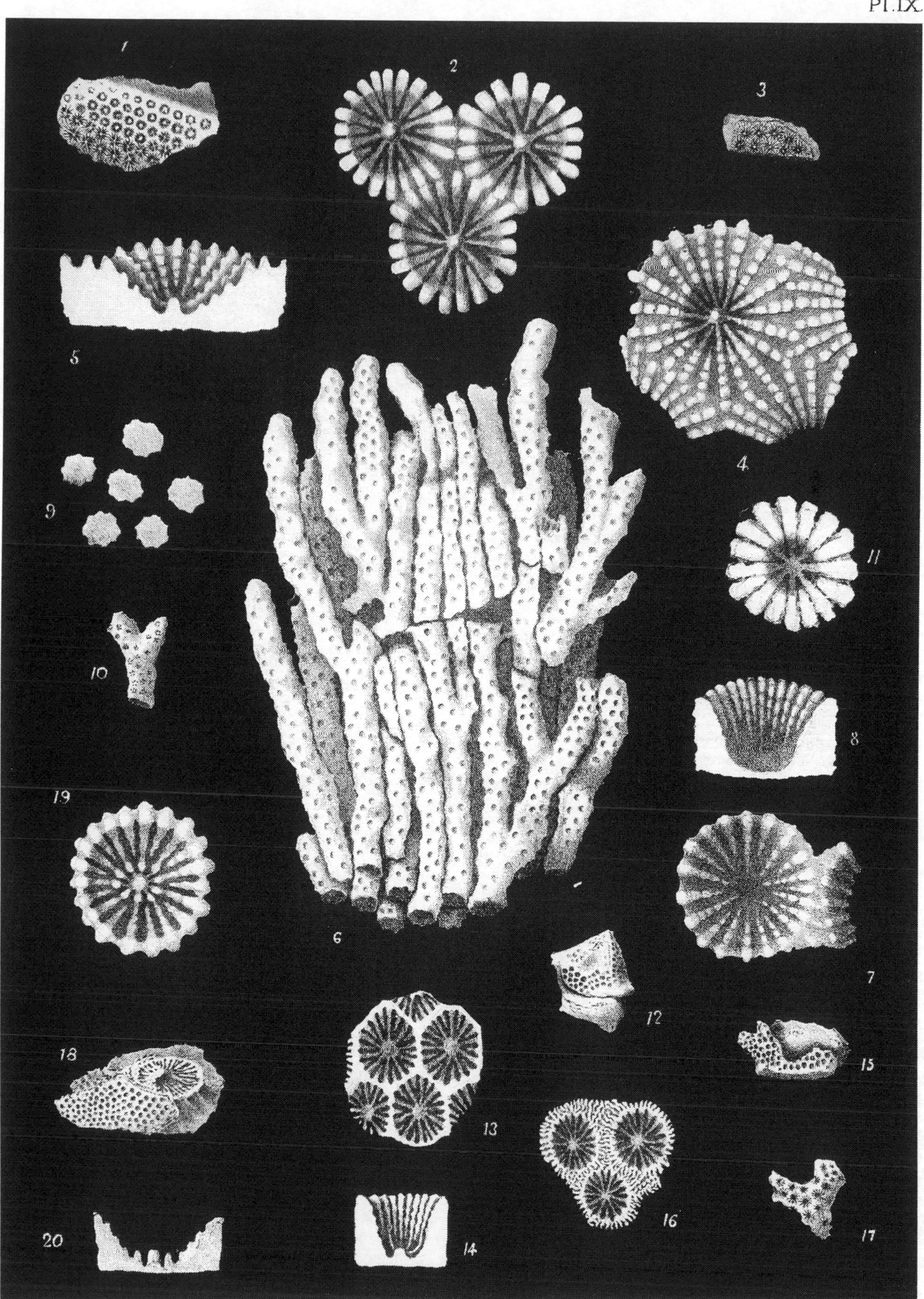

De Wilde lith.

M & N Hanhart imp

LIASSIC. CORALS.

PLATE X.

CORALS FROM BROCASTLE ; MARTON, NEAR GAINSBOROUGH ; NEWARK, IN NOTTINGHAMSHIRE ; AND FROM THE NORTH OF IRELAND.

Fig.

1. *Thecosmilia Brodiei*, Duncan. (P. 13.) The upper part of a corallite, natural size.
2. The calice, magnified.
3. A side view of a septum, magnified.
4. A septum seen from above, magnified.
5. Some corallites of *Thecosmilia irregularis*, Duncan. (P. 15.) Showing the gemmation from the calicular edge, and the rough and ridged epitheca.
6. A corallite of *Thecosmilia Martini*, E. de From. (P. 14.)

7 and 8. Views of its transverse section and calice, magnified.

9. Corallites, with strong epitheca.
10. A section of a cast of *Thecosmilia Michelini*, Terquem. (P. 14.) From Cowbridge, magnified.
11. The calicular end of a corallite.
12. The calice, magnified.
13. A corallite, showing the rounded ridges of the epitheca.
14. A calice.
15. The upper surface of *Montlivaltia papillata*, Duncan. (P. 36.)
16. A side view of the calice.
17. Septal dentations, magnified.
18. The base of the corallum.
19. *Montlivaltia papillata*, Duncan, variety. (P. 37.) Its calice.
20. The side of the calice.
21. The septa seen from above, magnified.
22. *Montlivaltia Hibernica*, Duncan. (P. 39.) Its calice.
23. The septa seen from above, magnified.
24. *Montlivaltia Haimei*, Chapuis et Dewalque. (P. 35.) The Irish form. View of the calice.
25. The septa seen from above, magnified.
26. A variety of *Montlivaltia Haimei*. The calice.

27 and 28. Views of the corallum.

29. A variety of *Montlivaltia Haimei*. The calice.
30. The side view of the corallum.

31 and 32. Unusual shapes of the corallum.

33. *Oppelismilia gemmans*, Duncan. (P. 39.) The calicular surface, showing calicular gemmation.
34. The side view of the corallum.

De Wilde lith. M & N Hanhart imp

LIASSIC. CORALS.

PLATE XI.

LIASSIC CORALS FROM LUSSAY, IN SKYE, AND MARTON, NEAR GAINSBOROUGH.

FIG.

1. *Isastræa Murchisoni*, Wright. (P. 41.)
2. Calices, magnified.
3. A calice, magnified.
4. A septum, magnified.
5. *Septastræa Fromenteli*, Terquem et Piette. (P. 37.) Some fissiparous calices, slightly magnified.

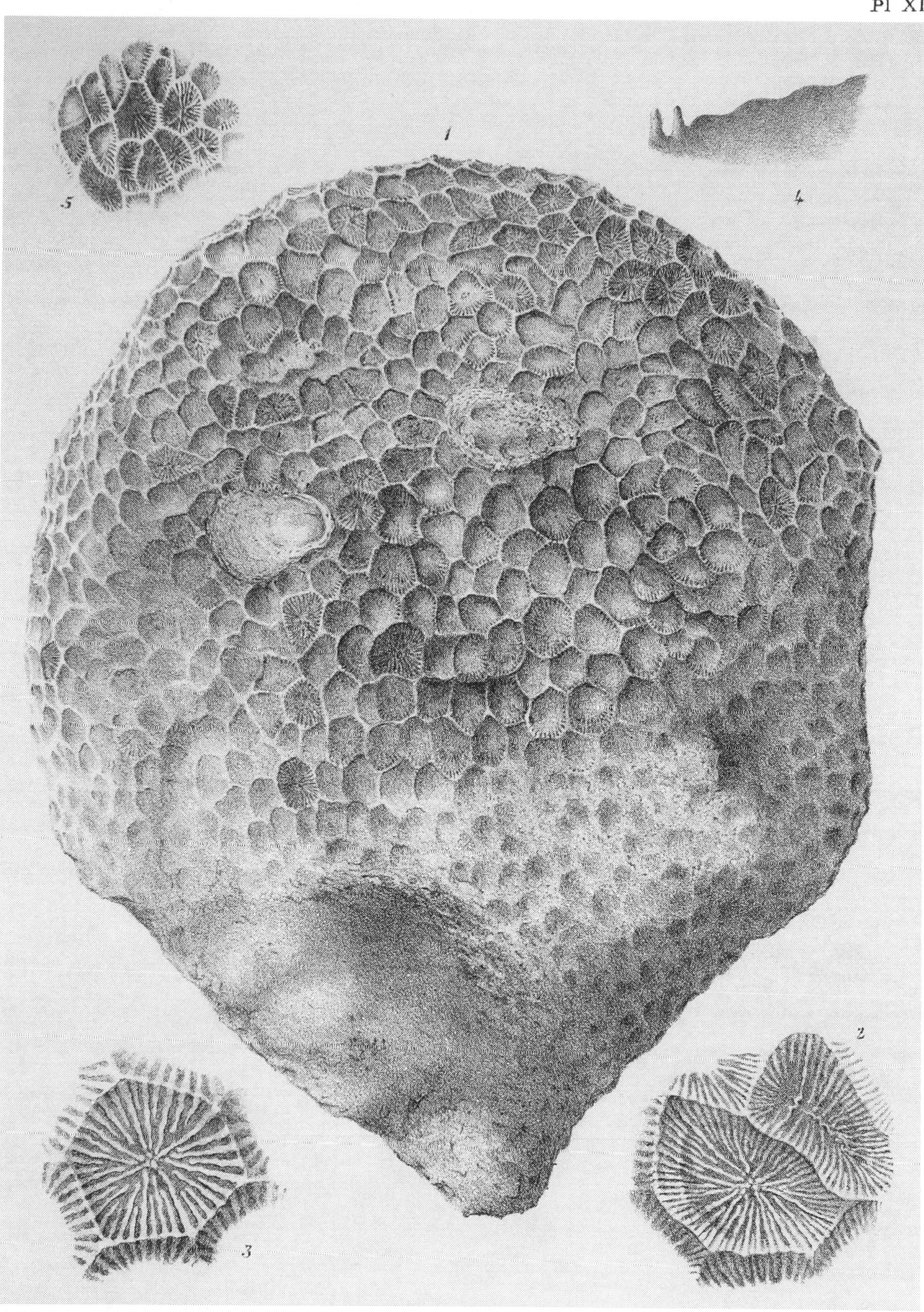

De Wilde lith.

M & N Hanhart imp

LIASSIC. CORALS.

PLATE XII.

CORALS FROM THE LOWER LIAS.

Fig.

1. *Thecosmilia Martini*, E. de Fromentel. (P. 45.)
2. Part of its epitheca magnified.
3. *Montlivaltia Ruperti*, Duncan. (P. 46.)
4. Its calice.
5. The calice, magnified.
6. A young specimen of *Montlivaltia Guettardi*, Blainville. (P. 51.)
7. The calice, magnified.

10., 11., 12. A full-grown specimen. (P. 51.)

13. Its calice, magnified.
14. The septa, magnified.
8. The calicular surface of a young *Thecosmilia Martini*.
9. The same, magnified.
15. *Lepidophyllia Stricklandi*, Duncan. (P. 53.)
16. A cast of *Thecosmilia Michelini*, Terquem. (P. 51.)
17. Part of the corallum of *Isastræa endothecata*, Duncan. (P. 53.)
18. A regular calice, magnified.
19. A side view of a magnified calice, showing the endotheca in the calice.
20. Oblique view of some calices, magnified.
21. Endothecal dissepiments connecting the septa, magnified.

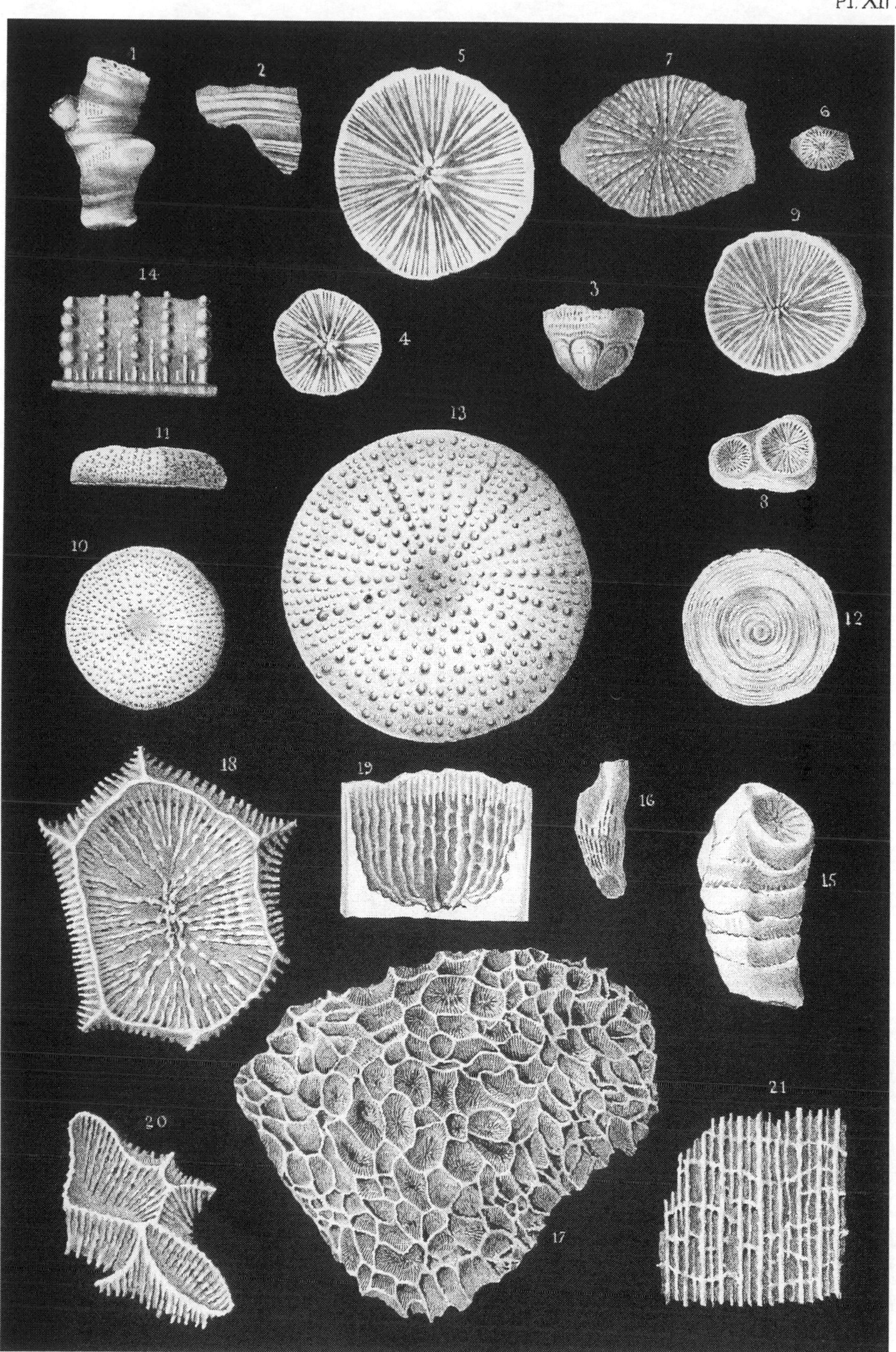

De Wilde lith

M & N Hanhart imp

LIASSIC. CORALS.

PLATE XIII.

CORALS FROM THE LOWER LIAS.

FIG.

1. The corallum of *Isastræa Stricklandi*, Duncan. (P. 54.)
2. A calice, magnified.
3. The dentations of a septum, magnified.
4. The strong dissepimental endotheca, magnified.
5. Part of the corallum of *Septastræa Eveshami*, Duncan. (P. 52.)
6. A calice, magnified.
7. An irregular and contorted calice, magnified.
8. The corallum of *Cyathocœnia globosa*, Duncan. (P. 55.)
9. Calices of *Cyathocœnia globosa*, magnified.
10. Part of the corallum of *Isastræa insignis*, Duncan. (P. 54.)
11. A system of four cycles of septa, magnified.

De Wilde lith. M & N Hanhart imp

LIASSIC. CORALS.

PLATE XIV.

CORALS FROM THE LOWER LIAS.

Fig.

1. A calice of *Montlivaltia rugosa*, Wright, magnified. (Pp. 58.)
2. An oblique view of some dentate septa of the same species, magnified.
3. A portion of a worn calice, magnified, showing the irregular septal arrangement of some specimens of this species.
4. A young specimen of *Montlivaltia mucronata*, Duncan. (Pp. 59.)
5. A variety of *Montlivaltia mucronata*.
6. One of its septa, magnified, showing the mucronate processes and the granular ornamentation.
7. One of the processes, magnified.
8. A side view of the corallum of a full-grown individual, rather enlarged.
9. A calice of a full-grown *Montlivaltia mucronata*, magnified. (Pp. 59.)
10. A deformed specimen of a variety of *Montlivaltia mucronata*.
11. A conical variety of *Montlivaltia mucronata*.
14. A variety of *Montlivaltia mucronata*, with a deeper calice than the type.
15. The calice, magnified.
16. One of its septa, magnified.
17. A process, magnified.
18. A variety of *Montlivaltia mucronata*.
12. The corallum of *Montlivaltia nummiformis*, Duncan. (P. 60.)
13. The basal epitheca and projecting septa, magnified.

Pl. XIV.

De Wilde lith

M & N Hanhart imp

LIASSIC. CORALS.

PLATE XV.

CORALS FROM THE LOWER LIAS.

Fig.

1. The upper calicular surface of *Montlivaltia radiata*, Duncan. (P. 61.)
2. The under surface.
3. A side view of the corallum, showing the central depression of the base.
4. The calice, magnified, showing the quaternary arrangement of the septa.
5. The base, magnified, showing the pellucid epitheca and the costæ.
6. The corallum of *Montlivaltia patula*, Duncan. (P. 56).
7. The calice, magnified.
8. Part of the wall and one of the septa, magnified, showing the direction of the teeth.
9. A very young *Montlivaltia*, of an unknown species. The calice magnified (P. 52.)
10. A cornute variety of *Montlivaltia mucronata*, Duncan. (P. 60.)
11. One of its septa, magnified.
12. A portion of the external surface of the type of *Montlivaltia mucronata*, Duncan, showing the dichotomous longitudinal bundles of costæ, magnified.
13. A conical variety of *Montlivaltia mucronata*, Duncan.
15. A section, magnified, of *Montlivaltia Ruperti*, Duncan.
16. A large and unusual shape of *Montlivaltia rugosa*, Wright.
17. A side view of its dentate septa.
14. A dentate process, magnified, showing the ornamentation.
18. The corallum of *Isastræa latimæandroidea*, Duncan. (P. 65.)
19. Its septa, magnified.
20. The corallum of *Isastræa Tomesii*, Duncan. (P. 46.)

De Wilde lith.

M & N Hanhart imp

LIASSIC. CORALS.

PLATE XVI.

CORALS FROM THE LOWER AND MIDDLE LIAS.

Fig.

1. *Lepidophyllia Hebridensis*, Duncan. Natural size. (P. 62.)
2. Calices, magnified, showing calicular gemmation.
3. Side view of corallites, showing the epitheca.
4. Septa, magnified.

5., 6., 7., 9. Common forms of { *Montlivaltia rugosa*, Wright, sp. (P. 58.) / *Thecocyathus rugosus*, Wright, MS. }

10., 11., 12., 13., 14. Unusual and young forms of the same species.

8. Septal ends (external) and intermediate endotheca (magnified).

15. A section magnified, showing the strong and arched endotheca between the septa.

De Wilde lith. M & N Hanhart imp

LIASSIC. CORALS.

PLATE XVII.

CORALS FROM THE MIDDLE LIAS.

Fig.

1. *Montlivaltia Victoriæ,* Duncan. Nat. size. A corallum, with a constriction near the calice. (P. 63.)

2. A magnified view of some septal (external) ends, with endotheca, simulating costæ and exotheca. The epitheca is shown covering these structures.

3. A specimen with a large calice.

4. A calice slightly magnified. The rudimentary septa are shown as faint white lines close to the thin margin.

5. Diagram of the septa.

6. Slightly magnified view of the thin epithecal wall and the curved endotheca.

7.
8. } Different forms of the species.
9.

10. A septum, magnified.

De Wilde lith.

M & N Hanhart imp.

LIASSIC. CORALS.

Printed in the United States
By Bookmasters